Dr. G

HANDBOOK OF FLUID SEALING

Other Books of Interest from McGraw-Hill

Avallone & Baumeister • MARKS' STANDARD HANDBOOK FOR MECHANICAL ENGINEERS
Bhushan & Gupta • HANDBOOK OF TRIBOLOGY
Brady & Clauser • MATERIALS HANDBOOK
Bralla • HANDBOOK OF PRODUCT DESIGN FOR MANUFACTURING
Brunner • HANDBOOK OF INCINERATION SYSTEMS
Corbitt • STANDARD HANDBOOK OF ENVIRONMENTAL ENGINEERING
Ehrich • HANDBOOK OF ROTORDYNAMICS
Elliot • STANDARD HANDBOOK OF POWERPLANT ENGINEERING
Freeman • STANDARD HANDBOOK OF HAZARDOUS WASTE TREATMENT AND DISPOSAL
Ganić & Hicks • THE MCGRAW-HILL HANDBOOK OF ESSENTIAL ENGINEERING INFORMATION AND DATA
Gieck • ENGINEERING FORMULAS
Grimm & Rosaler • HANDBOOK OF HVAC DESIGN
Harris • HANDBOOK OF ACOUSTICAL MEASUREMENTS AND NOISE CONTROL
Harris & Crede • SHOCK AND VIBRATION HANDBOOK
Hicks • STANDARD HANDBOOK OF ENGINEERING CALCULATIONS
Hodson • MAYNARD'S INDUSTRIAL ENGINEERING HANDBOOK
Jones • DIESEL PLANT OPERATIONS HANDBOOK
Juran & Gryna • JURAN'S QUALITY CONTROL HANDBOOK
Karassik et al. • PUMP HANDBOOK
Kurtz • HANDBOOK OF APPLIED MATHEMATICS FOR ENGINEERS AND SCIENTISTS
Mason • SWITCH ENGINEERING HANDBOOK
Nayyar • PIPING HANDBOOK
Parmley • STANDARD HANDBOOK OF FASTENING AND JOINING
Rohsenow et al. • HANDBOOK OF HEAT TRANSFER APPLICATIONS
Rohsenow et al. • HANDBOOK OF HEAT TRANSFER FUNDAMENTALS
Rosaler & Rice • STANDARD HANDBOOK OF PLANT ENGINEERING
Rothbart • MECHANICAL DESIGN AND SYSTEMS HANDBOOK
Schwartz • COMPOSITE MATERIALS HANDBOOK
Schwartz • HANDBOOK OF STRUCTURAL CERAMICS
Shigley & Mischke • STANDARD HANDBOOK OF MACHINE DESIGN
Townsend • DUDLEY'S GEAR HANDBOOK
Tuma • ENGINEERING MATHEMATICS HANDBOOK
Tuma • HANDBOOK OF NUMERICAL CALCULATIONS IN ENGINEERING
Wadsworth • HANDBOOK OF STATISTICAL METHODS FOR ENGINEERS AND SCIENTISTS
Woodruff, Lammers, & Lammers • STEAM-PLANT OPERATION
Young • ROARK'S FORMULAS FOR STRESS AND STRAIN

HANDBOOK OF FLUID SEALING

Robert V. Brink, P.E. Editor in Chief
Daniel E. Czernik, P.E.
Leslie A. Horve, Ph.D., P.E.

McGRAW-HILL, INC.
New York St. Louis San Francisco Auckland Bogotá
Caracas Lisbon London Madrid Mexico Milan
Montreal New Delhi Paris San Juan São Paulo
Singapore Sydney Tokyo Toronto

Library of Congress Cataloging-in-Publication Data

Brink, Robert V.
Handbook of fluid sealing / Robert V. Brink, Daniel E. Czernik, Leslie A. Horve.
p. cm.
ISBN 0-07-007827-0
1. Sealing (Technology)—Handbooks, manuals, etc. I. Czernik, Daniel E. II. Horve, Leslie A. III. Title.
TJ246.B648 1993
621.8′85—dc20 92-14548
CIP

1 2 3 4 5 6 7 8 9 0 DOC/DOC 9 8 7 6 5 4 3 2

ISBN 0-07-007827-0

The sponsoring editor for this book was Robert Hauserman, the editing supervisor was Nancy Young, and the production supervisor was Suzanne W. Babeuf. This book was set in Times Roman by McGraw-Hill's Professional Book Group composition unit.

Printed and bound by R. R. Donnelley & Sons Company.

CONTENTS

Part 3 Fluids and Lubricants

Part 4 Economic and Manufacturing Considerations

PREFACE

In the field of gaskets and seals, gaskets are generally associated with sealing mating flanges, and seals are generally associated with sealing rotating, reciprocating, or oscillating shafts, pistons, valves, and other types of moving parts. Some designers refer to gaskets as static seals and consider seals to be dynamic sealing components. The *Handbook of Fluid Sealing* is divided in this way. Part 1 covers gaskets, the products for static applications, and Part 2 covers seals, the products used in dynamic applications. Part 3 discusses fluids and lubricants, and Part 4 discusses the economics and manufacturing of seals and gaskets.

The Handbook of Fluid Sealing is the first and only comprehensive text to address the entire field of sealing products. It provides practical information for the most prevalent types of sealing products and offers workable solutions to many typical sealing problems, showing standard and accepted practices in current use today. In-depth theoretical information is presented for those readers who need additional background, whether they are students, researchers, or engineers with a need to know. For the first time the thermodynamic life equation of an oil seal is presented in useful handbook form. Economic and manufacturing considerations are discussed in separate chapters as individual topics so that the information is more accessible to the business person or user who doesn't need or want a lot of background seal technology to answer his or her questions.

The authors of this handbook, through their years of experience in sealing technology, have drawn extensively on their own personal research and technical publications. Without the contributions of so many others in the field, however, this handbook would have remained a mere abstract of papers. To Bob Hauserman and his excellent staff at McGraw-Hill must go the credit for recognizing the potential in what was most assuredly a very rough manuscript and turning it into another of McGraw-Hill's fine library of handbooks.

This book is dedicated in memory to the crew of the Space Shuttle Challenger.

Robert V. Brink
Editor in Chief

HANDBOOK OF FLUID SEALING

P • A • R • T • 1

SEALS FOR STATIC APPLICATIONS

CHAPTER 1

NONMETALLIC AND METALLIC GASKETS

DEFINITION

A gasket is a material or combination of materials clamped between two separable members of a mechanical joint. Its function is to effect a seal between the members (flanges) and maintain the seal for a prolonged period. The gasket must be capable of sealing the mating surfaces, impervious and resistant to the medium being sealed, and able to withstand the application temperature and pressure. Figure 1.1 shows the nomenclature associated with a gasketed joint.

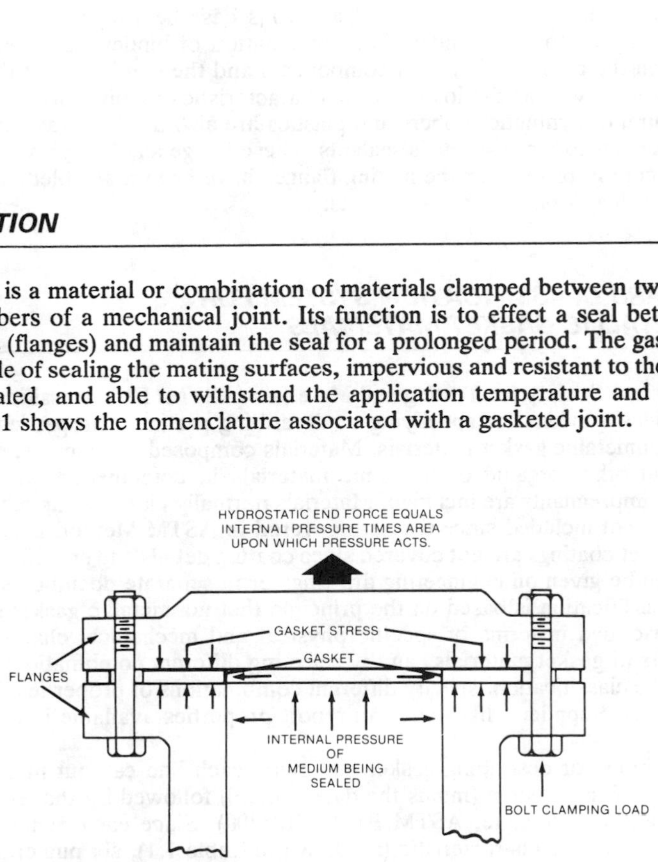

FIGURE 1.1 Nomenclature of a gasketed joint.

The range of environmental conditions for industrial gasketing is extremely large. Applications with a low clamp load, along with thin flanges such as those associated with washing machine water pumps, are on the one extreme. On the other extreme are the high clamp load and rigid joints associated with pressure vessel applications. The American Society of Mechanical Engineers (ASME) code for gasketing these latter applications is normally followed. The code specification, while being applicable for the latter, is not useful for the wide variety of industrial gasketing, which do not have specified flange thickness, bolt size, bolt

spacing, etc. This book contains information for the gasket engineer to consider when designing gaskets for industrial applications where the ASME code is usable and also for gasket design when the code is not applicable.

NONMETALLIC GASKETS

Most nonmetallic materials consist of a fibrous base held together or strengthened with an elastomeric binder. The combination of binders and base chosen depends on the compatibility with components and the conditions of the sealing environment as well as the load-bearing characteristics required for the application. Natural or synthetic rubbers and plastics are also used for gaskets.

Another category of gaskets is sealants. These are generally high-viscosity liquids that cure in place after the mating flanges have been assembled. They generally are called *formed-in-place* gaskets.

STANDARD CLASSIFICATION SYSTEM FOR NONMETALLIC GASKET MATERIALS

The American Society for Testing and Materials (ASTM) F104 classification system provides a means for specifying or describing pertinent properties of commercial nonmetallic gasket materials. Materials composed of asbestos, cork, cellulose, and other organic or inorganic materials in combination with various binders or impregnants are included. Materials normally classified as rubber compounds are not included since they are covered in ASTM Method D 2000 (SAE J200). Gasket coatings are not covered since coating details and specifications are intended to be given on engineering drawings or in separate documents.

This classification is based on the principle that nonmetallic gasket materials can be described in terms of specific physical and mechanical characteristics. Thus, users of gasket materials can, by selecting different combinations of statements in the classification, specify different combinations of properties desired in various parts. Suppliers, likewise, can report properties available in their products.

In specifying or describing gasket materials, each line call-out must include the number of this system (minus the date symbol) followed by the letter F and six numerals, for example, ASTM F104 (F125400). Since each numeral of the call-out represents a characteristic (as shown in Table 1.1), six numerals are always required. The numeral 0 is used when the description of any characteristic is not desired. The numeral 9 is used when the description of any characteristic (or related test) is specified by some supplement to this classification system, such as notes on engineering drawings.

It should be noted that although asbestos is listed in the ASTM classification system, it has been essentially eliminated from the vast majority of gasketing materials.

TABLE 1.1 Basic Physical and Mechanical Characteristics

Basic six-digit number	Basic characteristic
First numeral	Type of material (the principal fibrous or particulate reinforcement material from which the gasket is made) shall conform to the first numeral of the basic six-digit number as follows: 0 = not specified 1 = asbestos or other inorganic fibers (type 1) 2 = cork (type 2) 3 = cellulose or other organic fibers (type 3) 4 = fluorocarbon polymer 9 = as specified*
Second numeral	Class of material (method of manufacture or common trade designation) shall conform to the second numeral of the basic six-digit number as follows: When first numeral is 1, for second numeral: 0 = not specified 1 = compressed asbestos (class 1) 2 = beater addition asbestos (class 2) 3 = asbestos paper and millboard (class 3) 9 = as specified* When first numeral is 2, for second numeral: 0 = not specified 1 = cork composition (class 1) 2 = cork and elastomeric (class 2) 3 = cork and cellular rubber (class 3) 9 = as specified* When first numeral is 3, for second numeral: 0 = not specified 1 = untreated fiber—tag, chipboard, vulcanized fiber, etc. (class 1) 2 = protein treated (class 2) 3 = elastomeric treated (class 3) 4 = thermosetting resin treated (class 4) 9 = as specified* When first numeral is 4, for second numeral: 0 = not specified 1 = sheet PTFE 2 = PTFE of expanded structure 3 = PTFE filaments, braided or woven 4 = PTFE felts 5 = filled PTFE 9 = as specified*

TABLE 1.1 Basic Physical and Mechanical Characteristics (*Continued*)

Basic six-digit number	Basic characteristic
Third numeral	Compressibility characteristics, determined in accordance with 8.2, shall conform to the percentage indicated by the third numeral of the basic six-digit number (example: 4 = 15 to 25% maximum): 0 = not specified 5 = 20 to 30% 1 = 0 to 10% 6 = 25 to 40% 2 = 5 to 15%† 7 = 30 to 50% 3 = 10 to 20% 8 = 40 to 60% 4 = 15 to 25% 9 = as specified*
Fourth numeral	Thickness increase when immersed in ASTM no. 3 oil, determined in accordance with 8.3, shall conform to the percentage indicated by the fourth numeral of the basic six-digit number (example: 4 = 15 to 30%): 0 = not specified 5 = 20 to 40% 1 = 0 to 15% 6 = 30 to 50% 2 = 5 to 20% 7 = 40 to 60% 3 = 10 to 25% 8 = 50 to 70% 4 = 15 to 30% 9 = as specified*
Fifth numeral	Weight increase when immersed in ASTM no. 3 oil, determined in accordance with 8.3, shall conform to the percentage indicated by the fifth numeral of the basic six-digit number (example: 4 = 30% maximum): 0 = not specified 5 = 40% max. 1 = 10% max. 6 = 60% max. 2 = 15% max. 7 = 80% max. 3 = 20% max. 8 = 100% max. 4 = 30% max. 9 = as specified*
Sixth numeral	Weight increase when immersed in water, determined in accordance with 8.3, shall conform to the percentage indicated by the sixth numeral of the basic six-digit number (example: 4 = 30% maximum): 0 = not specified 5 = 40% max. 1 = 10% max. 6 = 60% max. 2 = 15% max. 7 = 80% max. 3 = 20% max. 8 = 100% max. 4 = 30% max. 9 = as specified*

*Specified on engineering drawings or other supplements to this classification system. Suppliers of gasket materials should be contacted to find out what line call-out materials are available. Refer to ANSI/ASTM F104 for further details.

IMPORTANT MATERIAL PROPERTIES FOR GASKETING

The following are properties of the gasket material that are important for sealing performance in the application:

- *Chemical compatibility:* To be resistant to the media being sealed
- *Heat resistance:* To withstand the temperature of the environment
- *Compressibility or macro-conformability:* To conform to the distortions and undulations of the mating flanges
- *Micro-conformability:* To "flow into" the irregularities of the mating flanges' surface finishes
- *Recovery:* To follow the motions of the flanges caused by thermal or mechanical forces
- *Creep relaxation:* To retain sufficient stress for continued sealing over an extended period of time
- *Erosion resistance:* To accommodate fluid impingement in cases where the gasket is required to act as a metering device
- *Compressive strength:* To resist crush and/or extrusion caused by high stresses
- *Tensile or radial strength:* To resist blow-out due to the pressure of the media
- *Shear strength:* To handle the shear motion of the mating flanges due to thermal and mechanical effects of the mating flanges
- *"Z" strength:* To result in easy used-gasket removal without internal fracture of the material
- *Antistick:* To ensure gasket removal without sticking
- *Heat conductivity:* To permit the desired heat transfer of the application
- *Acoustic isolation:* To provide the required noise isolation of the application
- *Dimensional stability:* To permit correct assembly

IMPORTANT GENERAL SHEET CHARACTERISTICS

Sheet materials for gaskets are made by two basic processes:

1. Beater addition
2. Calendering

In the beater addition process, a slurry of water, rubber latex, fibers, and fillers is deposited on a belt and the water is drawn off; a gasket material results. After the water is drawn off, there are voids in the material. These voids must be closed for the gasket to seal. Some enhancements added to gaskets to close these voids are identified later.

In the calendering process, uncured rubber, solvents, fibers, and fillers are squeezed together between two calendering rollers. The compound attaches to one of the rolls and is removed when the proper thickness of the material is achieved. Sheet materials made by the calendering process are called *compressed* products. They do not contain the void content of the beater sheets but many times still need sealing enhancements due to poor clamping load distribution.

To ensure that there will be no health problems for the fabricator and/or end user, the material must be nontoxic. This is a most important characteristic of the material. Another important characteristic of the gasket is sealability. This is not only dependent upon the gasket's material properties; it is also a function of the environmental conditions such as clamp load, bolt span, flange rigidity, etc.

There is another sheet material characteristic that is not necessarily application important, but it is required for general sealing performance. This characteristic is uniformity of the material in regard to consistency of formulation thickness, density, and surface finish:

- Consistency of formulation is vital for compatibility of the material with the medium being sealed, as well as for other reasons.
- Many gaskets have precise compressed thickness requirements since they are shims in addition to being gaskets. Thickness tolerance and density uniformity from lot to lot are important for compressed thickness control.
- Uniformity of density is also very critical for some of the fabricators' added enhancements such as saturating, laminating, and surface coating.
- Surface finish similarity within a lot and from lot to lot is important for surface sealing of the mating flanges.

IMPORTANT MATERIAL CHARACTERISTICS FOR PROCESSING AND/OR ASSEMBLY

There are a number of sheet material characteristics that are important to the fabricator for processing sheets into gaskets and/or for the end user who assembles the gasketed joint. These characteristics are:

Dust and sliver amounts: During blanking of the sheet material, lack of dust is desired since dust causes many problems during processing. It gets into bearings and other machine components, resulting in premature repair, downtime, and reduction of productivity. In addition, dust on the gasket results in adherence difficulties during subsequent operations such as coating and printing. Since the conversion away from asbestos, dust has become a major problem with gasket materials, caused by the replacement of asbestos fibers in the sheets with inorganics. Dust is also a major complaint of workers.

Slivers are also characteristic of nonasbestos material. Slivers are little pieces of the material that occur during blanking. They can cause dents in the gasket, resulting in potential leak paths. Being sliver free is very important to gasket fabricators because sliver removal is costly.

Tool wear and life: The conversion of the asbestos-based sheets to nonasbestos-based sheets has resulted in significant tool wear increases. Reports of decreased tool life abound. Some fabricators indicate a reduction of life by factors of one-third to one-twentieth. The exact reason or reasons for the increased wear are not known, but many gasket manufacturers are conducting investigations of tool wear and the Gasket Fabricators Association also has a study project in this regard.

Scuff resistance: The sheet material's ability to resist scuffing is important to the fabricator. The rough handling and transporting of materials during fabrication can result in scuffing of surfaces of the gaskets, thus rendering their sealing performance questionable.

Breaking strengths: The breaking strength of the material must be sufficient to resist fracture during processing. Some of the processing operations exert considerable tensile pull on the material.

Handling characteristics: The handling characteristics of the gasket material are important during processing, and these same characteristics are also important during assembly of the gasket. Rigidity of the gasket, for example, may be important for ease of assembly and to ensure proper installation. In some cases, the gasket may be installed by robots, and this type of handling has to be taken into account.

NONMETALLIC GASKET MATERIALS

The primary criterion for a material to be impervious to a fluid is to achieve a sufficient density to eliminate voids which might allow capillary flow of the fluid through construction. This requirement may be met in two ways: by compressing the material to fill the voids and/or partially or completely filling them during fabrication by means of binders and fillers. Also, to maintain its impermeability for a prolonged time, the constituents of the material must be able to resist degradation and disintegration resulting from chemical attack and/or temperature of the application.

Most gasket materials are composed of a fibrous or granular base material, forming a basic matrix or foundation, which is held together or strengthened with a binder. The choice of combinations of binder and base material depends on the compatibility of the components, conditions of the sealing environment, and load-bearing properties required for the application. Some of the major constituents and the properties which are related to impermeability are listed below.

Base Materials

Cork and cork-rubber: High compressibility allows easy density increase of the material, thus enabling an effective seal at low flange pressures. The temperature limit is approximately 120°C (250°F) for cork and 150°C (300°F) for cork-rubber compositions. Chemical resistance to water, oil, and solvents is good, but resistance to inorganic acids, alkalies, and oxidizing environments is poor. It conforms well to distorted flanges.

Cellulose fiber: Cellulose has good chemical resistance to most fluids except strong acids and bases. The temperature limitation is approximately 150°C (300°F). Changes in humidity may result in dimensional changes and/or hardening.

Asbestos fiber: This material has good resistance to 425°C (800°F) and is noncombustible. It is almost chemically inert (crocidolite fibers, commonly known as *blue asbestos,* resist even inorganic acids) and has very low compressibility. The binder determines the resistance to temperature and medium that can be sealed. This fiber is only permitted for use in special industrial applications where safety of the gasketed joint is involved and where no other fiber can reliably seal. As mentioned earlier, asbestos has been eliminated from most gasketing materials.

Nonasbestos fibers: A number of nonasbestos fibers are being used in gaskets. Some of these are cellulose, glass, carbon, polyaramids, acrylics, ceramics, and various inorganic fibers. Temperature limits from 400 to 1300°C (750 to 2400°F) are obtainable. Use of these fillers is an emerging field today, and suppliers should be contacted before these fibers are specified for use. The nonasbestos formulations differ considerably from the asbestos formulations that they have replaced.

Rubbers: In addition to natural rubber, there are a wide variety of synthetic rubbers which are used for gaskets. Some of the most common types that are used are:

- Natural, NR
- Buna N, nitrile, NBR
- Buna S, styrene butadiene, SBR
- Neoprene, polycloroprene, CR
- Butyl, isobutylene, IR
- Silicone, silicone and hydrocarbons, MQ
- Hypalon, CSM
- Ethylene propylene, EP and EPDM
- Viton, fluorinated rubber, FKM
- Polyurethane, AU and EU polymers

Plastics: This is a newer variety of gasket materials which generally resist temperatures and corrosive environments better than rubbers. Two of the popular materials are:

- Teflon—polytetrafluoroethylene—developed by DuPont de Nemours
- Kel-F—trifluorochloroethylene—developed by M. W. Kellog Co.

Rubber and plastic material properties are discussed thoroughly in Part 2 of this handbook.

Binders and Fillers

Rubber: Rubber binders provide varying temperature and chemical resistance depending on the type of rubber used. These rubber and rubberlike materials are used as binders and, in some cases, as gaskets:

- *Natural:* Good mechanical properties and is impervious to water and air. It has uncontrolled swell in petroleum oil, fuel, and chlorinated solvents. The temperature limit is 120°C (250°F).
- *Styrene:* Similar to natural rubber but has slightly improved properties. The temperature limit also is 120°C (250°F).
- *Butyl:* Excellent resistance to air and water, fair resistance to dilute acids, and poor resistance to oils and solvents. It has a temperature limit of 150°C (300°F).
- *Nitrile:* Excellent resistance to oils and dilute acids. It has good compression set characteristics and has a temperature limit of 150°C (300°F).
- *Neoprene:* Good resistance to water, alkalies, nonaromatic oils, and solvents. Its temperature limit is 120°C (250°F).

- *Ethylene propylene rubber:* Excellent resistance to hot air, water, coolants, and most dilute acids and bases. It swells in petroleum fuels and oils without severe degradation. The temperature limit is 150°C (300°F).
- *Hypalon:* Excellent resistance to oils and also has good resistance to flame. Its temperature limit is 150°C (300°F).
- *Polyurethane:* Excellent resistance to oils, solvents, and ozone. Its temperature limit is 100°C (212°F).
- *Silicone:* Good heat stability and low-temperature flexibility. It is not suitable for high mechanical pressure. Its temperature limit is 315°C (600°F).
- *Fluoroelastomer:* Good resistance to oils, fuel, and chlorinated solvents. It also has excellent low-temperature properties. Its temperature limit is 315°C (600°F).

Resins: These usually possess better chemical resistance than rubber. Temperature limitations depend on whether the resin is thermosetting or thermoplastic.

Tanned glue and glycerine: This combination produces a continuous gel structure throughout the material, allowing sealing at low flange loading. It has good chemical resistance to most oils, fuels, and solvents. It swells in water but is not soluble. The temperature limit is 95°C (200°F). It is used as a saturant in cellulose paper.

Fillers: In some cases, inert fillers are added to the material composition to aid in filling voids. Some examples of the fillers used are barytes, asbestine, and cork dust.

Sealing Enhancements

Often, a gasket's sealing requirements are such that the produced gasket sheet material cannot accommodate them. In these cases, the gasket fabricator relies on any number of sealing aids to improve the gasket to meet these requirements. Some of these sealing enhancements are:

1. *Saturation:* The voids in the gasket material can be filled with a number of chemical resins. The saturants, in addition to filling the voids, can add improved heat and chemical resistance to the gasket. They also can alter the physical properties of the material.
2. *Coating:* Coatings are applied to the gasket for a variety of reasons. Some of these reasons are:
   - Improve surface seal.
   - Improve antifret characteristics. Antifret is the reduction of scrubbing of the gasket due to flange shearing motion.
   - Reduce or eliminate sticking upon removal.
   - Provide sticking or tack for ease of assembly.
   - Provide the gasket with a barrier coat for subsequent printing (printing is defined below).
   - Reduce sticking of gaskets together during processing and shipping.
   - Provide color to the gasket for various identification reasons.
3. *Eyeletting:* Metal eyelets are used at port openings to protect the gasket material from the sealed media and/or to provide high sealing stress on the eyelet

(see Fig. 1.2). Eyelets can also be used at bolt holes to reduce distortion of the gasketed joint.

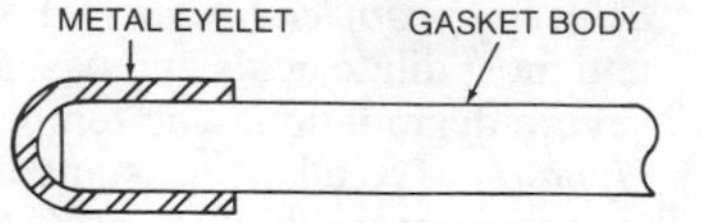

FIGURE 1.2 Eyeletted gasket.

4. *Printing:* A common sealing enhancement used today is printing on the gasket. This is usually done by a version of the silk screening process. In this process, elastomeric beads are deposited on the gasket in strategic locations (see Figs. 1.3 and 1.4). This results in improved sealing at these locations because of the higher resulting stress of the rubberlike sealing interface with the mating flange. In addition, the higher recovery properties of the beads also improve long-term sealing. Various elastomers are used, with silicone being the most popular. Sealing is accomplished by reducing sizes of voids and rubber to metal surface seal.

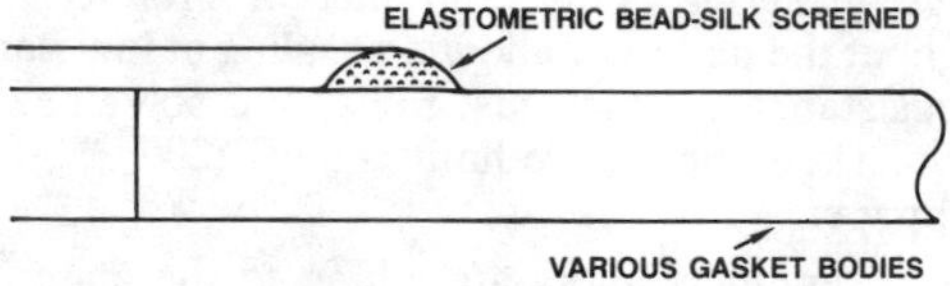

FIGURE 1.3 Printed gasket.

Rigid beads, such as those that result when rigid epoxies are deposited, are sometimes incorporated in gaskets for compression-limiting purposes or for reduction in joint distortion. Use of this technique has virtually changed the look of the engine gasketing industry. While the printing patents were in existence, there were more than 30 worldwide licensees of the technology.

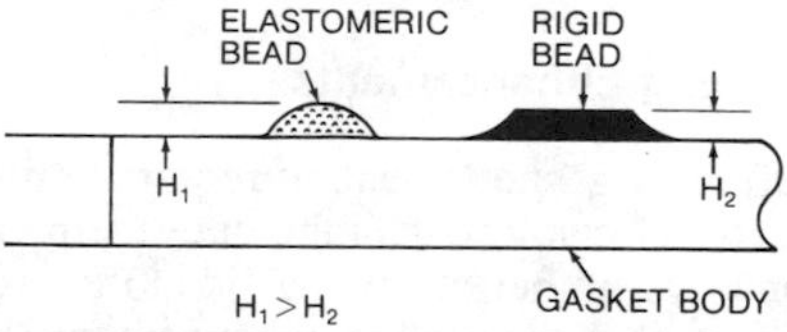

FIGURE 1.4 Double printed gasket.

An extension of this technique is to deposit the bead or beads using a tracer or robot. This permits utilization of more material and increases the thickness possibilities for the beads. Another extension of this technique is to combine it with embossing, where an elastomeric bead is deposited into the emboss, generally with a robot. This combination can be likened to a trapped O-ring. The end product has extensive recovery characteristics.

A variation of the printing process is called *MIP,* for mold in place. An elastomeric bead is molded to the gasket material or gasket. This provides for precision thickness variation of the bead, resulting in exacting three-dimensional gaskets. Liquid injection molded silicone is sometimes used in this regard.

5. *Grommeting:* Grommets in our industry are rubber parts and/or rubber parts which are reinforced with metal or plastic (see Fig. 1.5). They are molded products and are added to the gasket to provide improved sealing at difficult-to-seal locations. Their cross-sections are virtually unlimited and, therefore,

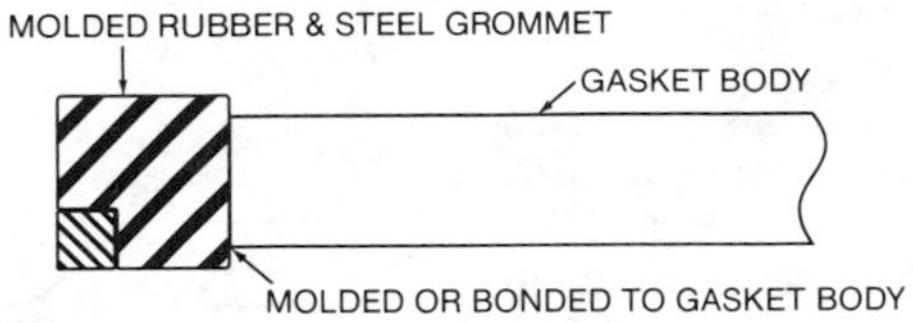

FIGURE 1.5 Grommeted gasket.

permit a large range of design possibilities. Many materials are used for grommets. Most common are nitrile, neoprene, polyacrylic, silicone, and fluoroelastomer.

Reinforcements

Some of the properties of nonmetallic gasket materials can be improved if the gaskets are reinforced with metal or fabric internal cores. Significant improvements in torque retention and blowout resistance are normally seen. Traditionally, perforated or upset metal cores have been used to support gasket facings (see Fig. 1.6). A number of designs have been used for production. The size of the perforations and their frequency in a given area are the usual specified parameters.

FIGURE 1.6 Perforated core.

Adhesives have been developed that permit the use of an unbroken metal core to render support to a gasket facing (see Fig. 1.7). Laminated composites of this type have certain characteristics that are desired in particular gaskets.

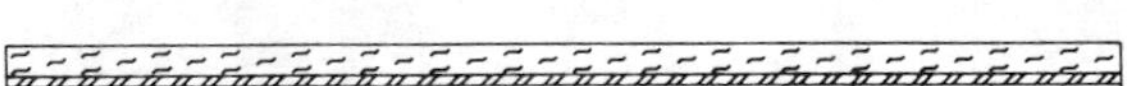

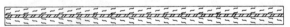

FIGURE 1.7 Solid core.

External metal reinforcements, both perforated skin and solid skin types, have been used (see Figs. 1.8 and 1.9).

FIGURE 1.8 Perforated skins.

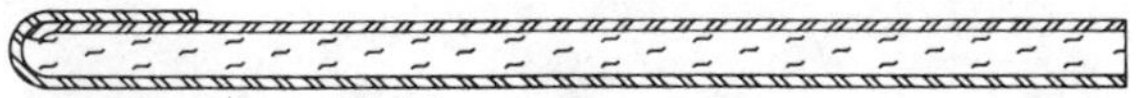

FIGURE 1.9 Solid skin.

Embossing

After a gasket material has been reinforced by combining and/or laminating, it can be embossed (see Fig. 1.10). This results in high stresses at the embossments for localized sealing purposes.

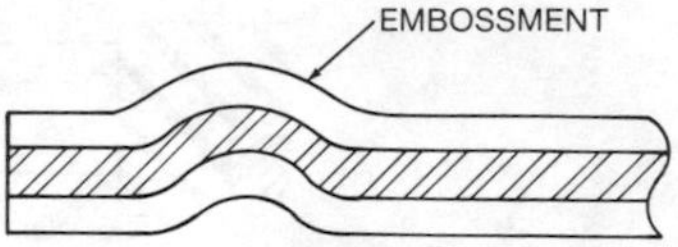

FIGURE 1.10 Unbroken metal core emboss.

METALLIC GASKETS

There are many metals used for gasketing purposes. Some of the most common range from soft varieties such as copper, aluminum, brass, and nickel to the high-alloyed steels. Noble metals such as platinum, silver, and gold also have been used to a limited extent.

Metallic gaskets are available in a wide array of designs that are both standard and custom designs. Some are to be used unconfined while others are used in a confined position. Both elastic and plastic sealing is used. Some metallic designs use the internal pressure to improve the sealing.

Since there is such a wide variety of designs and materials used for metallic gaskets, it is recommended that the reader directly contact metallic gasket suppliers for design and sealing information.

CHAPTER 2
GASKET PROPERTIES

IDENTIFICATION, TEST METHOD, AND SIGNIFICANCE OF GASKET PROPERTIES

Table 2.1 lists some of the most significant gasket properties which are associated with creating and maintaining a seal. This table also shows the test method as well as the significance of each property in a gasket application.

TABLE 2.1 Identification, Test Method, and Significance of Various Properties Associated with Gasket Materials

Property	Test method	Significance in gasket application
Sealability	Fixtures per ASTM F37-62T	Resistance to fluid leakage
Heat resistance	Exposure testing at elevated temperatures	Resistance to thermal degradation
Oil and water immersion characteristics	ASTM D-1170	Resistance to fluid attack
Antistick characteristics	Fixture testing at elevated temperatures	Ability to release from flanges after use
Stress versus compression and spring rates	Various compression test machines	Sealing pressure at various compressions
Compressibility and recovery	ASTM F36-61T	Ability to follow deformation and deflection
Creep relaxation and compression set	ASTM F38-62T and D-395-59	Related to torque loss and subsequent loss of sealing pressure
Crush and extrusion characteristics	Compression test machines	Resistance to high loadings and extrusion characteristics at room and elevated temperatures

LOAD-BEARING PROPERTIES

Conformability and Recovery

Since sealing conditions vary widely depending on the application, it is necessary to vary the load-bearing properties of the gasket elements in accordance with these conditions. Figure 2.1 illustrates stress compression curves for several gasket components and indicates the difference in the stress compression properties used for different sealing locations.

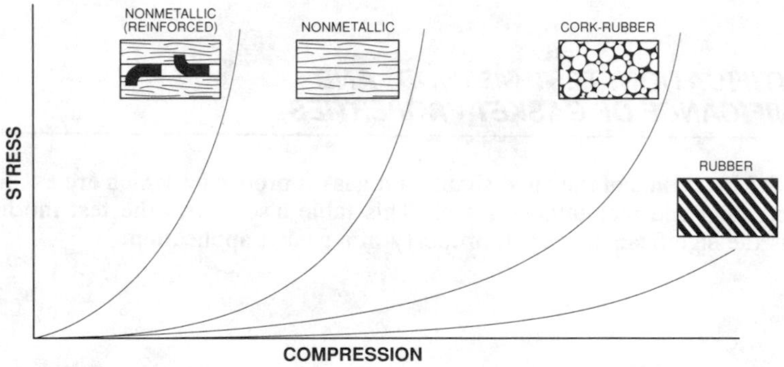

FIGURE 2.1 Stress versus compression for various gasket materials.

Gasket thickness and compressibility must be matched to the rigidity, roughness, and unevenness of the mating flanges. A complete seal can only be achieved if the stress level imposed on the gasket at clampup is adequate for the specific material. Minimum seating stresses for various gasket materials should be obtained from the material fabricator or gasket supplier. In addition, the load remaining on the gasket during operation must remain high enough to prevent blowout of the gasket. During operation, the hydrostatic end force, which is associated with the internal pressure, tends to unload the gasket. Figure 2.2 is a graphical representation of a gasketed joint; it shows the effect of the hydrostatic end force. The bolt needs to be capable of handling the maximum load imposed on it without yielding. The gasket must be capable of sealing at the minimum load resulting on it and resist blowout at this load level.

Gaskets fabricated from compressible materials should be as thin as possible. The thickness should be no greater than that which is necessary for the gasket to conform to the unevenness of the mating flanges. The unevenness is associated with surface finish, flange flatness, and flange warpage during use. It is important to use the gasket's unload curve in considering its ability to conform. Figure 2.3 shows typical load compression and unload curves for nonmetallic gaskets. The unload curve determines the recovery characteristic of the gasket which is required for conformance. Metallic gaskets will show no change in their load and unload curves unless yielding occurs. Load compression curves are available from gasket suppliers.

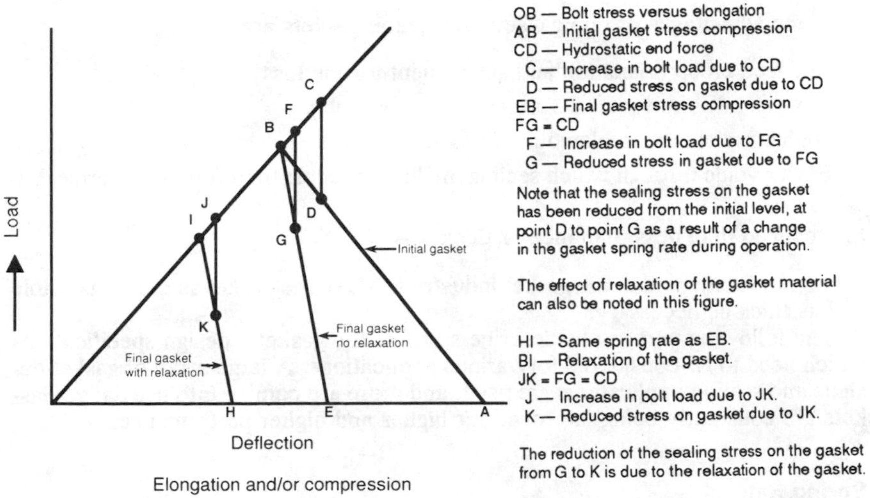
OB — Bolt stress versus elongation
AB — Initial gasket stress compression
CD — Hydrostatic end force
C — Increase in bolt load due to CD
D — Reduced stress on gasket due to CD
EB — Final gasket stress compression
FG = CD
F — Increase in bolt load due to FG
G — Reduced stress in gasket due to FG
Note that the sealing stress on the gasket has been reduced from the initial level, at point D to point G as a result of a change in the gasket spring rate during operation.
The effect of relaxation of the gasket material can also be noted in this figure.
HI — Same spring rate as EB.
BI — Relaxation of the gasket.
JK = FG = CD
J — Increase in bolt load due to JK.
K — Reduced stress on gasket due to JK.
The reduction of the sealing stress on the gasket from G to K is due to the relaxation of the gasket.
Load
Deflection
Elongation and/or compression
Initial gasket
Final gasket no relaxation
Final gasket with relaxation
A
B
C
D
E
F
G
H
I
J
K

FIGURE 2.2 Graphical representation of a gasketed joint and effect of hydrostatic end force.

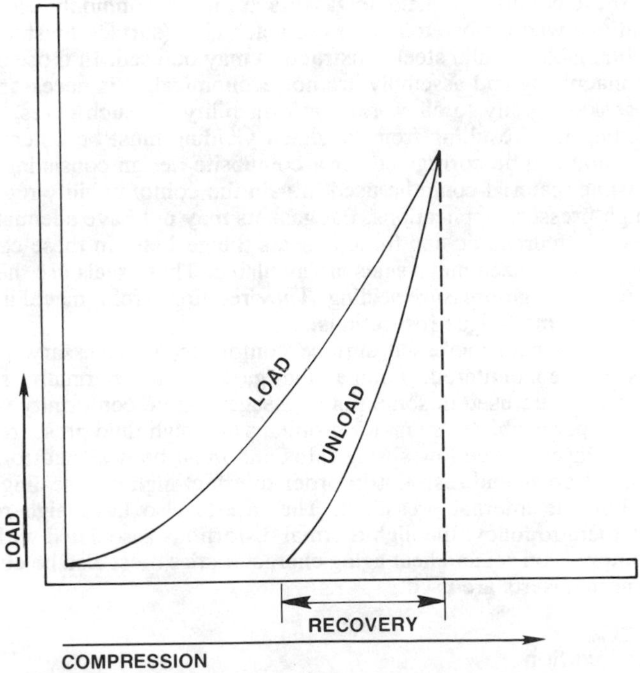
LOAD
UNLOAD
LOAD
RECOVERY
COMPRESSION

FIGURE 2.3 Load versus compression of a typical gasket material.

Some advantages of thin gaskets over thick gaskets are:

1. Reduced creep relaxation and subsequent torque loss
2. Less distortion of mating flanges during assembly
3. Higher resistance to blowout
4. Fewer voids through which sealing media can enter; therefore, less permeability
5. Reduced thickness tolerance or thickness

A common statement in the gasket industry is Make the gasket as thin as possible and as thick as necessary.

The following paragraphs describe some of the gasket's design specifications which need to be considered for various applications. A large array of gasket designs and sealing applications are used, and more are coming into use daily. Gaskets are constantly being improved for higher and higher performance.

Spring Rate

In high-pressure, clamp load, and temperature applications, a high spring rate (stress per unit compression) material is necessary to achieve high loading at low compression in order to seal the high pressures involved. These applications generally rely on sealing resulting from localized yielding under the unit loading. In addition to the high spring rate, high heat resistance is mandatory. To economically satisfy these conditions, metallic gaskets are most commonly used.

In applications where close tolerances in machining (surface finish and parallelism) are obtainable, a solid steel construction may be used. In those situations where close machining and assembly are not economical, it is necessary to sacrifice some gasket rigidity to allow for conformability. In such cases, conformability exceeding that resulting from localized yielding must be inherent in the design. The metal can be corrugated, or a composite design consisting of metal and compressible material could be used to gain the conformability required.

In very high-pressure applications, flat gaskets may not have adequate recovery to seal as the hydrostatic end force unseats the gaskets. In these cases, various types of self-energized metal seals are available. These seals use the internal pressure to achieve high-pressure sealing. They require careful machining of the flanges and have some fatigue restrictions.

In applications where increased surface conformity is necessary and lower temperatures are encountered, nonmetallic gaskets are normally specified. Elastomeric inserts are used in some fluid passages where conformity with sealing surfaces and permeability are major problems and high fluid pressures are encountered. The inserts have low spring rates and must be designed to have appropriate contact areas and restraint in order to effect high unit sealing stresses for withstanding the internal pressures. The inserts also have high recovery, which allows them to follow the high thermal distortions associated with the application. Compression set and heat aging characteristics must also be considered when elastomeric inserts are used.

Creep and Relaxation

After the initial sealing stress is applied to a gasket, it is necessary to maintain a sufficient stress for the designed life of the unit or equipment. All materials exhibit, in varying degrees, a decrease in applied stress as a function of time, commonly referred to as *stress relaxation.* The reduction of stress on a gasket is actually a combination of two major factors: stress relaxation and creep (compression drift). By definition:

Stress relaxation is a change in stress(s) on a specimen under constant strain *e* (*ds/dt*; *e* = constant).

Creep (compression drift) is a change in strain of a specimen under constant stress (*de/dt*; *s* = constant).

In a gasketed joint, stress is applied by tension in a bolt or stud and transmitted as a compressive force to the gasket. After loading, stress relaxation and creep occur in the gasket, causing corresponding lower strain and tension in the bolt. This process continues indefinitely as a function of time. The change in tension of a bolt is related to the often cited "torque loss" associated with a gasket application. Since the change in stress is due to two primary factors, a more accurate description of the phenomenon would be creep relaxation, hereafter called relaxation.

Bolt elongation, or stretch, is linearly proportional to bolt length. The longer the bolt, the higher the elongation. The higher the elongation, the lower the percentage loss for a given relaxation. Therefore, the bolts should be made as long as possible for best torque retention.

Relaxation in a gasket material may be measured by applying a load on a specimen by means of a strain gauged bolt-nut-platen arrangement as standardized by ASTM F38-62T. Selection of materials with good relaxation properties will result in the highest retained torque for the application. This results in the highest remaining stress on the gasket which is desirable for long-term sealing.

The amount of relaxation increases as thickness is increased for a given gasket material. This is another reason why the thinnest gasket that will work should be selected. Figure 2.4 shows the relaxation characteristics as a function of thick-

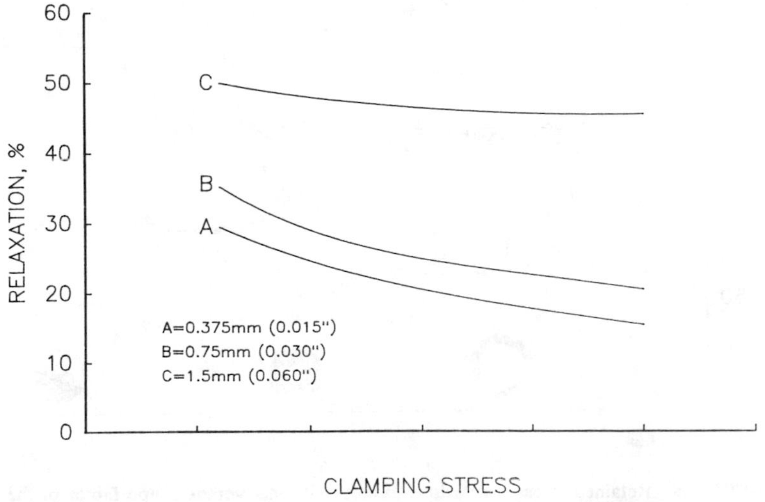

FIGURE 2.4 Relaxation versus stress for various thicknesses of gaskets.

ness and stress level for a particular gasket design. Note that as clamping stress is increased, relaxation is decreased. This is the result of more voids being eliminated as the stress level is increased.

Effect of Gasket Shape

The gasket's shape factor has an important effect on its relaxation characteristics. This is particularly true in the case of the highly compressible materials. Much of the relaxation of a material may be attributed to the releasing of forces through lateral expansion, or bulging. Therefore, the greater the area available for lateral expansion, the greater the relaxation. The shape factor of a gasket is the ratio of the area of one load face to the area free to bulge. For circular or annular samples, this may be expressed as:

$$\text{Shape factor} = \frac{1}{4t}(\text{O.D.} - \text{I.D.})$$

where t = thickness of gasket
O.D. = outside of diameter
I.D. = inside diameter

As the area to bulge increases, the shape factor decreases, the relaxation increases, and the retained stress decreases. Figure 2.5 shows the effect of shape factor on the gasket's ability to retain stress.

As can be noted in the shape factor equation, the shape factor decreases with increasing thickness. This is another reason why the gasket should be as thin as

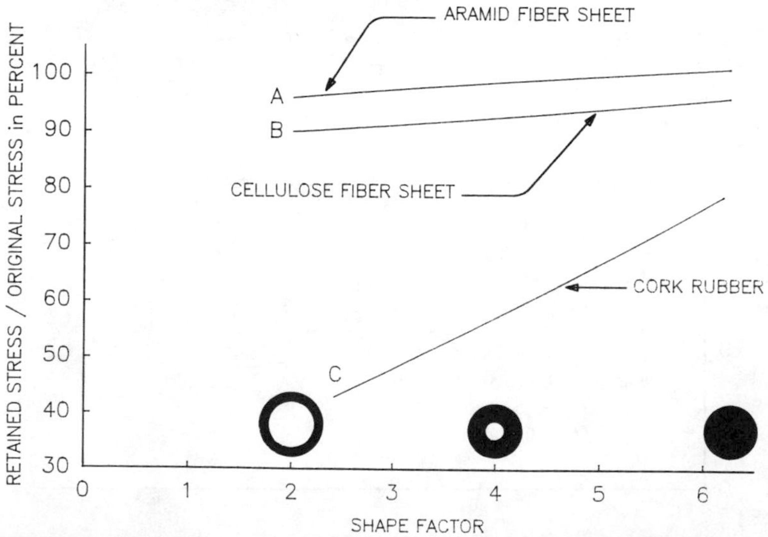

FIGURE 2.5 Retained stress for various gasket materials versus shape factor of the gasket.

possible. It must be thick enough, however, to permit adequate conformity. The clamp area should be as large as possible, consistent with seating stress requirements. Often designers reduce gasket width and subsequent clamp area, thereby increasing gasket clamping stress to obtain better sealing. Remember, however, that this reduction will decrease the gasket's shape factor, which may result in higher relaxation. Therefore, there is a compromise that must be made when a gasket's clamp area is reduced.

Environmental Conditions

Many environmental conditions and factors influence the sealing performance of gaskets. Flange design details, in particular, are most important. Design details such as number, size, length, and spacing of clamping bolts and flange properties such as thickness, modulus, surface finish, waviness, and flatness are important factors. In particular, flange bowing is a most common type of problem associated with the sealing of a gasketed joint. The amount of bowing can be reduced by reducing the bolt spacing. For example, if the bolt spacing were cut in half, the bowing would be reduced to one-eighth of its original value. Doubling the flange thickness could also reduce bowing to one-eighth of its original value. A method of calculating the minimum stiffness required in a flange is available.

Different gasket materials and types require different surface finishes for optimum sealing. Soft gaskets such as rubber can seal very rough surface service finishes in the vicinity of 12.5 micrometers, or 12.5 μm (500 microinches, or μin) whereas some metallic gaskets may require finishes in the range of 0.8 μm (32 μin) for best sealing. Most gaskets, however, will seal adequately in the surface finish range of 1.6 to 3.2 μm (63 to 125 μin). There are two main reasons for the surface finish differences:

1. The gasket must be able to conform to the roughness for surface sealing.
2. It must have adequate conformability into the mating flange to create frictional forces and thereby resist radial motion due to the internal pressure. This is necessary to prevent blow-out.

In addition, elimination of the radial motion will result in maintaining the initial clampup sealing condition. The radial motion is micro in amount but can result in localized fretting, and a leakage path may be created.

As a result of the gasket sealing complexity due to the wide variety of environmental conditions, some trial and error gasket designs for specific applications result. Information on gasket design and selection is supplied in Chap. 3. This information will enable a designer to minimize the chance for leaks. Since the gasketed joint is so complex, adherence to the procedure will not ensure adequate performance in all cases. When inadequate gasket performance occurs, gasket manufacturers should be contacted for assistance.

TEST PROCEDURES FOR GASKET MATERIALS

American Society of Testing Materials (ASTM)

The following is a list of the ASTM Standard Test Methods for gasket materials. Following the list are the ASTM Standard Classifications and the scopes of each of the test methods. The ASTM is located at 1916 Race Street in Philadelphia, PA 19103.

F36	Test Methods for Compressibility and Recovery of Gasket Materials
F37	Test Method for Sealability of Gasket Materials
F38	Test Method for Creep Relaxation of a Gasket Material
F112	Test Method for Sealability of Enveloped Gaskets
F145	Recommended Practice for Evaluating Flat-Face Gasketed Joint Assemblies
F146	Test Method for Fluid Resistance of Gasket Materials
F147	Test Method for Flexibility of Nonmetallic Gasket Materials Containing Asbestos or Cork
F148	Test Method for Binder Durability of Cork Composition Gasket Materials
F152	Method for Tension Testing of Nonmetallic Gasket Materials
F363	Method for Corrosion Testing of Enveloped Gaskets
F433	Recommended Practice for Evaluating Thermal Conductivity of Gasket Materials
F434	Method for Blow-Out Testing of Preformed Gaskets
F495	Test Method for Ignition Loss of Gasket Materials Containing Inorganic Substances
F586	Test Method for Leak Rates Versus "y" Stresses and "m" Factors for Gaskets
F607	Test Method for Adhesion of Gasket Materials to Metal Surfaces
F806	Test Method for Compressibility of Laminated Composite Gasket Materials
F1087	Linear Dimensional Stability of a Gasket Material to Moisture

The following are test methods for vulcanized elastomers:

D395	Compression Set, Method "B"
D412	Elongation
D412	Tensile Modulus @ 100%, 200%, 300%
D412	Tensile Strength Die "C" (common)
D429	Adhesion Bond Strength Method "A"
D430	Flex Resistance
D471	Fluid Resistance Aqueous Fuels, Oils and Lubricants
D573	Heat Age or Heat Resistance
D575	Compression Deflection, Method "A"
D624	Tear Strength, Die "B" or "C"
D813	Crack Growth
D865	Deterioration by Heating
D945	Resilience
D2632	Resilience
D1053	Low Temperature Torsional
D1171	Ozone or Weather Resistance

D1329 Low Temperature Retraction
D1418 Elastomer Classification
D2000 SAE J200
D2137 Low Temperature Resistance "A"
D2240 Hardness Durometer "A" and "D"

Standard Classifications

ASTM F104—Classification System for Nonmetallic Gasket Materials. This classification system provides a means for specifying or describing pertinent properties of commercial nonmetal gasket materials. Materials composed of asbestos, cork, cellulose, and other organic or inorganic materials in combination with various binders or impregnants are included. Materials normally classified as rubber compounds are not included, since they are covered in Method D2000. Gasket coatings are not covered, since details thereof are intended to be given on engineering drawings or in separate specifications.

Since all the properties that contribute to gasket performance are not included, use of the classification system as a basis for selecting materials is limited.

ASTM F868—Classification for Laminated Composite Gasket Materials. This classification system provides a means for specifying or describing pertinent properties of commercial laminate composite gasket materials (LCGM). These structures are composed of two or more chemically different layers of material. These materials may be organic or inorganic or combinations with various binders or impregnants. Gasket coatings are not covered since details thereof are intended to be given on engineering drawings or as separate specifications. Commercial materials designated as envelope gaskets are excluded from this standard.

Since all properties that contribute to gasket performance are not included, use of this classification system as a basis for selecting LCGM is limited.

Scope of Procedure—Standard Test Methods

ASTM F36—Test Method for Compressibility and Recovery of Gasket Materials. This test method covers determination of the short-time compressibility and recovery at room temperature of sheet-gasket material and, in certain cases, gaskets cut from sheets. It is not intended as a test for compressibility under prolonged stress application, generally referred to as *creep* or for recovery following such prolonged stress application, the inverse of which is generally referred to as *compression sets.* Also, it is not intended for tests at other than room temperature.

ASTM F37—Test Method for Sealability of Gasket Materials. These test methods provide a means of evaluating fluid sealing properties of gasket materials at room temperature. Method A is restricted to liquid measurements and method B may be used for both gas and liquid leakage measurements.

These methods are suitable for evaluating the seal characteristics of a gasket material under differing compressive flange load. When desired, the method may be used as an acceptance test when the following test conditions are agreed upon between a supplier and purchaser: fluid, internal pressure on fluid, and flange load on gasket specimen.

ASTM F38—Test Method for Creep Relaxation of a Gasket Material. These test methods provide a means of measuring the amount of creep relaxation of a gasket material at a stated time after a compressive stress has been applied:

Method A: Creep relaxation measured by means of a calibrated strain gauge on a bolt

Method B: Creep relaxation measured by means of a calibrated bolt with dial indicator

ASTM F112—Test Method for Sealability of Enveloped Gaskets. This test method covers the evaluation of the sealing properties of enveloped gaskets for use with corrosion-resistant process equipment. Enveloped gaskets are described as gaskets having some corrosion-resistant covering over the internal area normally exposed to the corrosive environment. The shield material may be plastic (such as polytetrafluoroethylene) or metal (such as tantalum). A resilient conformable filler is usually used inside the envelope.

ASTM F145—Recommended Practice for Evaluating Flat-Face Gasketed Joint Assemblies. This practice permits measurement of gasket compression resulting from bolt loading on a flat-face joint assembly at ambient conditions.

ASTM F146—Test Method for Fluid Resistance of Gasket Materials. These test methods cover the determination of the effect on physical properties of nonmetallic gasketing materials after immersion in test fluids. The types of materials covered are those containing asbestos and other inorganic fibers (type 1), cork (type 2), and cellulose or other organic fiber (type 3) as described in Classification F104. These methods are not applicable to the testing of vulcanized rubber, a method that is described in Test Method D471. It is designed for testing specimens cut from gasketing materials or from finished articles of commerce.

This standard may involve hazardous materials, operations, and equipment. This standard does not purport to address all of the safety problems associated with its use. It is the responsibility of whoever uses this standard to consult with appropriate authorities and establish appropriate safety and health practices and determine the applicability of regulatory limitations prior to use.

ASTM F147—Test Method for Flexibility of Nonmetallic Gasket Materials Containing Asbestos or Cork. This test method covers the determination of the flexibility of nonmetallic gasket materials. It is designed for testing specimens cut from sheet goods or from the gasket in the finished form, as supplied for commercial use. Materials normally classified as elastomeric compounds are excluded since they are covered in Classification D2000.

ASTM F148—Test Method for Binder Durability of Cork Composition Gasket Materials. This test method covers three procedures for determination of the binder durability of cork-containing materials.

This standard may involve hazardous materials, operations, and equipment. This standard does not purport to address all the safety problems associated with its use. It is the responsibility of whoever uses this standard to consult with appropriate authorities and establish appropriate safety and health practices and determine the applicability of regulatory limitations prior to use.

ASTM F152—Method for Tension Testing of Nonmetallic Gasket Materials. These test methods cover the determination of tensile strength of certain nonmetallic gasketing materials at room temperature. The types of materials covered are those pertaining to asbestos and other inorganic fibers (type 1), cork (type 2), and cellulose or other organic fiber (type 3) as described in Method F104. These methods are not applicable to the testing of vulcanized rubber (see Method D412) or for rubber O-rings (see Method D1414).

ASTM F363—Method for Corrosion Testing of Enveloped Gaskets. This test method covers the evaluation of gaskets under corrosive conditions at varying temperature and pressure levels. The test unit may be glass lined if the flanges are sufficiently plane (industry accepted), thus providing resistance to all chemicals, except hydrofluoric acid, from cryogenic temperatures to 260°C (500°F) at pressures from full vacuum to the allowable pressure rating of the unit, or they may be made of other suitable material. The test unit described in the standard has an internal design pressure rating of 1034 kPa (150 psi) at 260°C.

ASTM F433—Recommended Practice for Evaluating Thermal Conductivity of Gasket Materials. This practice covers a means of measuring the amount of heat transfer quantitatively through a material or system. This recommended practice is similar to the Heat Flow Meter System of Method C518, but it is modified to accommodate small test samples of higher thermal conductance.

ASTM F434—Method for Blow-Out Testing of Preformed Gaskets. This test method covers the determination of the resistance against blow-out of preformed gaskets. The test is conducted under ambient conditions and should be used for comparison purposes only to select suitable designs and constructions for specific applications.

This standard may involve hazardous materials, operations, and equipment. This standard does not purport to address all of the safety problems associated with its use. It is the responsibility of whoever uses this standard to consult with appropriate authorities and establish appropriate safety and health practices and determine the applicability of regulatory limitations prior to use.

ASTM F495—Test Method for Ignition Loss of Gasket Materials Containing Inorganic Substances. This test method covers the determination of gasket material weight loss upon exposure to elevated temperatures.

ASTM F586—Test Method for Leak Rates Versus "y" Stresses and "m" Factors for Gaskets. This test method covers the determination of leak rates versus *y* stresses and *m* factors for gaskets gripped by pressure-containing flanged connections.

ASTM F607—Test Method for Adhesion of Gasket Materials to Metal Surfaces. This test method provides a means of determining the degree to which gasket materials under compressive load adhere to metal surfaces.

The test method may be employed for the determination of adhesion. The test conditions described are indicative of those frequently encountered in gasket application. Test conditions may also be modified in accordance with the needs of specific applications as agreed upon between the user and the producer.

This standard may involve hazardous materials, operations, and equipment. This standard does not purport to address all of the safety problems associated with its use. It is the responsibility of whoever uses this standard to consult with appropriate authorities and establish appropriate safety and health practices and determinc the applicability of regulatory limitations prior to use.

ASTM F806—Test Method for Compressibility and Recovery of Laminated Composite Gasket Materials. This test method covers determination of the short-term compressibility and recovery at room temperature of laminated composite gasket materials.

This test method is not intended as a test for compressibility under prolonged stress application, that is creep, or for recovery following such prolonged stress application, the inverse of which is generally referred to as compression set. Also, it is only intended for tests at room temperature.

This standard may involve hazardous materials, operations, and equipment. This standard does not purport to address all of the safety problems associated with its use. It is the responsibility of whoever uses this standard to consult with appropriate authorities and establish appropriate safety and health practices and determinc the applicability of regulatory limitations prior to use.

ASTM F1087—Linear Dimensional Stability of a Gasket Material to Moisture. This method covers a procedure to determine the stability of a gasket material to linear dimensional change due to hydroscopic expansion and contraction. It subjects a sample to extremes (i.e., oven drying and complete immersion in water), which have shown good correlation to low and high relative humidities.

Scope of Procedure—Test Methods for Vulcanized Elastomers

ASTM D395—Compression Set, Method "B." These test methods cover the testing of rubber intended for use in applications in which the rubber will be subjected to compressive stresses in air or liquid media. They are applicable particularly to the rubber used in machinery mountings, vibration dampers, and seals. Two methods are covered as follows:

Method	Section
A—Compression Set Under Constant Force in Air	7.10
B—Compression Set Under Constant Deflection in Air	11.14

The choice of method is optional, but consideration should be given to the nature of the service for which correlation of test results may be sought. Unless otherwise stated in a detailed specification, Method B shall be used. Method B is not suitable for vulcanizates harder than 90 IRHD. The values stated in SI units are to be regarded as the standard.

This standard may involve hazardous materials, operations, and equipment. This standard does not purport to address all of the safety problems associated with its use. It is the responsibility of whoever uses this standard to consult with appropriate authorities and establish appropriate safety and health practices and determine the applicability of regulatory limitations prior to use.

ASTM D412—Elongation. These test methods cover tension testing of rubber at various temperatures. These methods are not applicable to the testing of ebonite and similar hard, low-elongation materials. The method appears as follows:

Method	Section
A—Dumbbell and Straight Specimens	9 to 13
B—Cut Ring Specimens	14 to 18

The values stated in either SI or non-SI units shall be regarded separately as standard. The values in each system may not be exact equivalents; therefore, each system must be used independently of the other, without combining values in any way.

This standard may involve hazardous materials, operations, and equipment. This standard does not purport to address all of the safety problems associated with its use. It is the responsibility of whoever uses this standard to consult with appropriate authorities and establish appropriate safety and health practices and determine the applicability of regulatory limitations prior to use.

ASTM D412—Tensile Modulus @ 100%, 200%, 300%. These test methods cover tension testing of rubber at various temperatures. The methods are not applicable to the testing of ebonite and similar hard, low-elongation materials. The methods appear as follows:

Method	Section
A—Dumbbell and Straight Specimens	9 to 13
B—Cut Ring Specimens	14 to 18

The values stated in either SI or non-SI units shall be regarded separately as standard. The values in each system may not be exact equivalents; therefore, each system must be used independently of the other, without combining values in any way.

This standard may involve hazardous materials, operations, and equipment. This standard does not purport to address all of the safety problems associated with its use. It is the responsibility of whoever uses this standard to consult with appropriate authorities and establish appropriate safety and health practices and determine the applicability of regulatory limitations prior to use.

ASTM D412—Tensile Strength, Die "C" (common). These test methods cover tension testing of rubber at various temperatures. The methods are not applicable to the testing of ebonite and similar hard, low-elongation materials. The methods appear as follows:

Method	Section
A—Dumbbell and Straight Specimens	9 to 13
B—Cut Ring Specimens	14 to 18

The values stated in either SI or non-SI units shall be regarded separately as standard. The values in each system may not be exact equivalents; therefore, each system must be used independently of the other, without combining values in any way.

This standard may involve hazardous materials, operations, and equipment. This standard does not purport to address all of the safety problems associated with its use. It is the responsibility of whoever uses this standard to consult with appropriate authorities and establish appropriate safety and health practices and determine the applicability of regulatory limitations prior to use.

ASTM D429—Adhesion Bond Strength, Method "A." These test methods cover procedures for testing the static adhesional strength of rubber to rigid materials (in most cases metals).

Method A—Rubber Part Assembled Between Two Parallel Metal Plates

Method B—90° Stripping Test—Rubber Part Assembled to One Metal Plate

Method C—Measuring Adhesion of Rubber to Metal with a Conical Specimen

Method D—Adhesion Test—Post-Vulcanization (PV) Bonding of Rubber to Metal

Method E—90° Stripping Test—Rubber Tank Lining—Assembled to One Metal Plate

While the method may be used with a wide variety of rigid materials, such materials are the exception rather than the rule. For this reason, we have used the word *metal* in the text rather than *rigid materials.*

ASTM D430—Flex Resistance. These test methods cover testing procedures that estimate the ability of soft rubber compounds to resist dynamic fatigue. No exact correlation between these test results and service is given or implied. This is due to the varied nature of service conditions. These test procedures do yield data that can be used for the comparative evaluation of rubber compounds or composite rubber-fabric materials for their ability to resist dynamic fatigue.

ASTM D471—Fluid Resistance—Aqueous Fuels, Oils, and Lubricants. This test method measures the comparative ability of rubber and rubberlike composition to withstand the effect of liquids. It is designed for testing specimens of elastomeric vulcanizates cut from standard sheets (see Recommended Practice D3182), specimens cut from fabric coated with elastomeric vulcanizates (see Method D751), or finished articles of commerce (see Practice D3183). The method is not applicable to the testing of cellular rubbers, porous compositions, and compressed asbestos sheet except as provided in Note 5 of the standard.

In view of the wide variations often present in service conditions, this accelerated test may not give any direct correlation with service performance. However, the method yields comparative data on which to base judgment about expected service quality and is especially useful in research and development work.

ASTM D573—Heat Age or Heat Resistance. This test method describes a procedure to determine the influence of elevated temperature on the physical properties of vulcanized rubber. The results of this test may not give an exact correlation with service performance since performance conditions vary widely. The test may, however, be used to evaluate rubber compounds on a laboratory comparison basis.

ASTM D575—Compression Deflection, Method "A." These test methods describe two test procedures for determining the compression-deflection characteristics or rubber compounds other than those usually classified as hard rubber and sponge rubber.

ASTM D624—Tear Strength, Die "B" or "C." This test method covers the determination of the tear resistance of vulcanized rubber. It does not apply to testing of hard rubber. The values stated in SI units are to be regarded as the standard.

This standard may involve hazardous materials, operations, and equipment. This standard does not purport to address all of the safety problems associated with its use. It is the responsibility of whoever uses this standard to consult with appropriate authorities and establish appropriate safety and health practices and determine the applicability of regulatory limitations prior to use.

ASTM D813—Crack Growth. This test method covers the determination of crack growth of vulcanized rubber when subjected to repeated bend flexing. It is particularly applicable to tests of synthetic rubber compounds which resist the initiation of cracking due to flexing when tested by Method B or Method D430. Cracking initiated in these materials by small cuts or tears in service may rapidly increase in size and progress due to complete failure even though the material is extremely resistant to the original flexing-fatigue cracking. Because of this characteristic of synthetic compounds, particularly those of the styrene (SBR) type, this method in which the specimens are first artificially punctured in the flex area should be used in evaluating the fatigue-cracking properties of this class of material.

ASTM D865—Deterioration by Heating. This test method describes a procedure to determine the deterioration induced by heating rubber specimens in individual test tube enclosures with circulating air. This isolation prevents cross-contamination of compounds due to loss of volatile materials (for example, antioxidants) and their subsequent migration into other rubber compounds (specimens). The absorption of such volatile materials may influence the degradation rate of rubber compounds.

ASTM D945—Resilience. These test methods cover the use of the Yerrzley mechanical oscillograph for measuring mechanical properties of elastomeric vulcanizates in the generally small range of deformation that characterizes many technical applications. These properties include resilience, dynamic modulus, static modulus, kinetic energy, creep, and set under a given dead load. Measurements in compression and shear are described.

The test is applicable primarily, but not exclusively, to materials having static moduli at the test temperature such that loads below 2 MPa (280 psi) in compression or 1 MPa (140 psi) in shear will produce 20 percent deformation and having resilience such that at least three complete cycles are produced when obtaining the damped oscillatory curve. The range may be extended, however, by use of supplementary masses and refined methods of analysis. Materials may be compared either under comparable mean stress or mean strain conditions.

ASTM D2632—Resilience. This test method covers the determination of impact resilience of solid rubber from measurement of the vertical rebound of a dropped mass. The method is not applicable to the testing of cellular rubbers or coated fabrics.

Note 1—A standard method of test for impact resilience and penetration of rubber by a rebound pendulum is described in ASTM Method D1054, Test for Rubber Property-Resilience Using a Rebound Pendulum.

ASTM D1053—Low Temperature Torsional. These test methods describe the use of a torsional apparatus for measuring the relative low-temperature stiffening of flexible polymeric materials and fabrics coated therewith. A routine inspection and acceptance procedure, to be used as a pass-fail test at a specified temperature, is also described.

These test methods yield comparative data to access the low-temperature performance of flexible polymers and fabrics coated therewith. The values stated in either SI or non-SI units shall be regarded separately as the standard. The values in each system may not be exact equivalents; therefore, each system must be used independently of the other, without combining values in any way.

This standard may involve hazardous materials, operations, and equipment. This standard does not purport to address all of the safety problems associated with its use. It is the responsibility of whoever uses this standard to consult with appropriate authorities and establish appropriate safety and health practices and determine the applicability of regulatory limitations prior to use.

ASTM D1171—Ozone or Weather Resistance. This test method permits the estimation of the relative ability of rubber compounds used for applications requiring resistance to outdoor weathering or ozone chamber testing.

This test method is not applicable to materials ordinarily classed as hard rubber, but it is adaptable to molded or extruded soft rubber material and sponge rubber for use in window weather stripping as well as similar automotive applications.

This standard may involve hazardous materials, operations, and equipment. This standard does not purport to address all of the safety problems associated with its use. It is the responsibility of whoever uses this standard to consult with appropriate authorities and establish appropriate safety and health practices and determine the applicability of regulatory limitations prior to use.

ASTD D1329—Low-Temperature Retraction. This test method covers a temperature-retraction procedure for rapid evaluation of crystallization effects and for comparing viscoelastic properties of rubber and rubberlike materials at low temperatures. This test is useful when employed in conjunction with other low-temperature tests for selection for materials suitable for low-temperature service. It is also of value in connection with research and development, but it is not yet considered sufficiently well established for use in purchase specification.

ASTM D1418—Elastomer Classification. This practice establishes a system of general classification for the basic rubbers both in dry and latex forms determined from the chemical composition of the polymer chain.

The purpose of this practice is to provide a standardization of terms for use in industry, commerce, and government and is not intended to conflict with but rather to act as a supplement to existing trade names and trademarks.

In technical papers or presentations the name of the polymer should be used if possible. The symbols can follow the chemical name for use in later references.

This standard may involve hazardous materials, operations, and equipment. This standard does not purport to address all of the safety problems associated with its use. It is the responsibility of whoever uses this standard to consult with appropriate authorities and establish appropriate safety and health practices and determine the applicability of regulatory limitations prior to use.

ASTM D2000, SAE J200—Common Elastomer Specification for Auto and Truck Components. This classification system tabulates the properties of vulcanized rubber materials (natural rubber, reclaimed rubber, synthetic rubbers, alone or in combination) that are intended for, but not limited to, use in rubber products for automotive applications.

Note 1—This classification system may serve many of the needs of other industries in much the same manner as SAE numbered steels. It must be remembered, however, that this system is subject to revision when required by automotive needs. It is recommended that the latest revision always be used.

This classification system is based on the premise that the properties of all rubber products can be arranged into characteristic material designations. These designations are determined by types, based on resistance to heat aging, and classes, based on resistance to swelling in oil. Basic levels are thus established which, together with values describing additional requirements, permit complete description of the quality of all elastomeric materials.

In all cases where the provisions of this classification system would conflict with those of the detailed specifications for a particular product, the latter shall take precedence.

ASTM D2137—Low Temperature Resistance "A." These test methods cover the determination of the lowest temperature at which rubber vulcanizates and rubber-coated fabrics will not exhibit fractures or coating cracks when subjected to specified impact conditions.

ASTM D2240—Hardness Durometer "A" and "D." This test method covers two types of durometers, A and D, and the procedure for determining the indentation hardness of homogeneous materials ranging from soft vulcanized rubber to some rigid plastics.

This test method is not applicable to the testing of coated fabrics. The values stated in SI units are to be regarded as the standard. This standard may involve hazardous materials, operations, and equipment. This standard does not purport to address all of the safety problems associated with its use. It is the responsibility of whoever uses this standard to consult with appropriate authorities and establish appropriate safety and health practices and determine the applicability of regulatory limitations prior to use.

Materials Technology Institute of the Chemical Process Industries (MTI) and the Pressure Vessel Research Council (PVRC)

Listed below are gasket tests that were developed by the MTI and PVRC. Each of these PVRC/MTI gasket tests have an established testing procedure. This guarantees consistency of results for comparison purposes. Several of these procedures are being reviewed by ASTM for establishing consistent, uniform national gasket test procedures. Similar testing is being considered in Europe (EEC) as a part of ISO Standards.

1. FIRS—*FIR*e Simulation *S*creen Test: Test procedure using ATRS fixture without springs to obtain relaxation and post-test tensile properties of specimens typically exposed to 650°C (1200°F) for 20 min.
2. FITT—*FI*re Simulation *T*ightness *T*est: Test procedure using the single gasket 1200°F HOTT fixture to obtain tightness properties specimens typically exposed to 650°C (1200°F) for 20 min.
3. ATRS—*A*ged *T*ensile/*R*elaxation *S*creen Test: Test procedure using spring-loaded fixture to obtain relaxation and post-test tensile properties and weight loss. Temperature to 400°C (750°F) with aging to 1000 hours (1000 h) (or more).
4. ARLA—*A*ged *R*elaxation *L*eakage *A*dhesion Test: A screening test procedure using a spring-loaded fixture with ASTM F38-size platens to obtain relaxation and post-test tightness properties, adhesion, and weight loss. Temperature to 400°C (750°F) with aging to 1000 h (or more).
5. HATR—*H*igh Temperature *A*ged *T*ensile/*R*elaxation Screen Test: The same procedure as ATRS, but the fixture is extended to 565°C (1050°F) via improved materials of construction.
6. ROTT—*RO*om *T*emperature *T*ightness Tests: Three pressure levels are used in seating sequences so as to determine the correlation between pressure and leakage. Gasket tightness parameters are obtained for ASME code design use.
7. HOTT—*H*igh *O*perational *T*ightness *T*est: Test procedure using a hydraulic fixture to evaluate and confirm the sealing performance of gaskets at elevated temperature. Pressure and temperature are maintained throughout the test, which may typically last 5 to 14 days while load, leak rate, and displacement are monitored. Double gasket fixture tests are to 425°C (800°F); single gasket fixture tests are to 650°C (1200°F). Two test temperatures are required, (*a*) maximum vendor recommended usage temperature (T_{max}) and (*b*) one-half the maximum vendor recommended usage temperature ($T_{max}/2$).
8. EHOTT—This test is a combination of the ROTT and HOTT tests and should be used any time both of these tests are needed unless you are performing multiple temperature tests, in which case you should use the HOTT tests after running either one EHOTT or one ROTT test.
9. AHOT—*A*ged *HO*t *T*ightness Test: This leak performance test is conducted at temperatures up to 425°C (800°F) on pairs of NPS 4-inch (4 in) gaskets after they have been oven aged in special platens for several weeks at temperatures to 650°C (1200°F). The AHOT test is used when the exposure times and temperatures are impracticable for the HOTT test.

NPS 4-in gaskets are tested in a three-part sequence consisting of initial compression, aging, and final leak testing. Mounted in special platens, the gaskets are hot compressed at temperatures up to 650°C (1200°F) in the short-term HOTT fixture and then aged in a still air oven. During aging, the gasket is maintained in an air purge stream or with helium or nitrogen under light pressure.

The platens are transferred to the HOTT apparatus and tightness tested for 2 days at 425°C (800°F). This allows a practical evaluation of response to blow-out conditions and controlled thermal disturbances. Test duration is 2 to 6 weeks; it can be longer, if desired.

The following are facilities that perform all or some of the PVRC/MTI gasket tests:

Dr. Andes Bazergui or Mr. Michel Derenne
Ecole Polytechnic Institute
P.O. Box 6079, Sta. A
Montreal, QE H3C 3A7 CANADA
Phone: 514-340-4943/514-340-4857
Fax: 514-340-4600/514-340-4176

Dr. Y. Birembaut
CETIM, Nawtes
BP 957-44067
NANTES CEDEX FRANCE
Phone: 40-74-03-38
Fax: 40-37-36-99

Dr. Benard Nau
BLTRA Group Ltd.
Cranfield
Bedford MK43 OAJ England
Phone: 0234-750422
Fax: 0234-750074

CHAPTER 3

GASKET DESIGN AND SELECTION PROCEDURE

INTRODUCTION

The first step in the selection of a gasket for sealing a specific application is to choose a material that is both chemically compatible with the medium being sealed and thermally stable at the operating temperature of the application. The remainder of the selection procedure is associated with the minimum seating stress of the gasket and the internal pressure involved. In these regards two methods are proposed:

1. The American Society of Mechanical Engineers (ASME) code method
2. The simplified procedure proposed by Whalen

Table 3.1 shows some typical gasket designs used for ASME code applications.

ASME CODE PROCEDURE

The ASME Code for Pressure Vessels, Section VIII, Div. 1, App. 2, is the most commonly used design method for gasketed joints. It should be noted that the ASME is currently evaluating the Pressure Vessel Research Council's method for gasket design. It is likely that a nonmandatory appendix to the code will appear first. This method is compared to the traditional ASME code method presented later in this chapter by James R. Payne in the section entitled "Traditional versus New Bolt Load Calculations."

An integral part of the ASME code centers on two gasket factors:

1. An m factor, often called the gasket maintenance factor, which is associated with the hydrostatic end force and the operation of the joint.
2. A y factor, which is the minimum seating stress associated with a particular gasket material. The y factor is only concerned with the initial assembly of the joint.

The m factor is essentially a multiplier on pressure to increase the gasket clamping load to such an amount that the hydrostatic end force does not unseat the

TABLE 3.1 Typical Gasket Designs and Descriptions

Type	Cross section	Comments
Flat		Basic form. Available in wide variety of materials. Easily fabricated into different shapes.
Reinforced		Fabric- or metal-reinforced. Improves torque retention and blowout resistance of flat types. Reinforced type can be corrugated.
Flat with rubber beads		Rubber beads located on flat or reinforced material afford high unit sealing pressure and high degree of conformability.
Flat with metal grommet		Metal grommet affords protection to base material from medium and provides high unit sealing stress. Soft metal wires can be put under grommet for higher unit sealing stress.
Plain metal jacket		Basic sandwich type. Filler is compressible. Metal affords protection to filler on one edge and across surfaces.
Corrugated or embossed	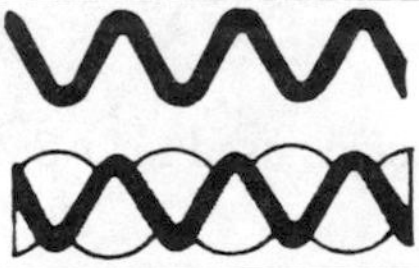	Corrugations provide for increased sealing pressure and higher conformability. Primarily circular. Corrugations can be filled with soft filler.
Profile		Multiple sealing surfaces. Seating stress decreases with increase in pitch. Wide varieties of designs are available.
Spiral-wound		Interleaving pattern of metal and filler. Ratio of metal to filler can be varied to meet demands of different applications.

gasket to the point of leakage. The factors were originally determined in 1937, and even though there have been objections to their specific values, these factors have remained essentially unchanged to date. The values are only suggestions and are not mandatory. Since the code is old and well known and its units are English, the next section will not be metricized. The reader is advised to use metric units as desired.

The ASME method uses two basic equations for calculating required bolt load, and the larger of the two calculations is used for design. The first equation is associated with W_{m2} and is the required bolt load to initially seat the gasket:

$$W_{m2} = (3.14)bGy \tag{3.1}$$

The second equation states that the required bolt operating load must be sufficient to contain the hydrostatic end force and simultaneously maintain adequate compression on the gasket to ensure sealing:

$$W_{m1} = \frac{3.14}{4} G^2P + 2b\,(3.14)GmP \tag{3.2}$$

where W_{m1} = required bolt load for maximum operating or working conditions, lb

W_{m2} = required initial bolt load at atmospheric temperature conditions without internal pressure, lb

G = diameter at location of gasket load reaction generally defined as follows:

When b_o is less than or equal to 1/4 in, G = mean diameter of gasket contact face, in; when b_o is greater than 1.4 in, G = outside diameter of gasket contact face less $2b$, in

P = maximum allowable working pressure, psi

b = effective gasket or joint contact surface seating width, in

$2b$ = effective gasket of joint contact surface pressure width, in

b_o = basic gasket seating width per Table 3.2 (The table defines b_o in terms of flange finish and type of gasket, usually from one-half to one-fourth gasket contact width.)

m = gasket factor per Table 3.1 (The table shows m for different types and thicknesses of gaskets ranging from 0.5 to 6.5.)

y = gasket or joint contact surface unit seating load, psi (per Table 3.1, which shows values from 0 to 26,000 psi)

Tables 3.2 and 3.3 are reprints of Tables 2-5-1 and 2-5-2 of the 1980 ASME code.

To determine bolt diameter based on required load and a specific torque for the grade of bolt, the following is used:

$$W_b = 0.17DT \qquad \text{(for lubricated bolts)} \tag{3.3}$$

or

$$W_b = 0.2DT \qquad \text{(for unlubricated bolts)} \tag{3.4}$$

where W_b = load per bolt, lb

D = bolt diameter, in

T = torque for grade of bolt selected, lb · in

Note that W_b is the load per bolt and must be multiplied by the number of bolts to obtain total bolt load.

To determine the bolt diameter based on the required load and the allowable bolt stress for a given grade of bolt, use

$$W_b = SA_b \tag{3.5}$$

where W_b = load per bolt, lb

S_b = allowable bolt stress for grade of bolt selected, psi

A_b = minimum cross-sectional area of bolt, in^2

TABLE 3.2 Gasket Materials and Contact Facings†

Gasket factors m *for operating conditions and minimum design seating stress* **y**

Gasket material	Gasket factor m	Minimum design seating stress y, psi	Sketches	Facing sketch and column to be used from Table 3.3
Self-energizing types (O-rings, metallic, elastomer, other gasket types considered as self-sealing)	0	0		
Elastomers without fabric or high percentage of asbestos fiber:			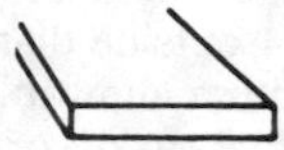	(1a), (1b), (1c), (1d), (4), (5); column II
Below 75A Shore Durometer	0.50	0		
75A or higher Shore Durometer	1.00	200		
Asbestos with suitable binder for operating conditions:			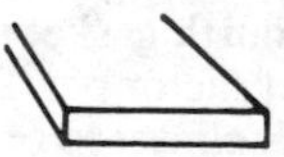	(1a), (1b), (1c), (1d), (4), (5); column II
$\frac{1}{8}$ in thick	2.00	1 600		
$\frac{1}{16}$ in thick	2.75	3 700		
$\frac{1}{32}$ in thick	3.50	6 500		
Elastomers with cotton fabric insertion	1.25	400	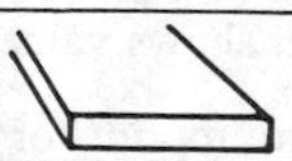	(1a), (1b), (1c), (1d), (4), (5); column II
Elastomers with asbestos fabric insertion (with or without wire reinforcement):			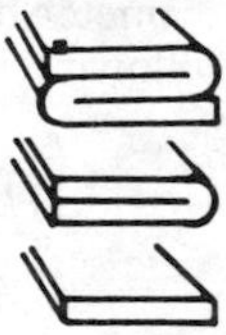	(1a), (1b), (1c), (1d), (4), (5); column II
3-ply	2.25	2 200		
2-ply	2.50	2 900		
1-ply	2.75	3 700		
Vegetable fiber	1.75	1 100	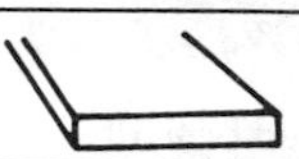	(1a), (1b), (1c), (1d), (4), (5); column II
Spiral wound metal, asbestos-filled:			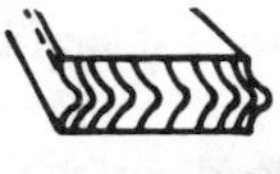	(1a), (1b); column II
Carbon	2.50	10 000		
Stainless or Monel	3.00	10 000		
Corrugated metal, asbestos inserted or corrugated metal, jacketed asbestos-filled:			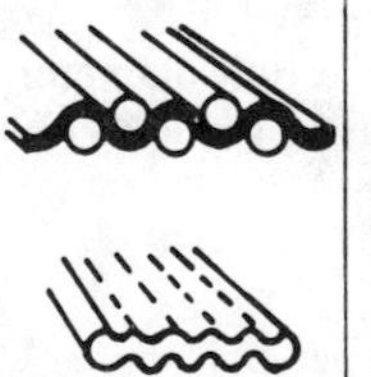	(1a), (1b); column II
Soft aluminum	2.50	2 900		
Soft copper or brass	2.75	3 700		
Iron or soft steel	3.00	4 500		
Monel or 4–6% chrome	3.25	5 500		
Stainless steels	3.50	6 500		

TABLE 3.2 Gasket Materials and Contact Facings† (*Continued*)

Gasket material	Gasket factor m	Minimum design seating stress *y*, psi	Sketches	Facing sketch and column to be used from Table 3.3
Corrugated Metal:				(1a), (1b), (1c), (1d); column
Soft aluminum	2.75	3 700		
Soft copper or brass	3.00	4 500		
Iron or soft steel	3.25	5 500		
Monel or 4–6% chrome	3.50	6 500		
Stainless steels	3.75	7 600		
Flat metal, jacketed asbestos-filled:				(1a), (1b), (1c),‡ (1d),‡ (2)‡: column II
Soft aluminum	3.25	5 500		
Soft copper or brass	3.50	6 500		
Iron or soft steel	3.75	7 600		
Monel or 4–6% chrome	3.50	8 000		
	3.75	9 000		
Stainless steels	3.75	9 000		
Grooved metal:				(1a), (1b), (1c), (1d), (2), (3): column II
Soft aluminum	3.25	5 500		
Soft copper or brass	3.50	6 500		
Iron or soft steel	3.75	7 600		
Monel or 4–6% chrome	3.75	9 000		
Stainless steels	4.25	10 100		
Solid flat metal:				(1a), (1b), (1c), (1d), (2), (3), (4), (5); column I
Soft aluminum	4.00	8 800		
Soft copper or brass	4.75	13 000		
Iron or soft steel	5.50	18 000		
Monel or 4–6% chrome	6.00	21 800		
Stainless steels	6.50	26 000		
Ring joint:			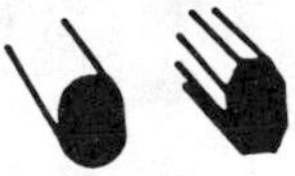	(6); column I
Iron or soft steel	5.50	18 000		
Monel or 4–6% chrome	6.00	21 800		
Stainless steels	6.50	26 000		

†This table gives a list of many commonly used gasket materials and contact facings with suggested design values of *m* and *y* that have generally proved satisfactory in actual service when using effective gasket seating width *b* given in Table 3.3. The design values and other details given in this table are only suggested and are not mandatory.

‡The surface of a gasket having a lap should not be against the nubbin.

TABLE 3.3 Effective Gasket Width†

	Facing sketch (exaggerated)		Basic gasket seating width b_o	
			Column I	Column II
(1a)	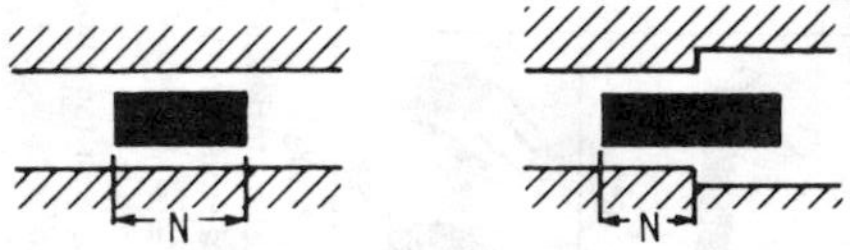		$\frac{N}{2}$	$\frac{N}{2}$
(1b)‡	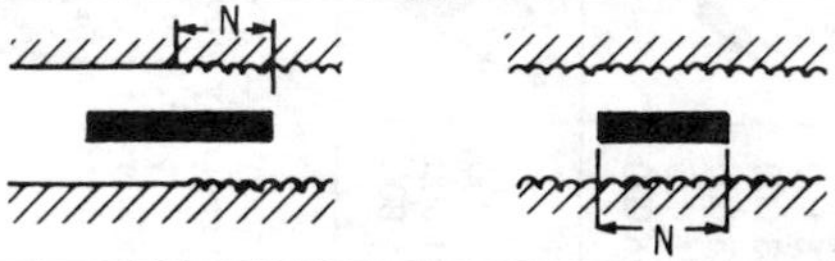			
(1c)	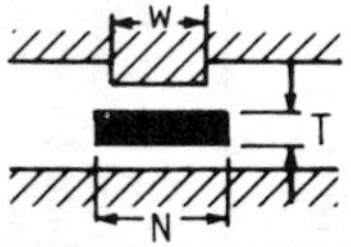	$w \leq N$	$\frac{w+T}{2}\left(\frac{w+N}{4} \text{ max.}\right)$	$\frac{w+T}{2}\left(\frac{w+N}{4} \text{ max.}\right)$
(1d)‡	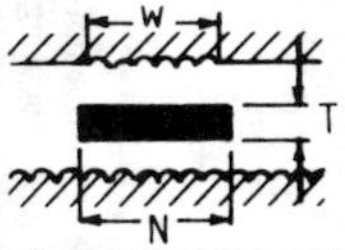	$w \leq N$		
(2) ⅟₆₄″ Nubbin	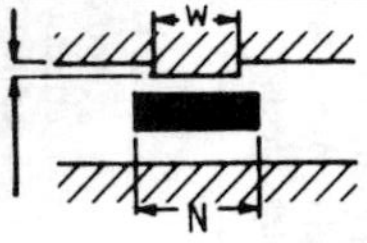	$w \leq \frac{N}{2}$	$\frac{w+N}{4}$	$\frac{w+3N}{8}$

(3) ¹⁄₆₄″ Nubbin	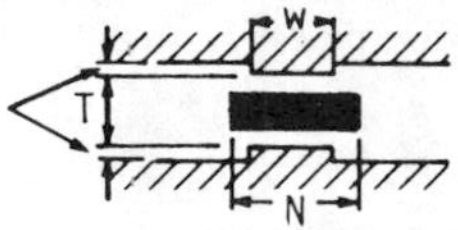	$w \leq \frac{N}{2}$	$\frac{N}{4}$	$\frac{3N}{8}$
(4)‡	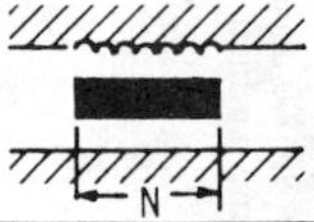		$\frac{3N}{8}$	$\frac{7N}{16}$
(5)‡	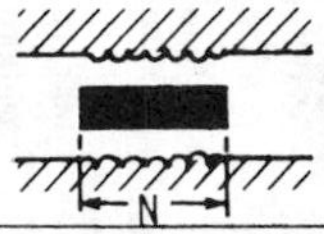		$\frac{N}{4}$	$\frac{3N}{4}$
(6)			$\frac{w}{8}$	

S81 Effective gasket seating width b:

$b = b_o$ when $b_o \leq \frac{1}{4}$ in $\quad b = 0.5\sqrt{b_o}$ when $b_o > \frac{1}{4}$ in

Location of gasket load reaction:

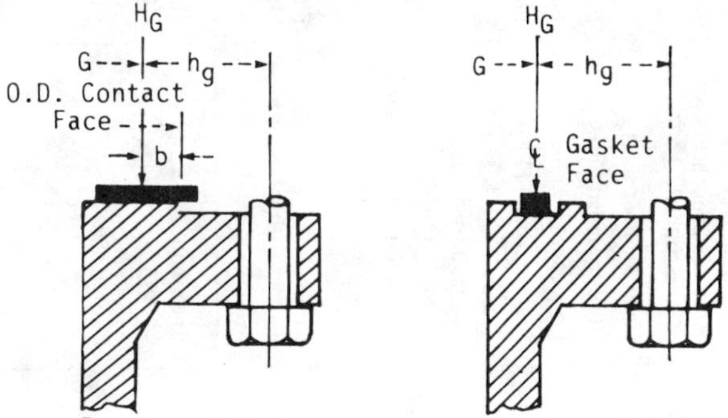

For b_o > 1/4 in. For $b_o \leqq$ 1/4 in.

†The gasket factors listed apply only to flanged joints in which the gasket is contained entirely within the inner edges of the bolt holes.
‡Where separations do not exceed ¹⁄₆₄-in-depth and ¹⁄₃₂-in-width spacing, sketches (1b) and (1d) shall be used.

TABLE 3.4 Minimum Recommended Seating Stresses for Various Gasket Materials

	Material	Gasket type	Minimum seating stress range (S_g), psi†
Nonmetallic	Asbestos fiber sheet	Flat	
	$\frac{1}{8}$ in thick		1400 to 1600
	$\frac{1}{16}$ in thich		3500 to 3700
	$\frac{1}{32}$ in thick		6000 to 6500
	Asbestos fiber sheet $\frac{1}{32}$ in thick	Flat with rubber beads	1000 to 1500 lb/in on beads
	Asbestos fiber sheet $\frac{1}{32}$ in thick	Flat with metal grommet	3000 to 4000 lb/in on grommet
	Asbestos fiber sheet $\frac{1}{32}$ in thick	Flat with metal grommet and metal wire	2000 to 3000 lb/in on wire
	Cellulose fiber sheet	Flat	750 to 1100
	Cork composition	Flat	400 to 500
	Cork-rubber	Flat	200 to 300
	Fluorocarbon (TFE)	Flat	
	$\frac{1}{8}$ in thick		1500 to 1700
	$\frac{1}{16}$ in thick		3500 to 3800
	$\frac{1}{32}$ in thick		6200 to 6500
	Nonasbestos fiber sheets (glass, carbon, aramid, and ceramics)	Flat	1500 to 3000 depending on composition
	Rubber	Flat	100 to 200
	Rubber with fabric or metal reinforcement	Flat with reinforcement	300 to 500
Metallic	Aluminum	Flat	10 000 to 20 000
	Copper	Flat	15 000 to 45 000 depending on hardness
	Carbon steel	Flat	30 000 to 70 000 depending on alloy and hardness
	Stainless steel	Flat	35 000 to 95 000 depending on alloy and hardness
	Aluminum (soft)	Corrugated	1000 to 3700
	Copper (soft)	Corrugated	2500 to 4500
	Carbon steel (soft)	Corrugated	3500 to 5500
	Stainless steel	Corrugated	6000 to 8000
	Aluminum	Profile	25 000
	Copper	Profile	35 000
	Carbon steel	Profile	55 000
	Stainless steel	Profile	75 000
Jacketed metal-asbestos	Aluminum	Plain	2 500
	Copper	Plain	4 000
	Carbon steel	Plain	6 000
	Stainless steel	Plain	10 000
	Aluminum	Corrugated	2000
	Copper	Corrugated	2500
	Carbon steel	Corrugated	3000
	Stainless steel	Corrugated	4000
	Stainless steel	Spiral-wound	3000 to 30 000

†Stresses in pounds per square inch except where otherwise noted.

SIMPLIFIED PROCEDURE

A simpler method of calculation has been suggested by Whalen. This method is also based on the seating stress S_g on the gasket, as shown in Table 3.4 and on the hydrostatic end force involved in the application. Basically, Whalen's equations accomplish the same thing as the code, but they are simplified since they use the full gasket contact width, regardless of the flange width and the surface finish of the sealing faces. This method is based on the total bolt load F_b being sufficient to:

1. Seat the gasket material into the flange surface
2. Prevent the hydrostatic end force from unseating the gasket to the point of leakage

In the first case, Table 3.4 lists a range of minimum seating stress values. The ranges shown were found in a search of the literature on gasket seating stresses. Gasket suppliers can be contacted to confirm these values.

As a starting point in the design procedure, the mean value of S_g could be used. Then, depending on the severity of the application and/or the safety factor desired, the upper and lower figures could be used. Two equations are associated with this procedure. The first is

$$F_b = S_g A_g \tag{3.6}$$

where F_b = total bold load, lb
S_g = gasket seating stress, psi (from Table 3.4)
A_g = gasket contact area, in^2

This equation states that the total bolt load must be sufficient to seat the gasket when the hydrostatic end force is not a major factor. The second equation associated with the hydrostatic end force is

$$F_b = KP_1A_m \tag{3.7}$$

where P_1 = test pressure or internal pressure if no test pressure is used
A_m = hydrostatic area on which internal pressure acts (normally) based on gasket's mid-diameter
K = safety factor (from Table 3.5)

The safety factors K from Table 3.5 are based on the joint conditions and operating conditions but not on the gasket type or flange surface finish. They are similar to the m factors in the ASME code. The equation using K states that the total bolt load must be more than enough to overcome the hydrostatic end force. The mid-diameter is used in A_m since testing has shown that just prior to leakage, the internal pressure acts up to the mid-diameter of the gasket.

After the desired gasket has been selected, the minimum seating stress, as given in Table 3.4, is used to calculate the total bolt load required by multiplying the seating stress and the gasket contact area [Eq. (3.6)]. Then the bolt load re-

TABLE 3.5 Safety Factors for Gasketed Joints

K factor	When to apply
1.2 to 1.4	For minimum-weight applications where all installation factors (bolt lubrication, tension, parallel seating, etc.) are carefully controlled; ambient to 120°C (250°F) temperature applications; where adequate proof pressure is applied.
1.5 to 2.5	For most normal designs where weight is not a major factor, vibration is moderate and temperatures do not exceed 400°C (750°F). Use high end of range where bolts are not lubricated.
2.6 to 4.0	For cases of extreme fluctuations in pressure, temperature, or vibration; where no test pressure is applied; or where uniform bolt tension is difficult to ensure.

quired to ensure that the hydrostatic end force does not unseat the gasket is calculated from Eq. (3.7). The total bolt load F_b calculated by Eq. (3.6) must be greater than the bolt load calculated in Eq. (3.7). If it is not, the gasket design must be changed, the gasket's area must be reduced, or the total bolt load must be increased.

Both the ASME procedure and the simplified procedure are associated with gasketed joints which have rigid, usually cast-iron flanges, have high clamp loads, and generally contain high pressures. There are many gasketed joints that have stamped-metal covers and splash or very low fluid pressure. In these cases, the procedures do not apply, and the stress distribution discussed in the next section should be considered by the designer.

CODE METHODS*

The Traditional ASME Code Method

The ASME Boiler and Pressure Vessel Code, Section VIII, Div. 1, Appendix 2 provides mandatory design "Rules for Bolted Flanged Connections with Ring Type Gaskets" where the gasket is within the circle defined by the bolt holes. Appendix 2 also suggests (but does not require) the gasket factor m and a minimum gasket seating stress y. These help establish a design bolt load for the joint which, in turn, is used to verify that the flange geometry is satisfactory for the governing design conditions. The design bolt load for the joint is calculated for operating and seating requirements from m and y as follows:

1. Determine the minimum bolt load required for operating conditions, W_{m1}:

$$W_{m1} = P(A_i) + S_g(A_g)$$

This equation means that the bolts must be designed for the sum of the pressure load (also called the hydrostatic end force), as represented by $P(A_i)$ and a gasket load sufficient to maintain a seal. P is the design pressure, psi. The code uses $A_i = 0.785G^2$ as the area against which the pressure acts and $A_g = 2b(3.14)G$ as the gasket area over which the minimum stress $S_g = m(P)$ must be maintained. G is the gasket O.D. less twice the effective width of the gasket which is defined by the code as b.

*This section was written by James R. Payne, JPAC Inc., Long Valley, NJ.

2. Determine the minimum bolt load required to seat the gasket, W_{m2}:

$$W_{m2} = \frac{(A_g)y}{2} = b(3.14)Gy$$

This equation means that the bolts must also be designed to exert a stress on the gasket that is sufficient to seat it.

3. Select the largest of W_{m1} or W_{m2} to determine the minimum required bolt area, A_m, as

$$A_m = \text{the greater of } A_{m1} \text{ or } A_{m2}$$

where

$$A_{m1} = \frac{W_{m1}}{S_b} \quad \text{and} \quad A_{m2} = \frac{W_{m2}}{S_a}$$

S_b and S_a are, respectively, the allowable operating and sealing design stresses for the bolts.

4. Define the seating design bolt load, W, from the average of the required and actual bolt areas:

$$W = 0.5(A_m + A_b)S_a$$

5. Define the operating design bolt load (also W) as $W = W_{m1}$.

From this point the ASME Code requires a stress analysis of the flange in question that will verify that the flange's stresses are adequate for these bolt loads.

A More Modern Method

Gasket factors similar to the ASME code m and y have been in use for around 50 years. Generally they have served industry well although one shortcoming has been that there is no standard way to replicate or verify the current factors or to get comparable new factors for new gasket products. Also, the ASME code m and y do not consider the leak rate of a gasketed joint. In these environmentally sensitive times there is a need for an approach to bolted joint design that considers leakage and makes the tightness of the joint a design criterion.

To solve this problem, the Pressure Vessel Research Committee (PVRC) has, for the past 10 years, directed a sponsor-funded gasket test program to provide the following:

- Gasket leakage data for more meaningful gasket constants
- The basis for a meaningful gasket tightness performance test

These tasks are nearly complete and an ASME Special Working Group (SWG/BFJ) is working to implement gasket constants derived from hundreds of PVRC-sponsored gasket tests. Also, an ASTM task group (F3.40.21) is evaluating the PVRC gasket tightness performance test as a draft ASTM Standard Test Method, which will permit manufacturers to get the new gasket constants for their new and improved gaskets.

The work of these industry committees is important to gasket suppliers, bolted joint designers, and users. As they complete their work, it is envisioned that two changes will take place:

1. A new appendix will appear in Section VIII, Div. 1 of the ASME Code, which will parallel the present Appendix 2 and eventually replace it. This appendix will contain tables with new gasket constants (called G_b, a, and G_s). There will be corresponding changes in the formulas that obtain the design bolt loads. The constants will be derived and condensed from a conservative interpretation of the new existing PVRC gasket test.
2. A new ASTM standard gasket test will appear which is designed to elicit the gasket constants G_b, a, and G_s for gasket materials.

While no one can exactly say what the final form of these documents will be, it is clear that the changes they represent will give the designer of gasketed bolted joints the opportunity of using gasket constants that have been certified by the gasket manufacturer or supplier in accord with meaningful new test standards. It is anticipated that the ASME table of gasket constants will be updated over the years as a more comprehensive array of gasket supplier-developed constants become available.

Bolt Load Calculations Based on Joint Tightness

The new gasket constants should be used for bolted joint designs where it is important that a minimum leak rate be considered. It will be seen from examples that getting a design bolt load from the new constants is fairly simple and straightforward once some new terminology is clarified. Tightness, the Tightness Parameter (T_p), the Standard Leak, and the new constants themselves need to be explained:

Tightness: Tightness may be thought of as the internal pressure needed to cause a small leak rate in a joint. If a tight joint requires 15,000 psi to cause a small leak rate, a pressure of 150 psi would cause the same small leak in a joint that is 100 times less tight. For a gasket, tightness is a measure of its ability to control the leak rate of the joint for a given load. With all other variables equal, a tighter gasket requires higher internal pressure to push the same rate of fluid through the joint. In other words, the tighter the seal, the lower the leak rate.

Tightness Parameter T_p: T_p is a measure of tightness that has been defined. T_p is proportional to pressure and inversely proportional to the square root of leak rate. More precisely, T_p is the pressure (in atmospheres) required to cause a helium leak of 1 mg/s for a 150-mm-O.D. gasket in a joint. Since this is about the same as the O.D. and NPS 4 joint, the pressure to cause a leak of 1 mg/s of helium for that joint is its tightness. A higher value of T_p means a tighter joint. Because of the square root, a joint that is 10 times tighter leaks 100 times less. More specifically,

$$T_p = \left(\frac{P}{p^*}\right)\left(\frac{1}{L_r}\right)^{0.5}$$

where P is the design pressure, p^* is atmospheric pressure, and L_r is the leak rate (mg/s) for a 150-mm-O.D. gasket. For design purposes, L_r must be defined in order to make use of T_p. The Standard Leak does this.

Standard Leak: In order to have joint designs that are based on leakage, some acceptable level of leakage must be defined. A leak of 1/2480 lb/h per inch gasket O.D. has been defined as a "standard" acceptable leak. For an NPS 10 joint this represents about 75 pints per day of nitrogen. This leak is related to a tightness of Class T2. When coupled with the standard leak, the required minimum tightness for design is

$$T_{p\min} = 0.1243\,P \qquad \text{(for } P \text{ in psi)}$$

Other tightness classes have been defined which in turn define corresponding values of $T_{p\min}$.

Constants G_b, a, *and* G_s: These are the constants used in formulas that give a design bolt load having the same meaning as the larger of W_{m1} or W_{m2} of the ASME code. G_b, a, and G_s are obtained by interpretation of leakage test data as plots of gasket stress (S_g) versus the tightness of parameter T_p on log-log paper. With reference to Figure 3.1, these constants are related to idealized gasket performance as follows:

G_b and a: These constants tell us what the gasket seating load should be because they are associated with the seating load sequence (part A) of a gasket test. Because they determine the seating stress, G_b and a are similar to the y stress of the present code. A log-log plot of part A data for gasket stress versus T_p gives, respectively, the intercept (for $T_p = 1$), which is G_b, and the slope, which is a.

Low values of G_b, a, and G_s are favorable. The value of $G_b(T_p)^a$ may be used to compare seating properties among gaskets for a constant value of T_p. Comparisons should be made at representative values of T_p, such as 100 and 1000. Such comparisons show the combined effect of G_b and a on the seating performance of a gasket (*see* Table 3.6).

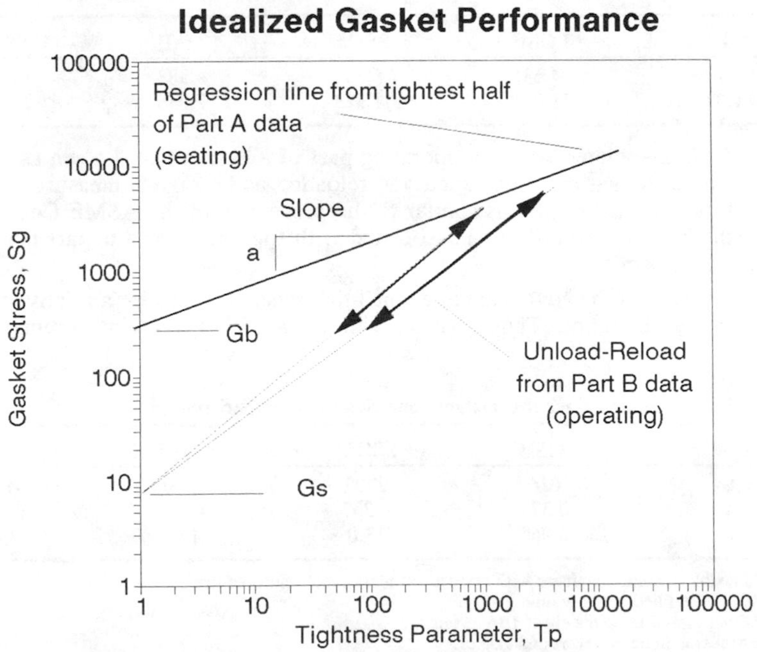

FIGURE 3.1 Gasket stress versus tightness (idealized).

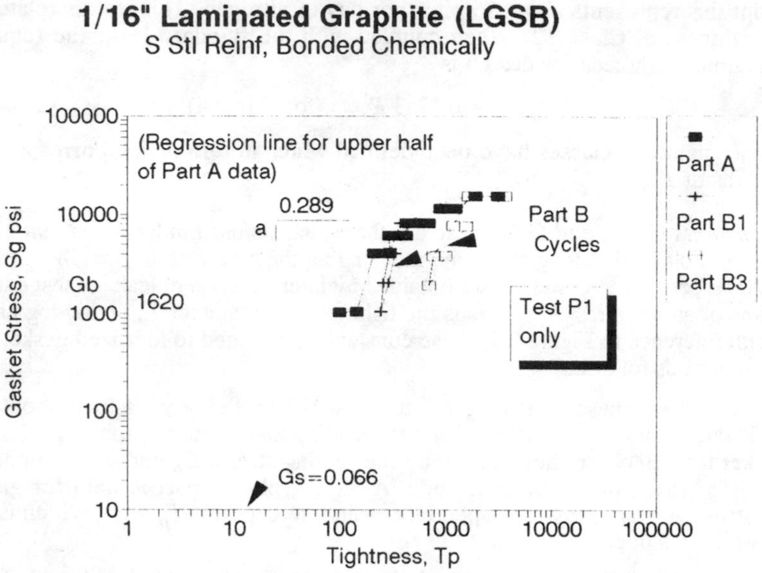

FIGURE 3.2 Gasket stress versus tightness.

The following shows a seating comparison for values of $G_b(T_p)^a$ (units are psi):

TP	LGSB	SSG6	CAP	SSA6
100	4,631	6,851	4,988	13,536
1,000	11,033	11,823	7,046	27,007

G_s Is associated with the operating part of a gasket test, known as part B, where the gasket is unloaded and reloaded as leakage is measured. It performs a function that is similar to the present m of the ASME Code. G_s is the intercept (for $T_p = 1$) associated with the data from the part B unload-reload sequences.

A low G_s has the advantage of a small tightness decrease for a relatively large bolt load reduction. Thus a gasket with a low G_s helps a joint to remain tight

TABLE 3.6 Typical Gasket Constants and Seating Comparisons

Constant	LGSB*	SSG6†	CAP‡	SSA6§
G_b	816	2300	2500	3400
a	0.377	0.237	0.15	0.30
G_s	0.066	13.0	117	7.0

*Flexible graphite laminate with bonded stainless steel reinforcement.
†Graphite filled spiral wound gasket.
‡Compressed asbestos sheet 1/16 in thick.
§Asbestos filled spiral wound gasket.

under circumstances that reduce the bolt load, for example, pressurization and thermal disturbances that cause flange rotation.

Figures 3.2 and 3.3 show the relationship of the constants to actual tightness test data for one and several tests, respectively. This data is from PVRC tests on a laminated graphite type of sheet gasket that is used in process piping.

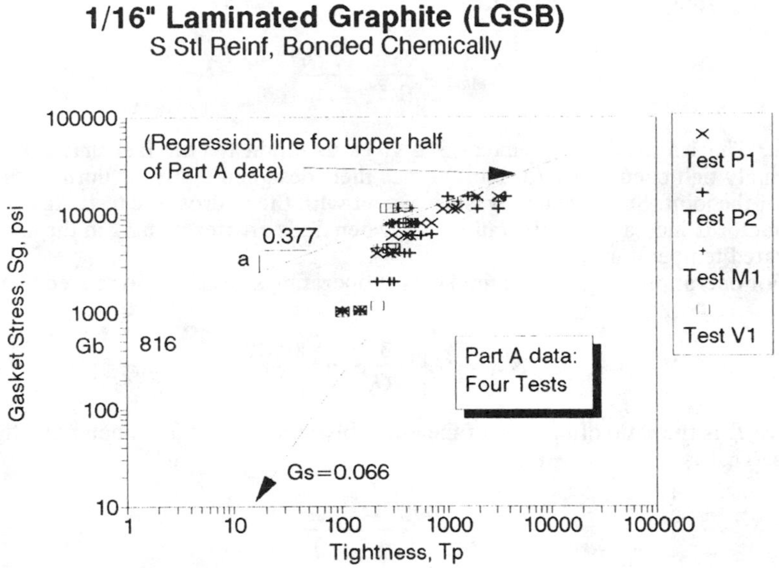

FIGURE 3.3 Gasket stress versus tightness.

Bolt Load Calculation Using T_{pmin}, G_b, a, and G_s

In a manner that is similar to the traditional ASME Code method, as explained above, the design bolt load for a joint is calculated for operating and seating requirements from the constants G_b, a, and G_s as follows:

1. Determine the minimum tightness required and the associated seating tightness:

$$T_{pmin} = 0.1243\, cP$$

Note: $c = 1.0$ for the standard tightness class ($T2$). For tightness class $T3$, $c = 1.0$.

2. Determine the required seating stress Sya to assure that T_{pmin} will be achieved. For manual joint boltup, when the operating and seating design stresses allowed by the code for the bolts are equal ($S_b = S_a$), the required seating stress is Sya, which is obtained from G_b and a and the tightness of T_{pmin}:

$$Sya = \frac{G_b}{e}(1.5\, T_{pmin})^a$$

Note: e = 3/4 for manual boltup. For other assembly methods, e has higher values. Also, note that the 1.5 factor is increased by the ratio S_a/S_b when S_a is greater than S_b in the case of elevated temperature joints.

3. Determine the minimum "design" gasket stress S_m that is required to maintain the specified minimum level of tightness (T_{pmin}) during operation. S_m is determined, from both seating and operating tightness requirements, as the greater of S_{m1}, S_{m2}, or $2P$.

For seating: S_{m2} is related to *Sya:*

$$S_{m2} = \frac{Sya}{1.5} - \frac{pA_i}{A_g}$$

Note: The 1.5 factor that reduces *Sya* gives us credit for the fact that bolts are normally tightened to 1.5 (or more) times their designed tightness during assembly of the joint. Such tightening is consistent with the hydrostatic test. Again, the 1.5 factor is increased by the ratio S_a/S_b when S_a is greater than S_b in the case of elevated temperature joints.

For operation: S_{m1}, the required gasket operating stress, is also related to *Sya:*

$$S_{m1} = G_s\left[\left(\frac{3}{4G_s}Sya\right)\right]^{(1/T_r)}$$

where T_r is the ratio of the logs of the assembly and the minimum operating tightness. That is

$$T_r = \frac{\log(1.5T_{pmin})}{\log(T_{pmin})}$$

Therefore S_m = the greater of S_{m1}, S_{m2}, or $2P$.

4. Determine the minimum bolt load required for operating conditions, W_m:

$$W_m = P(A_i) + S_m(A_g)$$

The equation for W_m means that the bolts are designed for the sum of the pressure load (also called the hydrostatic end force), as represented by $P(A_i)$, plus a gasket load that is sufficient to maintain a seal and adequate for seating. An example calculation of W_m follows.

Determination of Design Bolt Load with New Gasket Constants Example of the Convenient* Method for Flexible Graphite Laminate (LGSB)

The following data and Table 3.6 are used to determine the design bolt load. Use an assembly efficiency of e = 0.75 (0.75 is for manual boltup) and use c = 10.0 for a tightness of class T3-TIGHT.

*Convenient method assumes $T_{pn} = 1.5T_{pmin}$, except if the allowed bolt stress S_a is greater than S_b, $T_{pn} = 1.5\,(S_a/S_b)\,T_{pmin}$ may be used.

ASME code data	Gasket, in
Design pressure, P: 2000 psi	O.D., G_o: 10.63 Width, N: 0.5000 (assumes $b_o = N/2$) Eff. width: b': 0.250
Note: Press ratio, P_r: 136 Press area, in^2 $3.14/4(G_2) = A_i$: 80.52	Note: $N/b' = 2.000$ Eff. dia, G: 10.13 Area, in^2 AG: 15.90 ($3.14(G_o - N)N = A_g$) A_i/A_g = area ratio = 5.06

Tightness data	Gasket constants
$1.8257\,(cP_r) = T_{pmin} = 2486 = 0.1243(cP)$	$G_b = 816$ psi
$T_{pmin} \times 1.5 = T_{pn} = 3728$	$a = 0.377$
$\text{Log}T_{pn}/\log T_{pm} = T_r = 1.0519$	$G_s = 0.06640$ psi

Gasket design stresses are as follows:

$$G_b/e(T_{pn})^a = Sya = G_s[G_b/G_s(T_{pn})^a]^{1/T_r} = S_{m1} = G_s(eSya/G_s)^{1/T_r}$$

$$Sya/1.5 - P\,(A_i/A_g) = S_{m2} = 5982 \text{ psi}$$

For design select S_m as the largest of S_{m1} or S_{m2} or $2P$.
For the design bolt load, use $PA_i + S_m A_g = W_{m0} = 316{,}504$ lb (design value).

Determining the New Design Bolt Load (W_m)—A Summary

1. Get minimum tightness: $T_{pmin} = 0.1243cP$. *Note:* $c = 1.0$ for the standard tightness class.

2. Get seating stress S_{ya}: $S_{ya} = G_b/e\ (1.5T_{pmin})^a$. *Note:* $e = 3/4$ for manual boltup.

3. Get "design" stress, S_m to maintain T_{pmin}: S_m is the greater of S_{m1} or S_{m2} or $2P$.

   For seating: $S_{m2} = S_{ya}/1.5 - pA_i/A_g$
   For operation: $S_{m1} = G_s[(3/4GS)(Sya)]^{(1/T_r)}$
   $T_r = \log(1.5\ T_{pmin})/\log(T_{pmin})$

4. Calculate the design bolt load as: $W_m = P(A_i) + S_m(A_g)$, A_i = pressurized area, A_g = gasket area.

GASKET COMPRESSION AND STRESS-DISTRIBUTION TESTING

After a gasket has been selected and designed for a particular application, various tests can be performed to determine the gasket's compressed thickness and stress distribution. Inadequate compression or nonuniform stress distribution can result in a leaking joint. The tests can be performed to check for these possibilities and permit correction to ensure leak-tight joints.

1. *Lead:* In this test, lead pellets are used to accurately indicate the compressed thicknesses of a gasketed joint. The pellets, commonly called lead shot, are available from local gun supply stores. A diameter approximately twice the thickness of the gasket should be used. Lead solid-core solder can also be used if desired with the same size requirements. Pellets or solder are particularly well suited for doing this test since they exhibit no recovery after compression, whereas the actual gasketing material will almost always exhibit some recovery. The degree of nonuniform loading, flange bowing, or distortion will be indicated by the variations in the gasket's compressed thickness.

To begin, the original thickness of the gasket is measured and recorded at uniformly selected points across the gasket. At or near these points, holes are punched or drilled through the gasket. Care should be taken to remove any burrs. The punched holes should be approximately 1½ times the pellet diameter. Then the gasket is mounted on the flange. A small amount of grease can be put in the punched holes to hold the lead pellets if required. The pellets are mounted in the grease, and the mating flange is located and torqued to specifications.

Upon careful disassembly of the flange and removal of the pellets, their thicknesses are measured, recorded, and analyzed. Comparison of the pellets' compressed thicknesses to the gasket's stress-compression characteristics permits the desired stress-distribution analysis.

2. *Regular carbon paper:* This paper, once extensively used by secretaries in typing, is a two-piece system. One sheet is the carbon carrier and the other is a clean sheet of paper required for carbon transfer. Stress impressions done with this technique provide on-off or yes-no visual effect. That is, either sufficient stress was or was not available to transfer the carbon from the carbon paper to the clean sheet. This provides a very narrow range of stress information. In some instances, multiple layers of carbon paper are used for better measuring of the stresses. In such cases, compressibility of the paper stack has to be taken into account, particularly with gaskets that inherently possess little compressibility. Because of the limitations, this method is rarely used today.

3. Carbonless or no carbon required (NCR) paper is available as a single sheet that has pressure-sensitive chemicals within the paper. Stress applied to the paper crushes the encapsulated chemicals and colors the paper. The intensity of color is proportional to the stress on the paper, which is the same stress as is on the gasket.

This paper is an improvement over regular carbon paper since it is a single sheet and the impression density, or color intensity, is proportional to the stress applied. Therefore, elemental comparative impression testing at various stress levels, performed on load machines such as an Instron or MTS, can be used to calibrate the paper color impression density versus stress level. The paper is currently available in only one pressure range, and this places limitations on accurately quantifying the stress distribution data collected.

Stress-sensitive film, manufactured by Fuji Photo Film Co., Ltd., is available in one- or two-sheet systems and in three or four pressure ranges. It functions in a manner similar to the carbonless paper. This film is an improvement over the carbonless paper in that it permits "fine tuning" of the stress distribution because of the various ranges available. Also, with the use of a commercially available densitometer, impression color density can be directly converted to stress readings. Since the film is affected by time, temperature, and humidity, these conditions must be taken into account when analyzing the color intensities. Again, calibration of these films can be accomplished by conducting elemental stress compression and color intensity testing.

To perform stress distribution evaluations within a sealing joint, the regular carbon paper, NCR paper, or Fuji film is precut to the shape of the mating flanges, and the holes for the fasteners are punched out. The paper or film is then placed between the flanges and the fasteners are tightened to the specified torque. When the paper is removed from the actual gasketed joint, it shows the stress distribution pattern on the gasket. This is shown in Fig. 3.4.

A deficiency in each of the above techniques is that they all indicate a one-time maximum stress during the course of testing. Any reduction in stress during the torquing sequence or from external forces is not reflected in the impressions. The gasket designer needs to be aware of this, especially in low clamp load and nonrigid flange joints.

FIGURE 3.4 Gasket and stress impression.

There is new technology which can be used to measure stress distribution during both loading and unloading. It is described in the ASME/JSME paper "Measuring Real Time Static and Dynamic Gasket Stress Using a New Technique," June 1991, Book #H00644. The American Society of Mechanical Engineers is located at 345 East 47th St., New York, NY 10017.

FASTENERS IN THE GASKETED JOINT

The function of the fastener in a gasketed joint is to apply and maintain the load required to seal the joint.

Threaded nuts, bolts, and studs are the usual fasteners used in industrial gasketing to assemble mating parts. The fastener device must be able to produce a spring load on the gasket to compress it to its proper thickness and density for sealability. The fastener must also be able to maintain proper tension to maintain this compression of the gasket material throughout the life of the assembly. When flanges are made of dissimilar metals, bolting plays a most important part in obtaining and maintaining a satisfactory seal. The fasteners as well as the gasket must be able to compensate for the difference in expansion and contraction of the different flange materials.

Addressing the question of how many bolts or other fasteners can be used involves space available, economic limitations, and flange flexibility considerations as well as getting the required initial load. About 80 percent of the load applied by the fastener may be concentrated around the bolt area, leaving about 20 percent to be distributed out along the flange to the midpoint between the bolts. Exactly how this load is distributed is dependent upon the thickness, configuration, and rigidity of the flange. Therefore, it is difficult to provide useful rule of thumb guidelines, except, as noted earlier, that in a uniform flange, cutting the distance between bolts by half will reduce the bowing effect to one-eighth its original value, and conversely, perhaps, stiffening the flange is frequently more cost-effective than increasing the number of fasteners.

Washers spread the bolting load over a large area and prevent, or at least reduce, deformation of the flange area directly beneath the head of the bolt. The fit, flatness, and type of washer can have a significant effect on loading. The relationship of the shank clearance and washer I.D. chamfer is important since an increase in surface friction could result in a reduction in bolting loading.

When there is a reduction of clamp load on the gasket, there can be an increase in the movement of the sealing flange components. Therefore, a bolt-locking or thread-locking device may be beneficial.

In some cases star washers or spring washers or one of a variety of liquid or plastic thread locking compounds are used. Thread sealing compounds are also available for applications where the fastener threads into the cavity being sealed. Spring washers can also compensate somewhat for the compression set of the gasket and thus decrease the associated loss of clamp load.

It is essential to avoid overloading the gasket. Gasket materials will crush due to a combination of compression, shear, and extrusion-type displacements of the material. The maximum unit load is a function of the type of material, operating temperature, thickness, and section width, among the principal factors. The designer of the fastener system must know the crush limitations of the gasket material being considered for the application. This information is available from gasket suppliers. Deformation of the flange can also create higher than expected localized loading conditions that can aggravate crush and extrusion.

FLANGE AND GASKET DESIGN RECOMMENDATIONS

Materials

The following materials and combinations thereof are used for gasket flanges: steel, stainless steel, cast iron, aluminum, plastics, ceramics, and composites. Using dissimilar materials for the mating flanges can increase crush and extrusion problems due to the dissimilar thermal expansion characteristics of the different materials.

Finish

The flange surface roughness is important because of its effect on the degree of sealability (i.e., the rougher the surface, the higher the flange load required to provide an adequate seal). The smoother the finish, the lower the frictional resistance and the higher tendency to blow out. The flange surface may range from a rough casting to that produced by machine lapping.

The lay of the surface finish is also important since certain lays in specific applications may cause direct leakage paths while other lays would not.

Thickness

Flanges must have adequate thickness. Adequate thickness is required to transmit the load created by the bolt to the midpoint between the bolts. It is this midpoint that is the vital point of the design. Maintaining a seal at this location is important and should be kept constantly in mind. Adequate thickness is also required to minimize the bowing of the flange caused by the bolt loads. If the flange is too thin, the bowing will become excessive and no bolt load will be carried to the midpoint, as shown in Figs. 3.5 and 3.6.

Reinforcing flat sheet metal flanges by adding embossed metal strips or tapered risers or by turning up the edges of the flange to create a more rigid beam should be considered when working with thin flanges. See Figs. 3.7 and 3.8. Another recommendation for a sheet metal design is to provide a positive backdraft on the flange, concentrating the force on the inner edge of the gasket, as shown in Fig. 3.9. This prevents oil from escaping through the bolt holes, which can happen if the load is concentrated at the outer edge of the gasket.

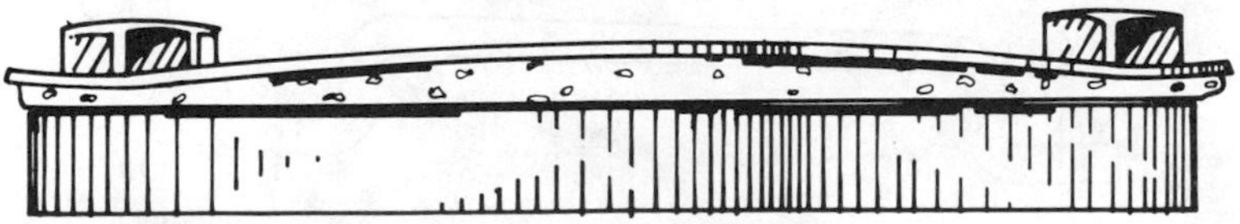

FIGURE 3.5 Illustration of flange bending or bowing.

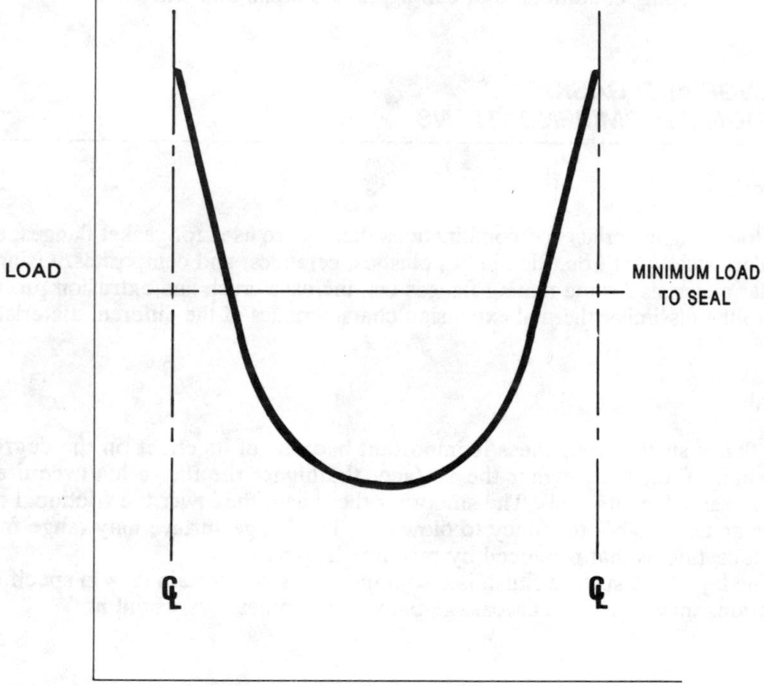

FIGURE 3.6 Midpoint loading.

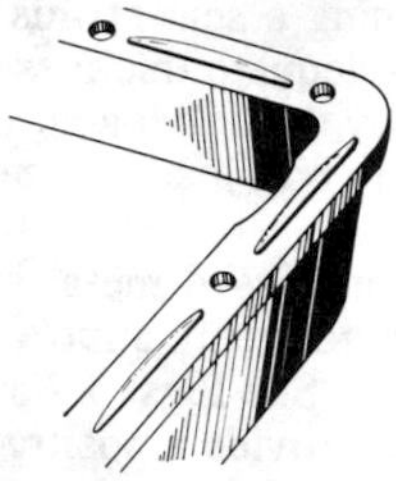

FIGURE 3.7 Tapered riser.

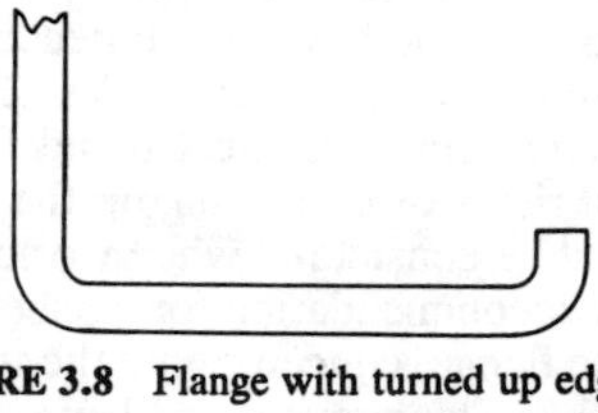

FIGURE 3.8 Flange with turned up edge.

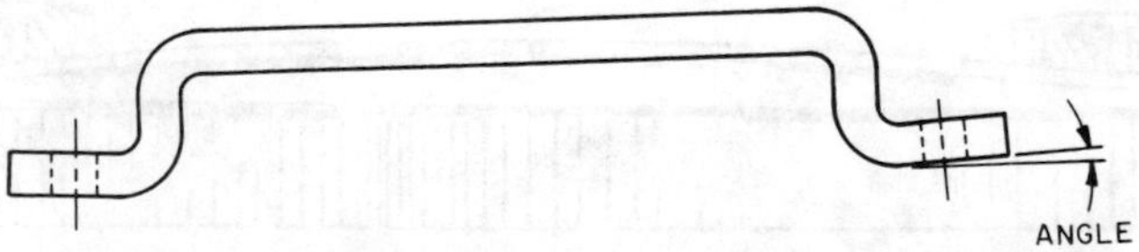

FIGURE 3.9 Flange with positive back draft.

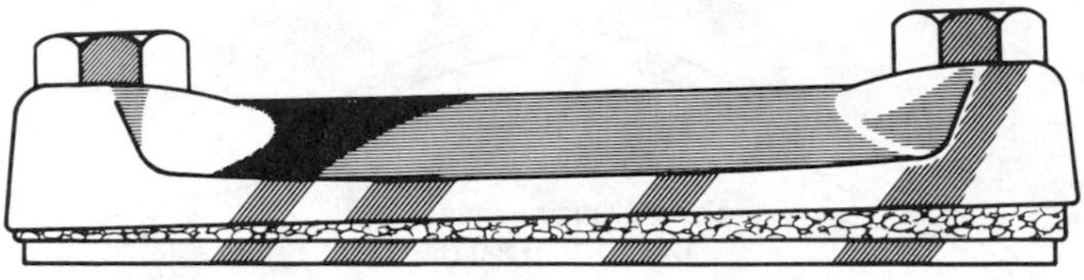

FIGURE 3.10 Cocking or nonparallelism in a flange.

A flange may also cock, or become nonparallel, during the clamping process. See Fig. 3.10. In most cases, cocking does not give rise to serious performance problems until the gasket compression loads in the area of low compression falls below the minimum load required to seal.

Internal pressure also can create loads on both bolts and flanges to create another type of distortion. If such loads are high enough, the bolts and flanges will exhibit distortion. For instance, bolts might elongate. This could very well be an elongation in addition to that caused by the initial tightening torques. And the flange might deflect or reveal a bowing in addition to that caused by the imposition of initial bolt loads. One other force or load created by internal pressure is blow-out. Figure 3.11 shows blow-out as acting on the inner edge of the gasket, tending to push it from between the flanges.

Bolting Pattern

The number and distribution of the bolts significantly affects the loading pattern between the sealing surfaces of the two flanges. The best clamping pattern is invariably a combination of the maximum practical number of bolts, even spacing, and optimum positioning.

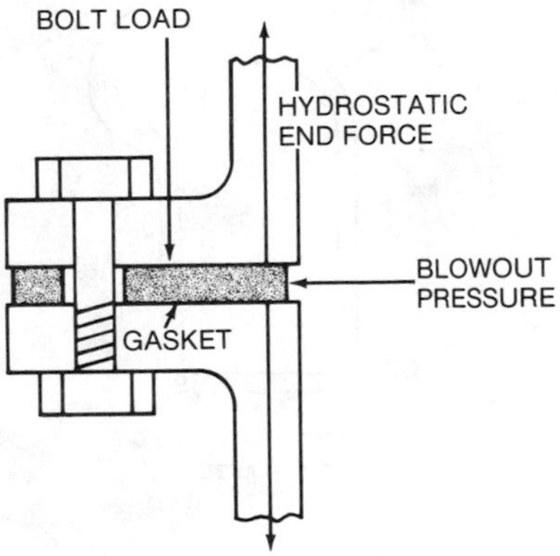

FIGURE 3.11 How internal pressure acts on a gasket joint.

Distribution

The loading pattern is a series of straight lines drawn from bolt to adjacent bolt until the circuit is completed. If the sealing area lies either side of this pattern, it may be a potential leakage spot, and the further the sealing area deviates from this pattern, the more perilous is the design. An example of these various conditions is illustrated in Fig. 3.12.

It is desirable to have an adequate number of bolts, properly spaced, at a joint. Sometimes in certain areas this is not possible. Where the spacing of the bolts must be wide, and larger diameter bolts cannot be used, the designer can help the sealing of the gasket by changing the flange face and gasket contours from a uniform width to a sculptured design. Such a revision is shown in Fig. 3.13. Because the width of the gasket is gradually reduced to the midpoint, the unit loading is increased and the midpoint load is better maintained than it would have been if the width of the gasket had not been reduced.

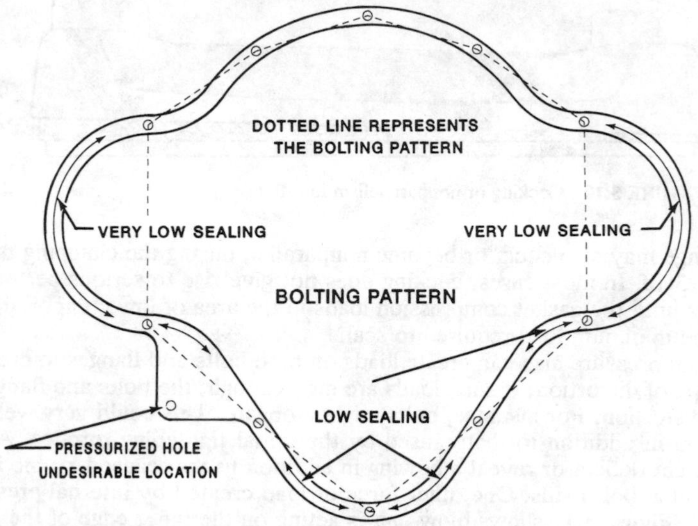

FIGURE 3.12 Good versus poor sealing areas.

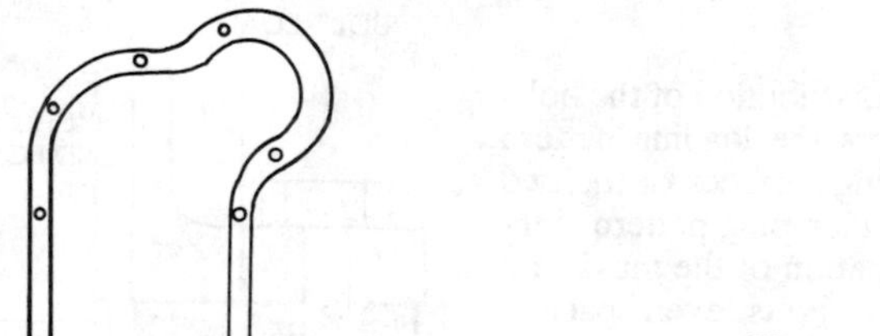

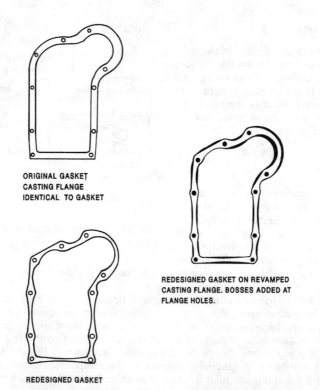

FIGURE 3.13 Original versus redesigned gasket for improved sealing.

Bolt Holes

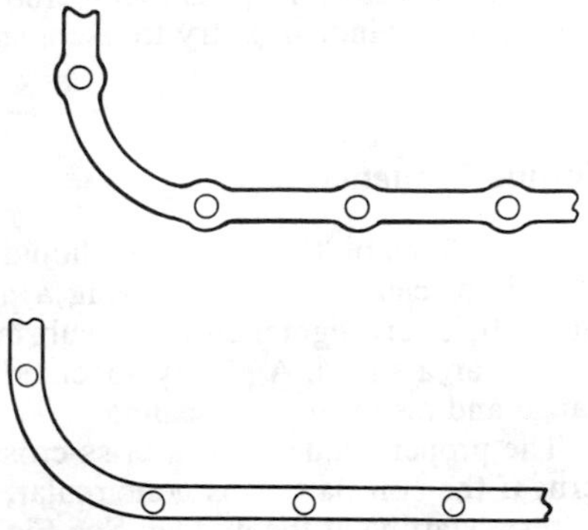

FIGURE 3.14 Example of revised flange and gasket to prevent rupture of gasket at bolt holes.

About 80 percent of the bolt load is concentrated around the area of the bolt. Because of the necessity of incorporating a bolt hole in a gasket, the amount of material left at this location may then be inadequate to carry the high bolt load, and, as a result, the gasket may rupture. This is especially true with thin sheet metal flanges where distortion at the bolt holes is common. To alleviate this condition, the flange face and the gasket should be enlarged as shown in Fig. 3.14. This will help compensate for gasket material removed to create the bolt hole.

Very small bolt holes or small noncircular openings can build up costs. The centers from such holes will probably require hand picking and, because small holes are easy to miss, extra inspection may be needed. It is best to

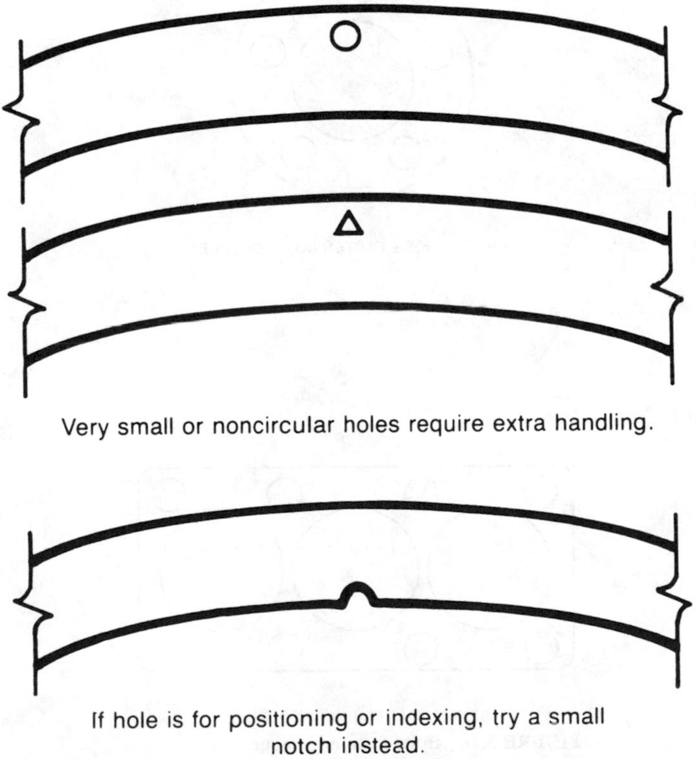

Very small or noncircular holes require extra handling.
If hole is for positioning or indexing, try a small notch instead.

FIGURE 3.15 Avoid small bolt holes.

avoid holes sizes under 2.5 mm (0.100 in) in diameter. If the hole is used just for positioning or indexing, try to use a small notch instead. See Fig. 3.15.

Torque Sequence

Ideally, all the bolts of the joint should be torqued simultaneously. Alternately, a good joint can be achieved using a proper torque sequence. The sequence in which bolts are tightened has a substantial bearing upon the distribution of the contact area stress. A poorly specified bolt sequence will cause distortion of the flange and result in poor sealing.

The proper sequence is a criss-cross pattern if the bolts are in a circular pattern. If the bolt pattern is noncircular, the fastening sequence is usually a spiral pattern, starting at the middle. See Fig. 3.16. Tightening of bolts is accomplished by hand or with multiheaded automatic torque wrenches. Tightening specifications can be in terms of a specified torque plus an additional angle of turn, torque applied for a specific time, tightening to the yield point of the bolt, or tightening to a specific torque value or load value.

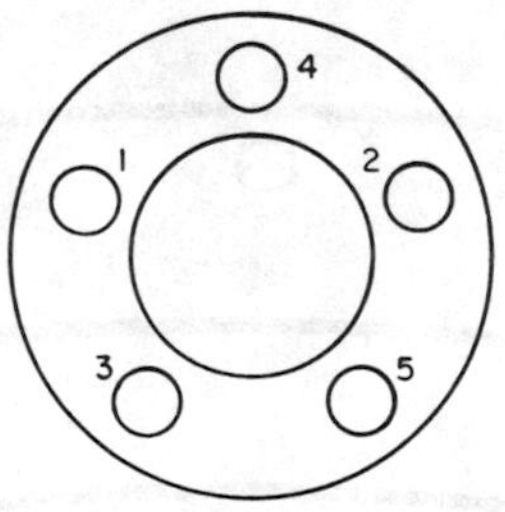

CRISS-CROSS FASTENING SEQUENCE

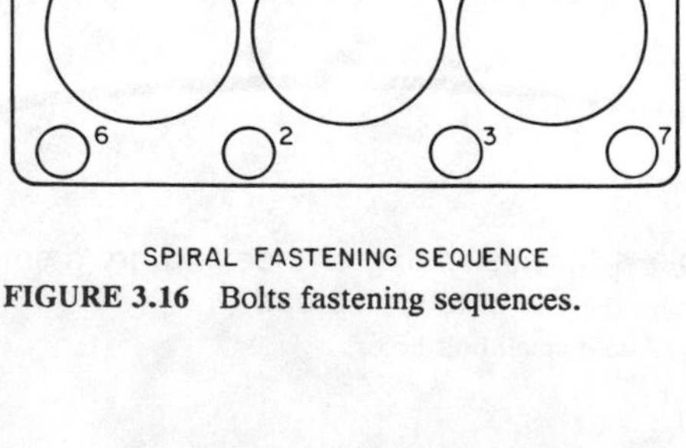

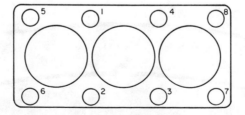

SPIRAL FASTENING SEQUENCE

FIGURE 3.16 Bolts fastening sequences.

GASKET INSTALLATION SPECIFICATIONS

An installation is only as good as its gasket; likewise, a gasket is only as good as its installation. The following are some recommendations associated with gasket installation:

- Be sure that mating surfaces are clean and in specification with regard to finish.
- Clean the bolt holes in mating flanges.
- Check the gasket for damage before installing it.
- Make certain the gasket fits the application.
- Specify lubricated bolts. Bolt threads and the underside of the bolt head should be lubricated.
- Make certain that the bolts do not bottom out in the mating flange.
- Specify the torque level and use of a torque wrench.
- Specify the torquing sequence. In addition to the sequence, two or three stages of torque before reaching specified level are recommended.

GASKET FABRICATION

Cutting Tools

Gasket materials are fabricated into parts with the following typical tools and machines:

1. *Steel rule dies:* These dies are generally used when cutting nonmetallic gasket materials where gasket dimension tolerances are greater ±0.3 mm (0.015 in). They can be used for low- or high-volume parts when cutting nonmetallic materials. The tooling is economical and can be run in platen presses, roll presses, mechanical presses, or rotary die machines. These dies also can be used on metals of light gauge. For metallic gaskets, this type of tooling is generally used for prototypes or low-volume parts. The dies are generally made as follows:
   a. *Standard jig cut steel rule die:* It is laid out by hand on the base board or onto a mylar which is then adhered to board. The holes are drilled and contour sawed on a jig saw. The rule is bent and fitted into the slots.
   b. *Standard jig cut and precision bored rule die:* The operations are the same as above except the holes are drilled in a precision machine such as a mill, jig borer, or N/C machine.
   c. *Most precise rule die:* The part is programmed into a laser cutting machine. The machine cuts out the holes and the contour to a precise dimension. The rule is bent by hand and pressed into the base board.
2. *Steel cutting dies for punch press:* These dies are either progressive dies or compound blank dies consisting of male and female die components. The tooling is considerably more expensive than rule dies, but it is more accurate and can hold tolerances of ±0.05 mm (0.002 in). These dies are used for high-volume parts, tight tolerance parts, and for metals 0.25 mm (0.010 in) and thicker.

The compound die produces the most accurate part and is generally less expensive than the progressive die. Sometimes because of weak walls or processing problems this type of tool cannot be used and a progressive die is used.

3. *Electrical discharge machining:* This is done with an Electrical Discharge Traveling Wire Machine. One can only cut ferrous metals using this machine. It is a highly accurate method of cutting parts. The parts are stacked [generally 25 to 50 mm high (1 to 2 in)] and machined together. This method is good for prototypes or very low-volume parts, but it is rather slow and expensive.
4. *Laser cutting:* This is an inexpensive method to machine parts. The cutting speeds are extremely fast (up to 7.5 m/min, or 8.3 yd/min). The parts are generally cut out of a sheet and can be nested for best material utilization. The laser can also hold tight tolerances ±0.05 mm (0.002 in) but does leave a slight slag on edges from the burning operation. You can cut most materials on this machine (it is not limited to ferrous metals as is the case with wire EDM).
5. *Water jet cutting:* This is another inexpensive method to machine parts, but is not as accurate as the laser [tolerance ±0.25 mm (0.010 in)]. The cutting speeds are fast, and one can cut materials that are stacked, thereby producing higher production rates. Here again nesting of parts for best material can be done.

Gasket Dimensions

The dimensions and tolerances associated with gaskets are determined by the cutting tools and the gasket materials. It is important that the critical dimensions of the gasket be clearly identified. Generally dowel and bolt hole locations are easier to control than are contours. Contours and part holes are dependent on hand jigging and hand rule bending. The gasket thickness also has an effect on dimensional tolerances. Gasket manufacturers should be consulted in regard to the plan dimensions and tolerances.

BIBLIOGRAPHY

Armstrong Gasket Design Manual, Armstrong Cork Co., Lancaster, Pa., 1978.

Bazergui, A., and L. Marchand, "Development of a Production Test Procedure for Gaskets," *Welding Research Council Bulletin No. 309,* Nov. 1985.

"Chemicals: Wonder Drugs or Cure-Alls?," *Import Car,* Feb. 1983.

Code for Pressure Vessels, The American Society of Mechanical Engineers, Sec. VIII, Div. 1, App. 2, 1980.

Czernik, D. E., "Gasketing the Internal Combustion Engine," SAE paper 800073, Feb. 1980

———, "Recent Developments and New Approaches in Mechanical and Chemical Gasketing," SAE paper 810367, Feb. 1981.

———, "Sealing Today's Engines," *Fleet Maintenance and Specifying,* Irving-Cloud, July 1977.

———, and F. L. Miszczak, "Measuring Real Time Static and Dynamic Gasket Stresses Using a New Technique," ASME/JSME Paper, June 1991.

———, J. C. Moerk, Jr., and F. A. Robbins, "The Relationship of a Gasket's Physical Properties to the Sealing Phenomena," SAE paper 650431, May 1965.

Faires, V. M., *Design of Machine Elements,* MacMillan, New York, 1955.

Freudenberger, R., "Gaskets or Glue?," *Motor Service,* April 1982.

"Gasket and Joint Design Manual for Engine and Transmission Systems," SAE publication, AE-13, 1988.

Hsu, K. H., J. R. Payne, J. B. Bickford, and G. F. Leon, "The US PVRC Elevated Temperature Bolted Flange Research Program," for presentation at the *2d Intl. Symp. on Fluid Sealing of Static Gasketed Joints,* La Baule, Fr. Sept. 18–20, 1990.

McDowell, D. J., "Choose the Right Gasket Material," *Assembly Engineering,* Oct. 1978.

Oren, J. W., "Creating Gasket Seals with Rigid Flanges," SAE paper 810362, Feb. 1981.

Payne, J. R., A. Bazergui, and G. F. Leon, "Getting New Gasket Design Constants from Gasket Tightness Data," Special Supplement, *Experimental Techniques,* Nov. 1988, pp. 22–27.

Rogers, R., R. Foster, and K. Wastler, "Factors Affecting the Formulation of Nonasbestos Gasket Materials," SAE paper 910206, Feb. 1991.

Rothbart, H. A., *Mechanical Design and Systems Handbook,* 2d ed., McGraw-Hill, New York, 1985, Sec. 27.4.

Standard Classification System for Nonmetallic Gasket Materials ANSI/ASTM F104-79a, American Society for Testing and Materials.

Standard Handbook of Machine Design, McGraw-Hill, New York, 1986, Chap. 26, Part 1—Gaskets, 1986.

Whalen, J. J., "How to Select the Right Gasket Material," *Product Engineering,* Oct. 1960.

Winter, J. R., "Gasket Selection—A Flowchart Approach," for presentation at the *2d Intl. Symp. on Fluid Sealing of Static Gasketed Joints,* La Baule, Fr. Sept. 18–20, 1990.

CHAPTER 4
ELASTOMERIC STATIC SEALS

AXIAL AND RADIAL APPLICATIONS

Axial and radial applications are the two most common types of installations. Static seals perform in the same manner as gaskets in that they provide a barrier to fluid leakage between two surfaces that for all practical purposes do not move relative to each other (see Fig. 4.1).

Application types include sewer pipe joints, mason jar seals, car door and window weather strip moldings, oil filter seals, drain plug seals, hydraulic end cap seals, house foundation barrier seals, toilet wax ring seals, tire valve seals, and many others too numerous to mention. The seals have nothing in common with one another except that their mating faces do not move relative to the seal surface.

MATERIAL CHOICES

The choice of seal material can be rubber (the most common), plastic (as in cans and bottle caps), rope, leather, metal, wax, treated paper, wood, and various combinations of material. This section of the handbook will deal only with the more common elastomeric types. Elastomers have an advantage over other material selections because of their ability to be molded and extruded into various shapes and usually manufactured to obtain physical properties conducive to sealing the various environments. Because rubber can be easily molded into various shapes there are a great variety of cross-sections (see Fig. 4.2).

Extruded elastomeric seals do not have the excellent physical properties molded seals have due to die swell, state of vulcanization, and the accompanying lack of density, strength, resistance to compression set, and attack by various fluids.

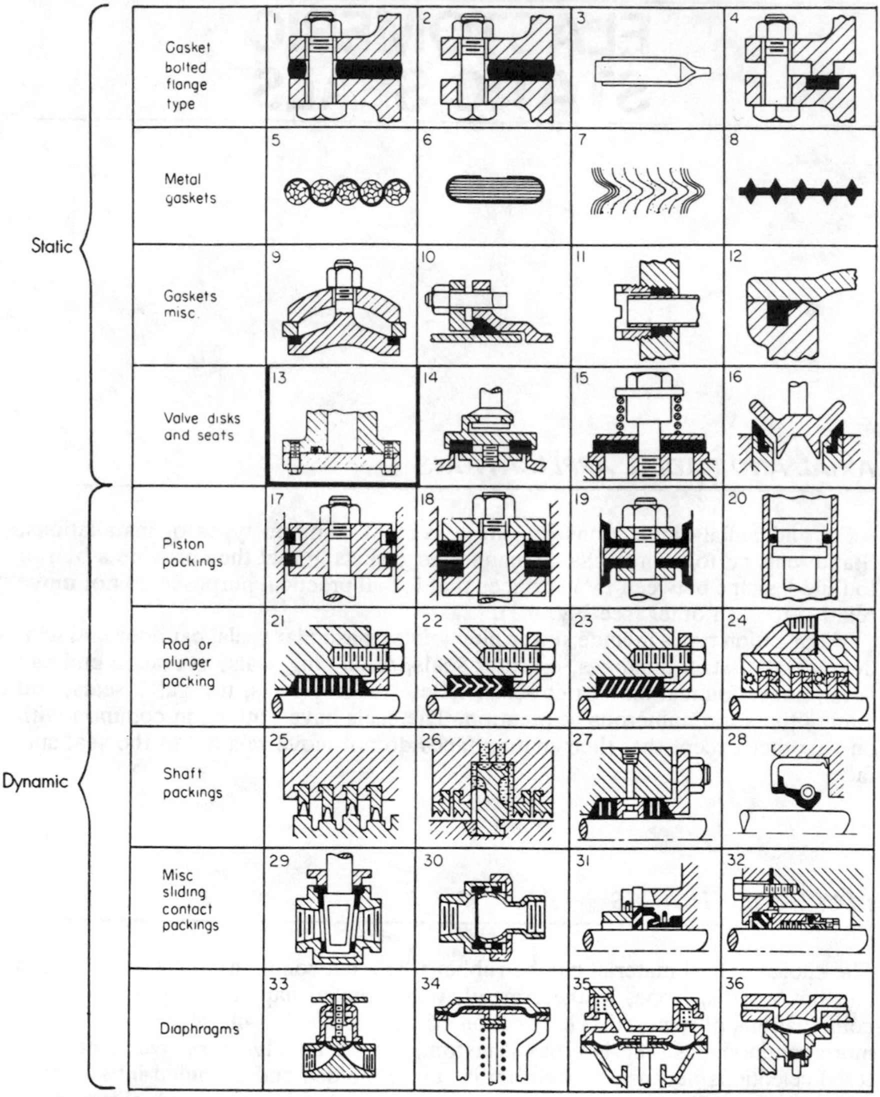

FIGURE 4.1 Typical seal applications.

APPLICATION DICTATES SEAL DESIGN

The application requirements for static sealing are not as stringent as for dynamic sealing as in the case of O-rings; for example, groove finish can be rougher and amount of squeeze over a minimum is not as critical since friction is not a factor.

There is a static seal cross-sectional shape for every letter of the alphabet (including several foreign languages), although the most prevalent cross-sections are

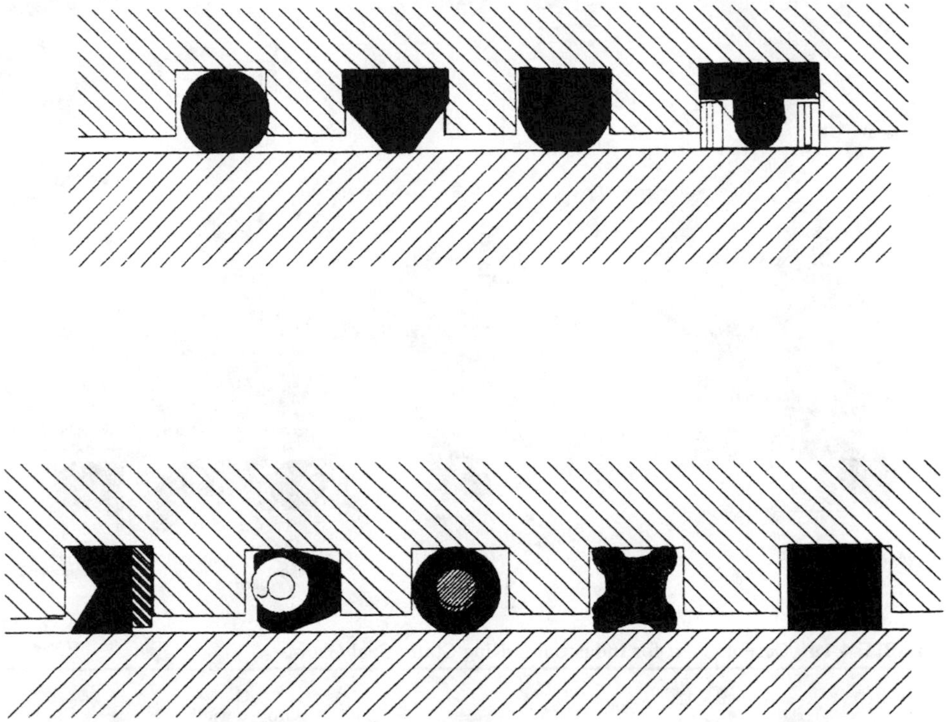

FIGURE 4.2 Typical seal cross-sections.

square (lathe cut rings) or round, such as the common O-ring. There are X-rings, D-rings, C-rings, U-rings, V-rings, delta-rings, and T-rings to name but a few.

SEAL STANDARDS

O-ring standards are specified in SAE J120A and SAE J515A and AS568 is an SAE aerospace industrial standard. Numerous military standards and Army-Navy standards such as MIL-G-5514F, AN6227, MS28775, M25988, M83248, MS9020, MS9355, MS29512, and MS28900 are also available to assist the user of O-rings. It is most important to follow the manufacturer's selection and installation recommendations (see Tables 4.1 and 4.2).

A typical O-ring groove is shown in Figs. 4.3 and 4.4, an axial installation is shown in Fig. 4.5, and a radial installation in Fig. 4.6. Keep in mind that only by closely working with the manufacturer's data can one be assured of proper fluid compatibility, specific groove size, and tolerances, temperature, and pressure limitations for a specific application.

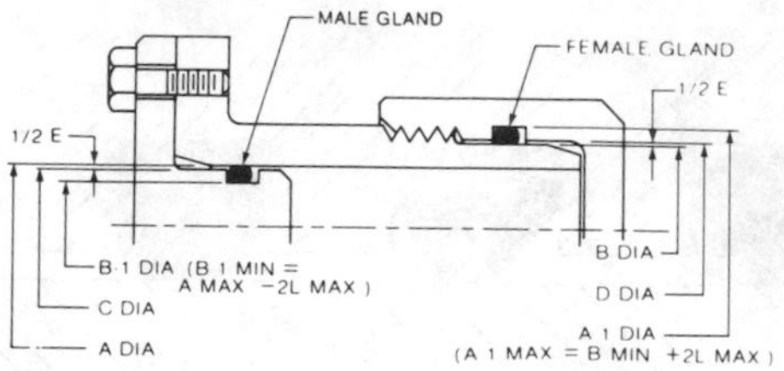

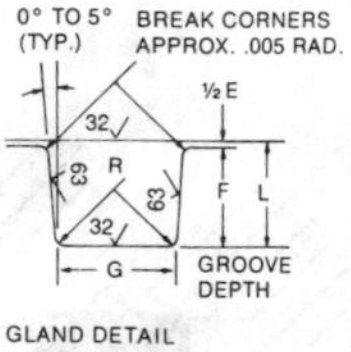

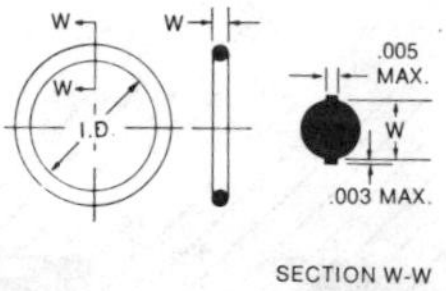

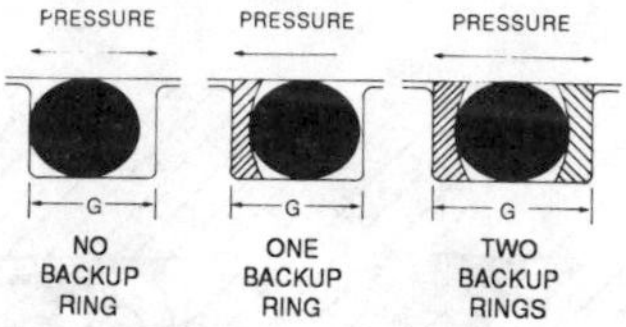

TABLE 4.1*a* Design Chart for Industrial O-Ring Static Seal Glands

O-ring size	W Cross-section Nominal	W Cross-section Actual	L* Gland depth	Squeeze Actual	Squeeze %	E†‡ Diametral clearance	G Groove width No backup rings	G Groove width One backup ring	G Groove width Two backup rings	R Groove radius	Eccentricity max.§
004 through 050	1/16	0.070 ±0.003	0.050 to 0.052	0.015 to 0.023	22 to 32	0.002 to 0.005	0.093 to 0.098	0.138 to 0.143	0.205 to 0.210	0.005 to 0.015	0.002
102 through 178	3/32	0.103 ±0.003	0.081 to 0.083	0.017 to 0.025	17 to 24	0.002 to 0.005	0.140 to 0.145	0.171 to 0.176	0.238 to 0.243	0.005 to 0.015	0.002
201 through 284	1/8	0.139 ±0.004	0.111 to 0.113	0.022 to 0.032	16 to 23	0.003 to 0.006	0.187 to 0.192	0.208 to 0.213	0.275 to 0.280	0.010 to 0.025	0.003
309 through 395	3/16	0.210 ±0.005	0.170 to 0.173	0.032 to 0.045	15 to 21	0.003 to 0.006	0.281 to 0.286	0.311 to 0.316	0.410 to 0.415	0.020 to 0.035	0.004
425 through 475	1/4	0.275 ±0.006	0.226 to 0.229	0.040 to 0.055	15 to 20	0.004 to 0.007	0.375 to 0.380	0.408 to 0.413	0.538 to 0.543	0.020 to 0.035	0.005

*For ease of assembly when backup rings are used, gland depth may be increased up to 5 percent.
†Clearance gap must be held to a minimum consistent with design requirements for temperature range variation.
‡Reduce maximum diametral clearance 50 percent when using silicone or fluorosilicone O-rings.
§Total indicator reading between groove and adjacent bearing surface.

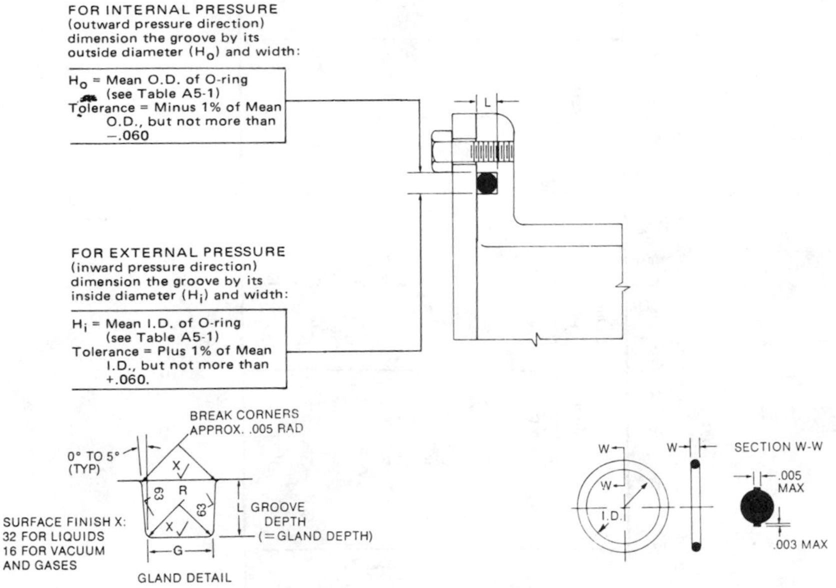

TABLE 4.1*b* Design Chart for O-Ring Face Seal Glands*

O-ring size	W Cross-section Nominal	W Cross-section Actual	L Gland depth	Squeeze Actual	Squeeze %	G Groove Width Liquids	G Groove Width Vacuum and gases	R Groove radius
004 through 050	1/16	0.070 ±.003	0.050 to 0.054	0.013 to 0.023	19 to 32	0.101 to 0.107	0.083 to 0.088	0.005 to 0.015
102 through 178	3/32	0.103 ±0.003	0.074 to 0.080	0.020 to 0.032	20 to 30	0.136 to 0.142	0.118 to 0.123	0.005 to 0.015
201 through 284	1/8	0.139 ±0.004	0.101 to 0.107	0.028 to 0.042	20 to 30	0.177 to 0.187	0.0157 to 0.163	0.010 to 0.025
309 through 395	3/16	0.210 ±0.005	0.152 to 0.162	0.043 to 0.063	21 to 30	0.270 to 0.290	0.236 to 0.241	0.020 to 0.035
425 through 475	1/4	0.275 ±0.006	0.201 to 0.211	0.058 to 0.080	21 to 29	0.342 to 0.362	0.305 to 0.310	0.020 to 0.035
Special	3/8	0.375 ±0.007	0.276 to 0.286	0.082 to 0.108	22 to 28	0.475 to 0.485	0.419 to 0.424	0.030 to 0.045
Special	1/2	0.500 ±0.008	0.370 to 0.380	0.112 to 0.138	22 to 27	0.638 to 0.645	0.560 to 0.565	0.030 to 0.045

*These dimensions are intended primarily for face type seals and low temperature applications.

TABLE 4.2 Properties of Elastomers Commonly Used for Gasket Materials

Properties	Styrene-butadiene	Ethylene propylene	Polychloroprene	Silicone	Nitrile-butadiene	Chlorosulfonated polyethylene	Fluorocarbon	Polyacrylate
Useful temperature range	−70 to +250°F	−65 to +350°F	−50 to +250°F	−120 to +600°F	−65 to +300°F	−60 to +250°F	−40 to +600°F	−30 to +400°F
Water	E	T	T	T	E	E	E	F
Acid	G	E	G	G	G	E	E	P
Alkali	E	G	E	F	G	E	G	P
Gasoline	P	P	F	P	E	F	E	G
Petroleum oil	P	P	G	P	E	G	G	E
Animal and vegetable oil	P-G	G	G	G	E	F	G	E
Hydrocarbon solvents	P	P	G (except aromatics)	P	G-E	F	E	G
Oxygenated solvents	F	G	P	P	P	F	F	P
Ozone	P	E	G	E	P	E	E	G

Key: E = excellent; G = good; F = fair; P = poor.

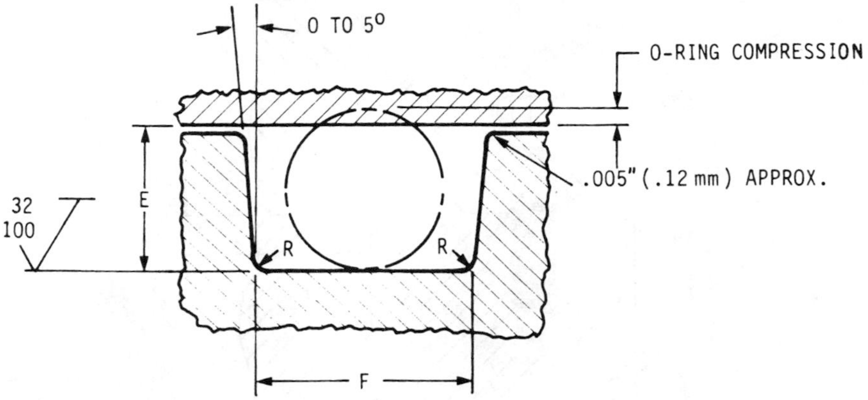

FIGURE 4.3 Typical O-ring groove.

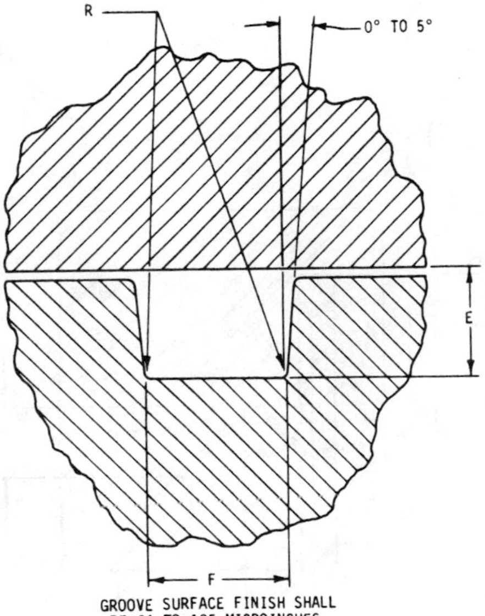

FIGURE 4.4 Typical groove for lathe cut rectangular rings.

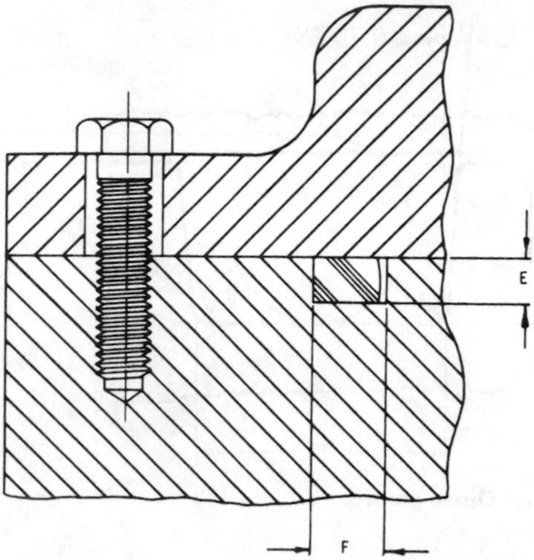

FIGURE 4.5 Typical static face-type seal application.

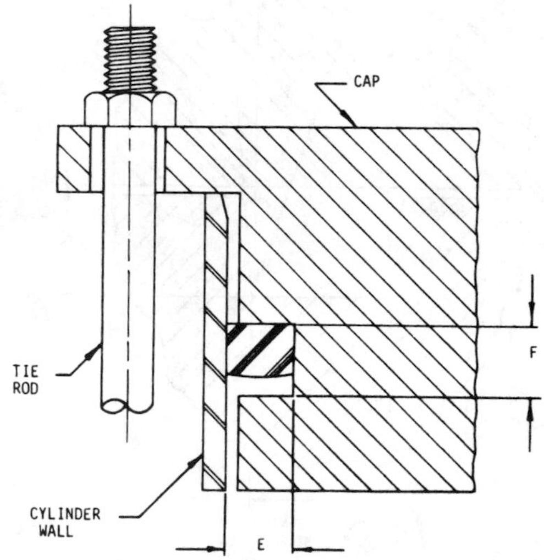

FIGURE 4.6 Typical static radial seal installation.

NITRILE, THE WORKHORSE ELASTOMER AND SOME IMPORTANT PHYSICAL PROPERTIES

The most common elastomer used in the manufacture of O-rings is nitrile (commonly called buna-N). A wide selection of elastomers is available to the user, and they are listed in Chap. 9 of this handbook. Selection of materials for elastomeric static seals depends on the operating temperature, the internal pressure, and the fluid to be sealed. The elastomer's physical properties, such as compression set, tensile set, oil swell, thermal expansion and contraction, hardness, shrinkage tolerances, and parting line flash (Fig. 4.7, Table 4.3), must be considered for each application.

Static seal leakage can be caused by a variety of reasons ranging from installation damage to extrusion of the seal. Extrusion and "nibbling" results when pulsating pressure forces the rubber into the clearance between the metal faces. Increasing the hardness of the elastomer will increase resistance to extrusion and permit greater clearances to be used between metal surfaces.

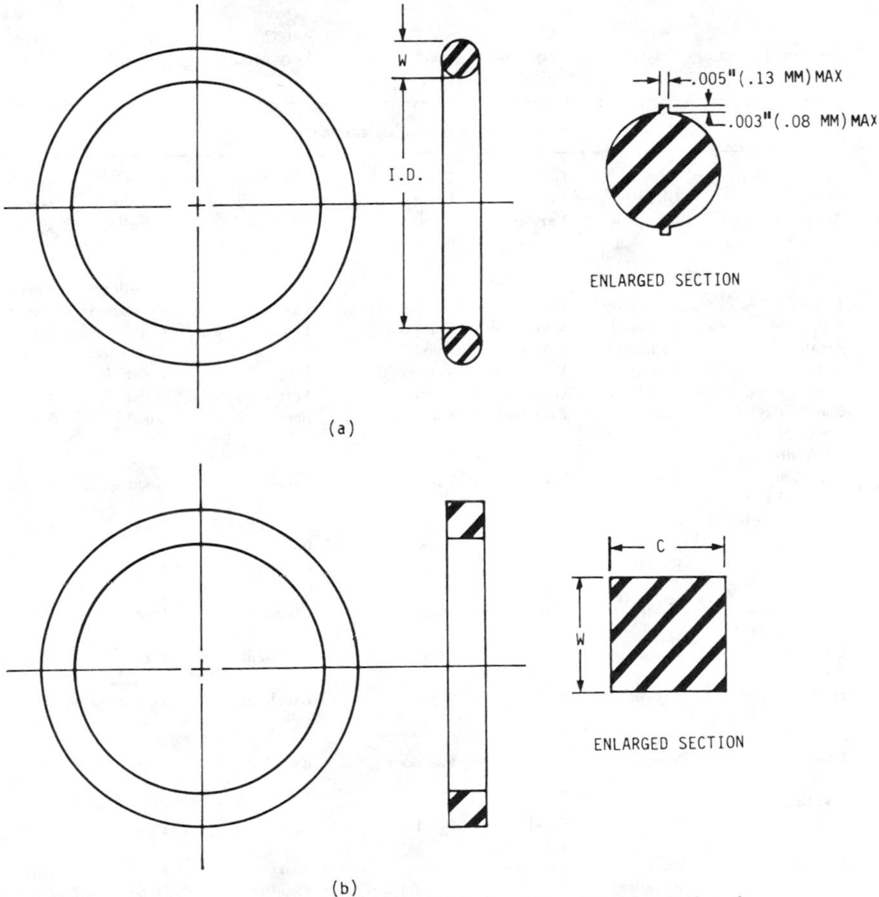

FIGURE 4.7 Cross-sections of typical seals. (*a*) O-ring; (*b*) rectangular-section ring.

TABLE 4.3 Application Compatibility

	Nitrile (Buna-N)	Ethylene-propylene	(Chloroprene) neoprene	Fluorocarbon (Viton, Fluorel)	Silicone	Fluoro-silicone
General						
Hardness range, A scale	40–90	50–90	40–80	70–90	40–80	60–80
Relative static ring cost	Low	Low	Low/Moderate	Moderate/High	Moderate	High
Continuous high-temperature limit	257°F 125°C	302°F 150°C	284°F 140°C	437°F 225°C	482°F 250°C	347°F 175°C
Low-temperature capability	−67°F −55°C	−67°F −55°C	−67°F −55°C	−40°F −40°C	−103°F −75°C	−85°F −65°C
Compression set resistance	Very good	Very good	Good	Very good	Excellent	Very good
Fluid compatibility summary						
Acid, inorganic	Fair	Good	Fair/good	Excellent	Good	Good
Acid, organic	Good	Very good	Good	Good	Excellent	Good
Aging (oxygen, ozone, weather)	Fair/poor	Very good	Good	Very good	Excellent	Excellent
Air	Fair	Very good	Good	Very good	Excellent	Very good
Alcohols	Very good	Excellent	Very good	Fair	Very good	Very good
Aldehydes	Fair/poor	Very good	Fair/poor	Poor	Good	Poor
Alkalis	Fair/good	Excellent	Good	Good	Very good	Good
Amines	Poor	Very good	Very good	Poor	Good	Poor
Animal oils	Excellent	Good	Good	Very good	Good	Excellent
Esters, alkyl phosphate (Skydrol)	Poor	Excellent	Poor	Poor	Good	Fair/poor
Esters, aryl phosphate	Fair/poor	Excellent	Fair/poor	Excellent	Good	Very good
Esters, silicate	Good	Poor	Fair	Excellent	Poor	Very good
Ethers	Poor	Fair	Poor	Poor	Poor	Fair
Hydrocarbon fuels, aliphatic	Excellent	Poor	Fair	Excellent	Fair	Excellent
Hydrocarbon fuels, aromatic	Good	Poor	Fair/poor	Excellent	Poor	Very good
Hydrocarbons, halogenated	Fair/poor	Poor	Poor	Excellent	Poor	Very good
Hydrocarbon oils, high aniline	Excellent	Poor	Good	Excellent	Very good	Excellent
Hydrocarbon oils, low aniline	Very good	Poor	Fair/poor	Excellent	Fair	Very good
Impermeability to gases	Good	Good	Good	Very good	Poor	Poor
Ketones	Poor	Excellent	Poor	Poor	Poor	Fair/poor
Silicone oils	Excellent	Excellent	Excellent	Excellent	Good	Excellent
Vegetable oils	Excellent	Good	Good	Excellent	Excellent	Excellent
Water/steam	Good	Excellent	Fair	Fair	Fair	Fair

TABLE 4.3 Application Compatibility (*Continued*)

Styrene-butadiene (SBR)	Poly-acrylate	Poly-urethane	Butyl	Polysulfide (Thiokol)	Chloro-sulfonated poly-ethylene (Hypalon)	Epichloro-hydrin (Hydrin)	Phospho-nitrilic fluoro-elastomer (PNF)
			General				
40–80	70–90	60–90	50–70	50–80	50–90	50–90	50–90
Low	Moderate	Moderate	Moderate	Moderate	Moderate	Moderate	High
212°F	347°F	212°F	212°F	212°F	257°F	257°F	347°F
100°C	175°C	100°C	100°C	100°C	125°C	125°C	175°C
−67°F	−4°F	−67°F	−67°F	−67°C	−67°C	−67°C	−85°F
−55°C	−20°C	−55°C	−55°C	−55°C	−55°C	−55°C	−65°C
Good	Fair	Fair	Fair/good	Fair	Fair/poor	Fair/good	Good
			Fluid compatibility summary				
Fair/good	Poor	Poor	Good	Poor	Excellent	Fair	Poor
Good	Poor	Poor	Very good	Good	Good	Fair	Fair
Poor	Excellent	Excellent	Very good	Excellent	Very good	Very good	Excellent
Fair	Very good	Good	Good	Good	Excellent	Good	Excellent
Very good	Poor	Poor	Very good	Fair/good	Very good	Good	Fair
Fair/poor	Poor	Poor	Good	Fair/good	Fair/good	Poor	Poor
Fair/good	Poor	Fair/good	Excellent	Poor	Excellent	Fair	Good
Fair	Poor	Poor	Good	Poor	Poor	Poor	Good
Poor	Excellent	Good	Good	Poor	Good	Good	Fair
Poor	Poor	Poor	Very good	Poor	Poor	Poor	Poor
Poor	Poor	Poor	Excellent	Good	Fair	Poor	Excellent
Poor	Fair/poor	Poor	Poor	Fair/poor	Fair	Good	Excellent
Poor	Fair/poor	Fair	Fair/poor	Good	Poor	Good	Poor
Poor	Very good	Good	Poor	Excellent	Fair	Very good	Excellent
Poor	Poor	Fair/poor	Poor	Good	Fair/poor	Very good	Excellent
Poor	Fair/good	Fair	Poor	Good	Fair	Excellent	Fair
Poor	Excellent	Excellent	Poor	Very good	Excellent	Excellent	Excellent
Poor	Excellent	Very good	Poor	Good	Very good	Excellent	Excellent
Fair/good	Very good	Fair	Excellent	Very good	Very good	Excellent	Fair
Poor	Poor	Poor	Excellent	Good	Fair	Fair	Poor
Excellent	Excellent	Excellent	Excellent	Excellent	Excellent	Excellent	Excellent
Poor	Good	Fair	Good	Poor	Good	Excellent	Fair
Fair	Poor	Poor	Excellent	Fair	Fair	Good	Fair

SOME INSTALLATION CONSIDERATIONS

Surface finish is important but not as important as with dynamic seals where abrasive wear can be a problem. A 0.13- to 0.40-μm (32- to 63-μin) roughness is acceptable for groove finish.

Many users tend to overfill the seal groove, ignoring SAE and manufacturer's specifications, but keep in mind that the O-ring is neither compressible nor very strong and will split and crack if too much compressive force is used. So use recommended groove designs and allow for expansion of the O-ring. A minimum squeeze on the O-ring of 0.15 mm (0.006 in) should be used to account for the differences in the shrinkage of rubber and metal as temperatures vary. The maximum squeeze permissible is 35 percent of the O-ring cross-section diameter. Most manufacturers permit standard groove tolerances to be used up to 10.3 N/mm^2 (1500 psi). It is recommended, however, that if unusual conditions of fluid, temperature, and pressure are involved, you should by all means contact the manufacturer.

Installing O-rings over shafts can sometimes cause them to twist and suffer spiral leakage failure, hence the use of T-rings with backup rings to prevent leakage. A T-ring seal will usually fit into the same groove as an O-ring. T-rings have been used up to 137 N/mm^2 (20,000 psi) and are used frequently in aircraft applications. The rectangular lathe-cut rings also can minimize twisting at installation but are usually limited to 10.3 N/mm^2 (1500 psi). It is important to remember also that molded seals have much greater density and strength and better overall physical properties than extruded seals. Window seals are usually extruded, for example, whereas cylinder head seals and flange seals would be molded. It's also important in highly engineered applications to avoid using a spliced seal. The joint creates a point of differential swell and thermal expansion, and the bond itself quite often is made imperfectly. A seal formed or molded into its final shape (usually circular) is the preferable type of sealing product. Seals up to 5 m (15 ft) in size are routinely molded in full round condition.

TESTING CAN AID THE SELECTION PROCESS

Contrary to popular notion, one test is not worth a thousand expert opinions, and quite often many tests are required to statistically substantiate the design of choice. Properly designed elastomeric static seal applications for flanges and flange fittings as high as 171.7 N/mm^2 (25,000 psi) can be made, although coated metal seals are more commonly used in high-pressure applications.

An important factor to keep in mind when sealing static applications with soft sealing materials under high pressure is to have the two mating metal surfaces touching each other to prevent extrusion of the seal (Fig. 4.8). The maximum limiting pressure is not due to the seal design only but rather to the design of the entire installation (Fig. 4.9). Temperature is always a modifying factor as is the type of fluid, so always refer to the manufacturer's technical data when considering a new application.

When cutting the installation grooves for seals, the Vee groove is the cheapest but permits easier extrusion, and if there is some movement of the sealing faces, the Vee groove creates the most friction (Fig. 4.10). The dovetail, or undercut groove, is best at preventing extrusion and reduces friction if there is motion. It

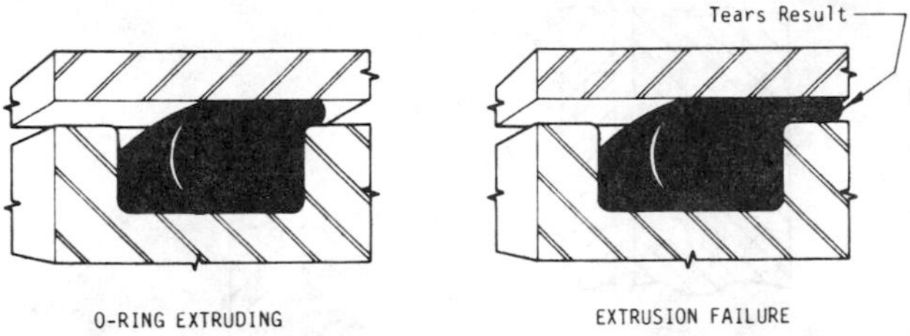

FIGURE 4.8 Tears can result from nibbling if extrusion occurs.

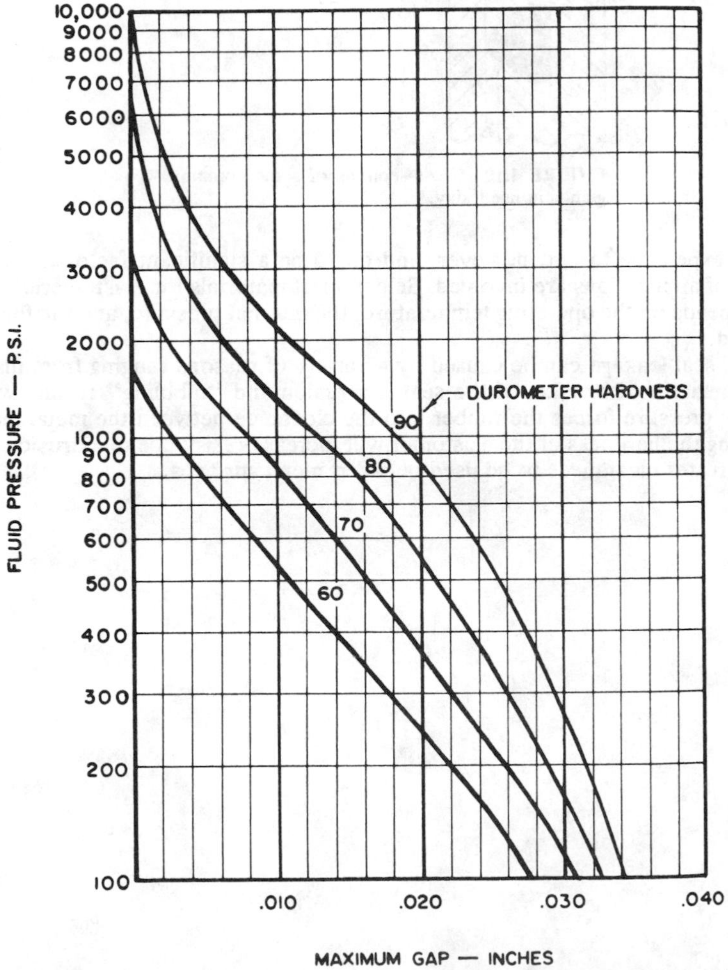

FIGURE 4.9 Maximum permissible clearance at a given pressure for various rubber hardness values. (*Courtesy, Hydraulics & Pneumatics.*)

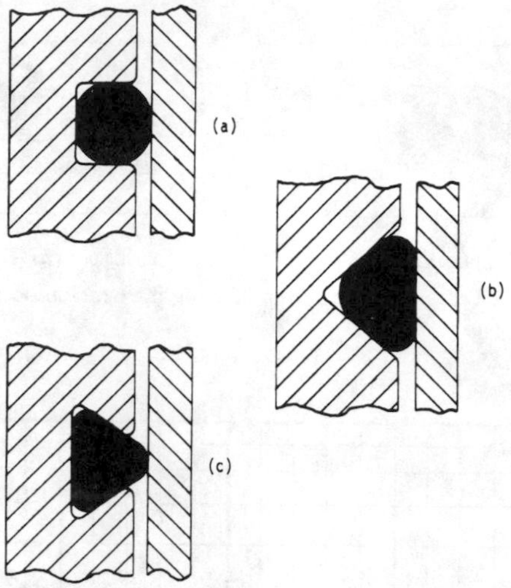

FIGURE 4.10 Cross-sections of some common glands in use today.

is more expensive to cut, however, and could be a significant factor if a large number of applications are involved. Selection of materials for elastomeric static seals depends on the operating temperature, the internal pressure, and the fluid to be sealed.

Static seal leakage can be caused by a variety of reasons ranging from installation damage to extrusion of the seal. Extrusion and "nibbling" results when pulsating pressure forces the rubber into the clearance between the metal faces. Increasing the hardness of the elastomer will increase resistance to extrusion and permit greater clearances to be used between metal surfaces.

CHAPTER 5
CHEMICAL GASKETS AND SEALANTS

Chemical gaskets and sealants are available in liquids of varying viscosity, pastes, puttylike mastics, and tapes dispensed manually or automatically from bottles, tubes, and cartridges. Chemical gaskets have achieved considerable acceptance and adoption in industrial gasketing.

CHEMICAL GASKET SEALING

Two classes of materials dominate the "chemical" gasket field. They are the room-temperature vulcanizing (RTV) silicones and anaerobics. The RTV silicones cure by absorbing moisture in the air and giving off acetic acid; the newer types give off amines. Silicone RTV having a low content of silicone volatiles are available. These volatiles may foul sensors used in engine emission systems. The anaerobics cure in the absence of air when in contact with an active metal (i.e., when air is excluded such as in a clamped gasketed joint).

Both of the classes of materials contain 100 percent solids, and this ensures that no voids or shrinkage occur during drying or curing. Since the mating flanges are assembled when the material is uncured or wet, shrinkage during drying and curing could result in leak paths. There are other types of sealant materials available, but they are restricted to limited applications and therefore are not covered in this handbook. In some industries, the chemical gaskets are called formed-in-place gaskets.

The theory of sealing chemical gaskets differs substantially from that of typical mechanical gasketing. In the latter case, the material is compressed, and it exerts a sealing force on the flange proportional to its stress-retention and recovery characteristics. Due to the support of the load, the gasket tends to follow the motion of the mating flanges whether they are thermal or mechanically induced. In some applications bolt stretch is a factor in maintaining joint tightness.

In the case of chemical gaskets, metal-to-metal or flange-to-flange contact occurs, and the gasket does not support the load or contribute to tolerance stack-up. Chemical gaskets seal by filling the gaps between the flanges and adhering to the flanges. The high extensibility of silicone RTVs allows them to follow flange motion. To ensure adequate silicone gasket thickness to follow motion, it is a common practice to include gasket stops designed into one of the mating flanges. In some cases, the flange is grooved to permit a thick silicone RTV bead.

In the case of the anaerobics, they seal by completely filling the gap and adhesively stopping the flange motion by bonding the flanges together. Anaerobics have little extensibility but do possess very high shear and tensile strength. Bond strength is a controlled variable. Ease of disassembly and flange clean-up are a function of bond strength. They are used for heavy cast flanges and tight fitting cases, such as oil seal cases.

When metal-to-metal contact occurs with chemical gaskets, torque loss is normally negligible. In addition, flange motion is generally less than that in the case of a mechanical gasketed joint since the fastener torque is retained. Also, flange distortion is minimized since the flanges are assembled while the chemicals are uncured and metal-to-metal contact occurs. Chemical gaskets generally do not contribute to tolerance stackup.

CHEMICAL GASKET—CURED AND UNCURED PROPERTIES

With chemical gaskets, one must be concerned with not only the cure properties of the material but with the uncured properties as well. We have already mentioned the importance of 100 percent solids content. One concern is the viscosity of the chemical gaskets. The chemicals must fill the gap as well as the unevenness of the mating flanges and remain in place until cured. Therefore, the viscosity must be high enough to prevent flowing out. In the case of the anaerobics, for example, intimate contact with the mating flanges is required to ensure exclusion of air, curing, and adhesion. Therefore, the viscosity of the compounds will be dictated by the gap thickness that needs to be filled.

To completely fill the gap, an excess amount of material is required. This, however, can result in the material flowing into the cavity to be sealed and mixing with the sealed media. This possibility must be considered when specifying the chemical gaskets. There have been cases where the excess material has caused problems by blocking passageways in the assembly.

The most popular silicone RTVs on the market cure by absorbing moisture from the air; therefore, their curing characteristics and rate of cure are a function of relative humidity. Figure 5.1 is a graph showing the rate of cure versus percentage of relative humidity for an RTV compound. Note that 20 percent and higher relative humidity is required for rapid cures. Figure 5.2 is a graph that shows the rate of cure of an anaerobic versus gap size. Note that there is a limit to the size gap that can be filled and still result in the cure of the anaerobic. The activity level of the flange metal also affects the rate of cure. In addition, depending upon the internal pressure and the time permitted for curing, there are limitations on the size of the gap permitted for both the silicone RTVs and the anaerobics.

Concerning storage stability, the RTVs must be kept tightly sealed until just prior to use. Anaerobics conversely do require exposure to oxygen in order to prevent cure; therefore, the storage containers must take this into account. Most chemical gaskets have about a 1-year shelf life.

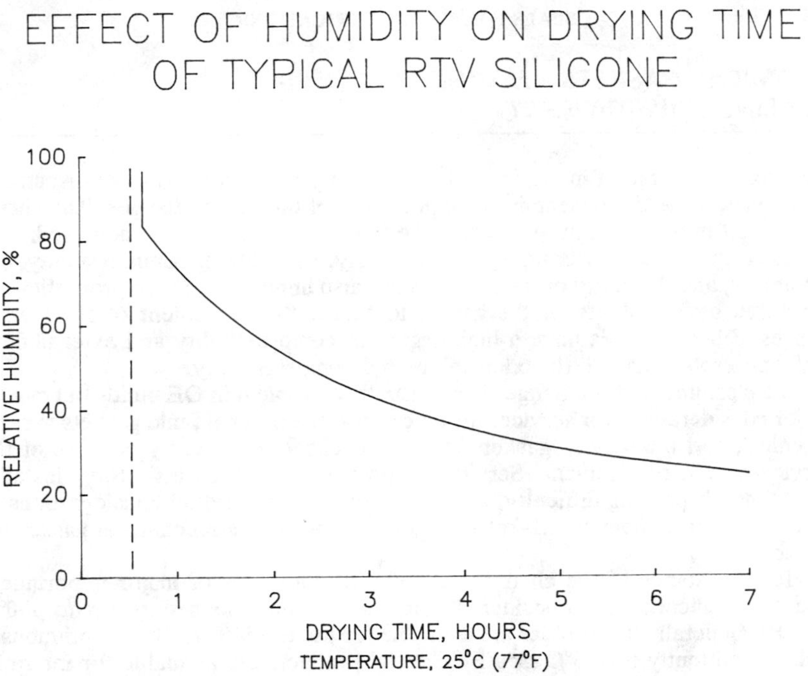

FIGURE 5.1 RTV cure time for various percentages of relative humidities.

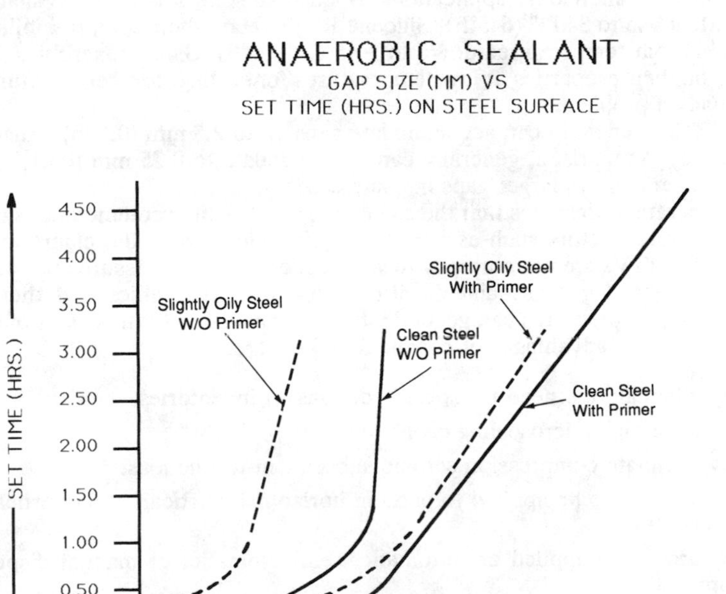

FIGURE 5.2 Anaerobic cure versus size of gap.

CHEMICAL GASKETS AND THE SEALING ENVIRONMENT

Another consideration in sealing with chemical gaskets concerns the dispensing of the material and the cleanliness requirement of the mating flanges. The chemical gasket must be firmly bonded to the flanges for a seal to be achieved. The adhesive strength of the bond is a key property. In addition, compressibility, extensibility, and flexibility of the products are also important. These properties are measured by the ability of the gasket to follow the movement of the mating flanges. Silicone RTVs have a high degree of compressibility and extensibility, while anaerobics have little extensibility and compressibility.

The cleanliness of the flanges is not usually a problem in OE builds but can be major considerations for service. In some cases, the original build gaskets may be chemical, and the service gaskets may be mechanical primarily because of the service-related requirement. Service requirements, such as inventory, installation, time, dispensing difficulty, shelf life, and so forth, are all considerations in determining if a chemical gasket should be changed to a mechanical gasket for service.

Heat resistance of the chemical gaskets, of course, is of major importance. The RTV silicones are considered suitable for continuous duty up to 190°C (375°F). Generally, anaerobics are available for use to 154°C (310°F) continuously and intermittently to 175°C (350°F). Special products are available for intermittent use up to 230°C (450°F). One needs to refer to the manufacturer's recommendations for individual applications. Special silicone RTVs are available for intermittent use to 340°C (645°F). Silicone RTVs retain their compressibility and extensibility at temperatures down to −45°C (−50°F). Since anaerobics do not depend on their properties at low temperatures for sealing, low temperatures are not usually a problem.

The RTVs generally can accommodate gaps up to 2.5 mm (0.1 in). Anaerobic sealants, by comparison, generally can accommodate to 0.25 mm (0.01 in). The use of primers allows larger gaps in some cases.

The operating pressures that the chemical gaskets can accommodate is dependent on several factors such as cure time, joint width, gap, and clamp load. In general, the RTVs are not normally recommended for high-pressure applications. Due to the small gap associated with sealing with anaerobics and their high strength, higher pressures can generally be accommodated by this class of chemical gaskets. The advantages of chemical gaskets are:

1. They eliminate the need for specific cut gasket inventories.
2. They have high microsealing capabilities.
3. They eliminate compression set and subsequent torque loss.
4. They are able to be applied to parts in horizontal, vertical, and overhead positions.
5. They are easily applied on automatic, semiautomatic, or manual dispensing equipment.
6. They do not require cutting tools.

The disadvantages of chemical gaskets are:

1. Some of them create by-products which can cause corrosion of the mating flanges and/or the assembly.

2. In some cases, the working time is dependent on temperature and humidity.
3. They have a limitation of gap filling ability.
4. They have limited flange motion following ability.
5. Some chemical gaskets can cause some skin irritations.

SUMMARY

Chemical gaskets definitely have a place in industrial gasketing applications. Their sealing ability, however, varies greatly with the chemical compound itself. It is therefore recommended that the material manufacturers be consulted when specifying a chemical gasket for a specific application. In addition, one can consult *Design Guide for Formed-in Place Gaskets,* SAE Publication J-1497 published in May 1988. It contains information on the types of sealants available, cure systems, cured and uncured properties of the chemical gaskets, and application techniques for the initial seal and for surface resealing.

P • A • R • T • 2

SEALS FOR DYNAMIC APPLICATIONS

CHAPTER 6

RECIPROCATION, OSCILLATION, SLOW ROTATION

Seals are used in reciprocating and rotating applications to prevent fluid leakage and/or to exclude contaminants. Referred to as packings, they can be classified into four basic categories: (1) Compression packings, (2) molded packings, lathe cut, and O-rings, (3) floating seals, and (4) diaphragms (see Fig. 6.1).

Compression packings are used primarily in rotary or reciprocating shaft applications. They are squeezed between the throat of a box and a gland to effect a seal. The packing flows outward to seal against the bore of the box and inward to seal against the moving shaft or rod. Compression packings require periodic tightening to compensate for compression set, wear, and loss of volume. Automatic or molded packings rely on operating pressure to create a seal; therefore, very little gland adjustment is required. Flexible lips seal against one or both surfaces in a stuffing box. One lip seals against the stationary bore and the other lip

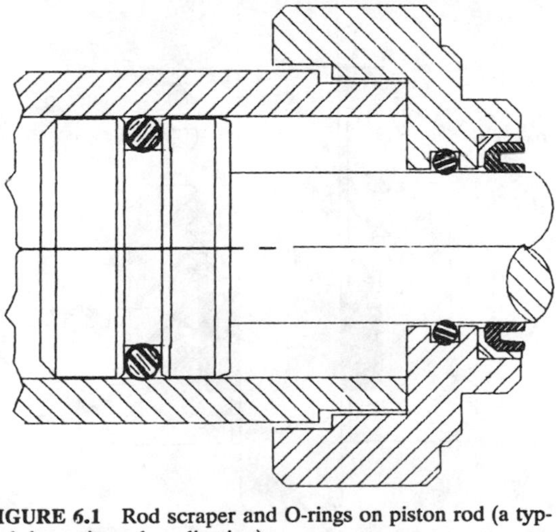

FIGURE 6.1 Rod scraper and O-rings on piston rod (a typical dynamic seal application).

against the moving shaft or piston. In some cases, O-rings, quad rings, and other molded static seals are used in dynamic applications. These seals are called *squeeze* seals and depend on internal pressure to effect a seal. More molded packings are used in reciprocating applications. Floating packings include piston rings and segmental rod packings that fit into grooves. These rings are used in rotary and reciprocating motion. Diaphragms are used primarily in reciprocating piston applications.

COMPRESSION PACKINGS

Compression packings create a seal by being squeezed between the throat of a stuffing box and its adjustable gland (Fig. 6.2). The squeeze force pushes the material against the throat of the box and the reciprocating or rotating shaft. When leakage occurs, the gland is tightened further.

Materials are extremely important when selecting the proper packing for an application. The packing must be able to withstand the conditions outlined in Table 6.1. Packing material is made of fabric or metal. The packing cross-section can be round, square, or rectangular.

Fabric packings are made from plant fibers, mineral fibers, animal products, or artificial fibers. Plant fibers include flax, jute, ramie, and cotton. Cotton is usually used in cloth form, and the other materials are braided to form high-strength, water-resistant packings. All plant fibers are restricted to low-temperature applications. Asbestos is the most versatile mineral fiber. It is resistant to strong mineral acids. Asbestos loses strength at about 480°C (800°F).

Leather strips are used to make braided packings. Leather is restricted to applications of 93°C (200°F) or less. Artificial fibers are used in applications to seal corrosive liquids at temperatures up to 260°C (500°F). PTFE and graphite are

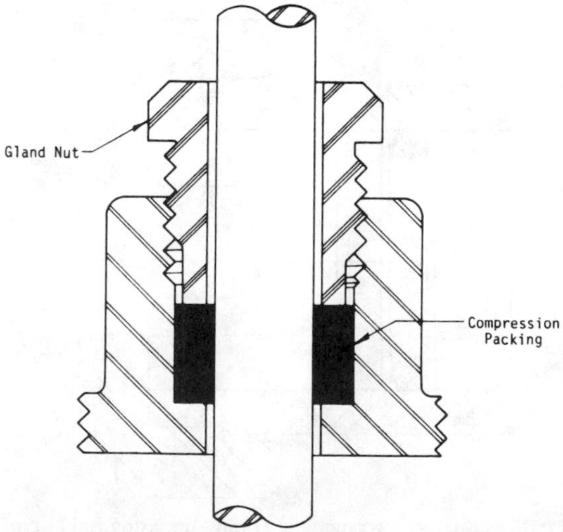

FIGURE 6.2 Compression packing installed in stuffing box.

TABLE 6.1 Compression Packings for Various Service Conditions

Fluid medium	Service condition			
	Reciprocating shafts	Rotating shafts	Pistons or cylinders	Valve stems
Acids and caustics	Asbestos (blue) Metallic Semimetallic TFE fluorocarbon resins and yarns	Asbestos (blue) Semimetallic TFE fluorocarbon resins and yarns	TFE fluorocarbon resins	Asbestos (blue) Semimetallic TFE fluorocarbon resins and yarns Graphite yarn
Air	Asbestos Metallic Semimetallic	Asbestos Semimetallic	Leather Metallic	Asbestos Semimetallic
Ammonia	Duck and rubber Metallic Semimetallic	Asbestos Semimetallic	Duck and rubber	Asbestos Duck and rubber Semimetallic
Gas	Asbestos Metallic Semimetallic	Asbestos Semimetallic	Leather Metallic	Asbestos Semimetallic
Cold gasoline and oils	Asbestos Semimetallic	Asbestos Semimetallic	Leather	Asbestos Semimetallic
Low-pressure steam	Asbestos Duck and rubber Metallic Semimetallic	Asbestos Metallic Semimetallic	Duck and rubber Metallic	Asbestos Duck and rubber Semimetallic
High-pressure steam	Asbestos Metallic Semimetallic	Asbestos Metallic Semimetallic	Metallic	Asbestos Metallic Semimetallic
Cold water	Duck and rubber Flax, jute, or ramie Leather Semimetallic	Asbestos Flax, jute, or ramie Semimetallic	Duck and rubber Semimetallic	Asbestos Flax or cotton
Hot water	Duck and rubber Leather Semimetallic	Asbestos Semimetallic	Duck and rubber	Asbestos Duck and rubber Semimetallic

(*Courtesy,* Machine Design.)

used to make yarns for packings. The simplest yarn-type packing consists of strands of material twisted together to form a rope (Fig. 6.3). Braided packings can be obtained in round, square, or interlocked configurations (Fig. 6.4).

The packings are impregnated with mineral oil, grease, or graphite. This lubricates the moving parts of the application to reduce wear, retain packing flexibility, and help effect a seal. Cloth packings are made from sheets of asbestos or cotton duck that are rolled, folded, or laminated with rubber to form the desired cross section (Fig. 6.5).

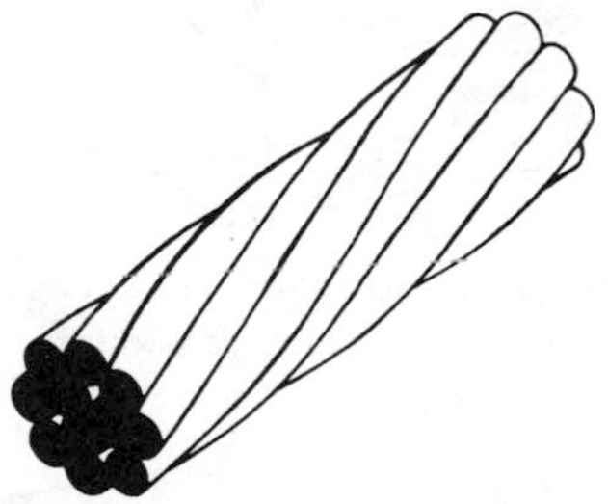

FIGURE 6.3 Twisted yarn fabric packing.

Metallic packings are used in high-temperature applications. Lead is used in applications up to 232°C (450°F). Soft pure copper is flexible; it is used in hot oil, tar, or asphalt pump applications. Aluminum foil is more flexible than copper and is resistant to sour crude oil. Shafts for copper and aluminum packings must be hardened to 500 Brinell hardness number (Bhn); copper and aluminum can handle 538°C (1000°F) application temperatures. Pure nickel is used to handle steam or caustic alkali materials at temperatures of up to 815°C

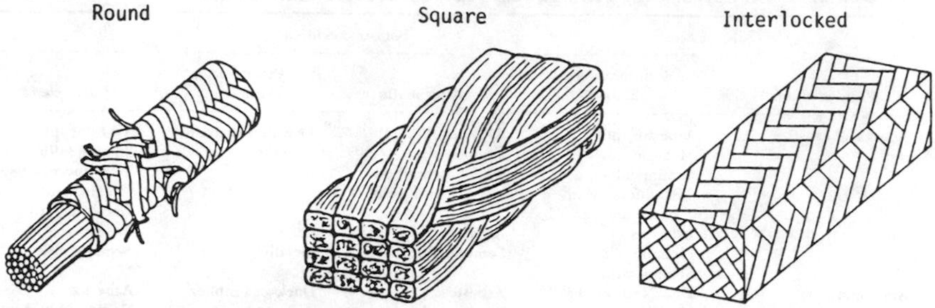

FIGURE 6.4 Braided yarn fabric packing types.

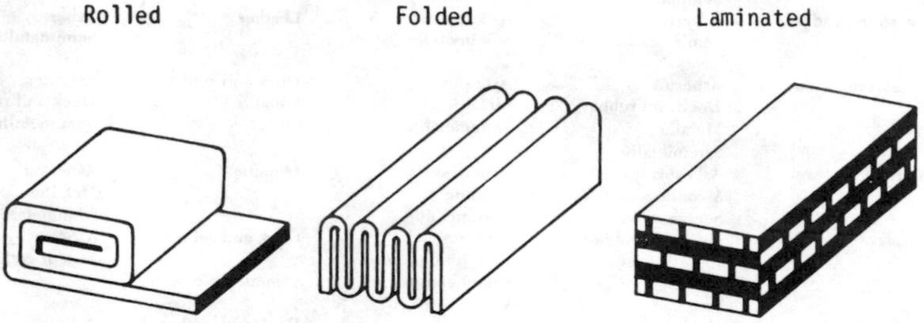

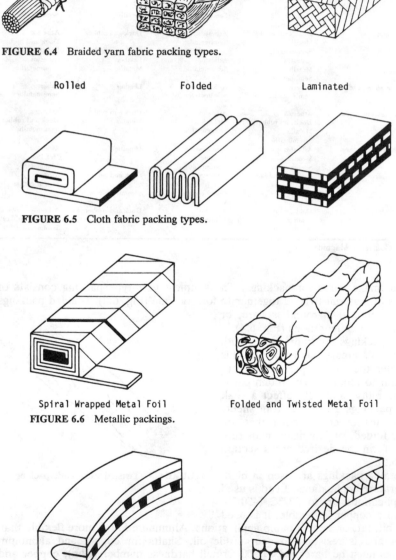

FIGURE 6.5 Cloth fabric packing types.

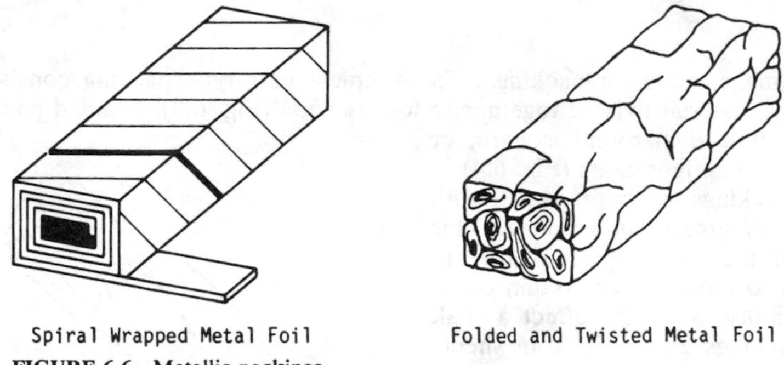

FIGURE 6.6 Metallic packings.

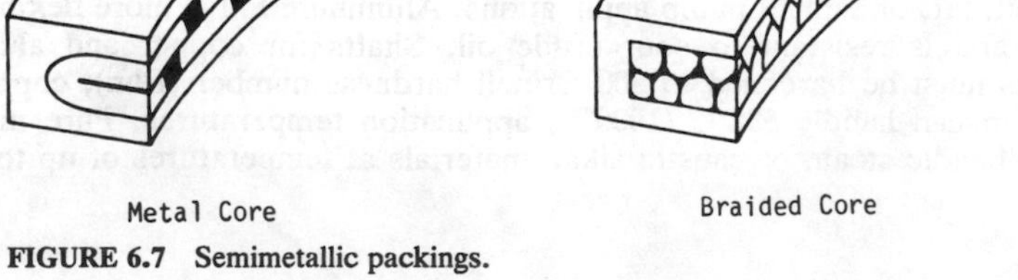

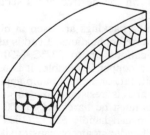

FIGURE 6.7 Semimetallic packings.

(1500°F). Metal packing designs are shown in Fig. 6.6. Metal is sometimes combined with other materials to form semimetallic designs (Fig. 6.7).

The types of compression packings that should be used for various service and fluid conditions are summarized in Table 6.1.

MOLDED PACKINGS

Molded packings are sometimes referred to as automatic, hydraulic, or mechanical packings. They rely on the fluid pressure of the application to press the packing material against the wear surfaces. Lip-type packings are used primarily for reciprocating shafts. V- and U-ring packings can be used for both high- (50,000 psi, or 345 N/mm^2) and low-pressure applications. Multiple V-rings are often installed in sets. They are primarily packed on the outside of a reciprocating rod (Fig. 6.8). U-rings are usually used to seal a piston (Fig. 6.9). Cup packings have a single lip and are used to seal pistons (Fig. 6.10).

Flanged packings are designed to seal on the inner diameter (Fig. 6.11) and are used for low-pressure, outside packed installations where there is not enough room for V- or U-rings. Exclusion seals are used in conjunction with packings to keep out solid and liquid contaminants. Rod wipers (Fig. 6.12) are usually molded from tough materials that resist abrasion. Carboxylated nitriles and polyurethanes are usually used. They are hard (85 to 95 Shore A) with high (6000 psi, or 41 N/mm^2) tensile strength. V- and U-cup packings are sometimes combined with a rod wiper to form a unitized seal (Figs. 6.13 and 6.14).

Wiper scraper seals (Fig. 6.15) have a metal ring made from copper, aluminum, bronze, or brass that scrapes debris from the shaft. The metal used must have low-corrosion and low-friction characteristics and must be soft enough to resist scoring the shaft. Boots (Fig. 6.16) are used to prevent contamination from getting to the packings. Many shapes of boots are available. Specialized boots for hydraulic and pneumatic cylinders accommodate various stroke lengths.

Packings are made from rubber, fabric reinforced rubber, and leather. Leather

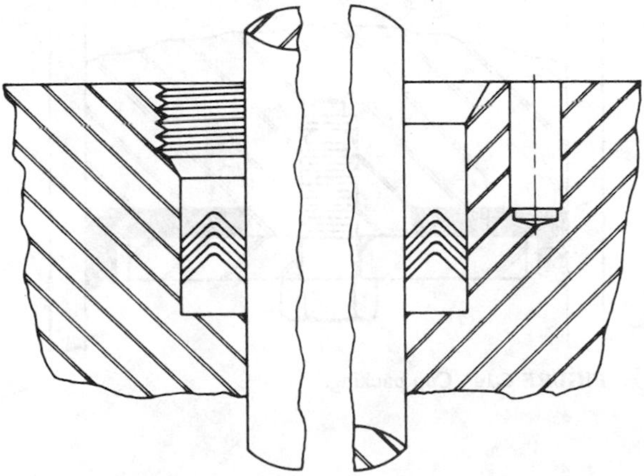

FIGURE 6.8 V-ring packing in a gland.

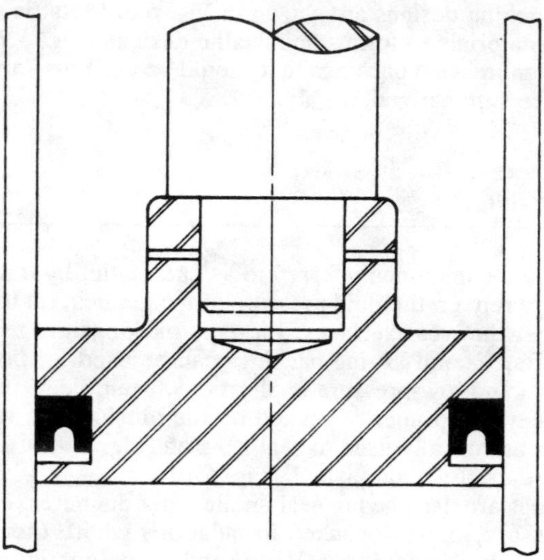

FIGURE 6.9 U-packing used to seal a piston.

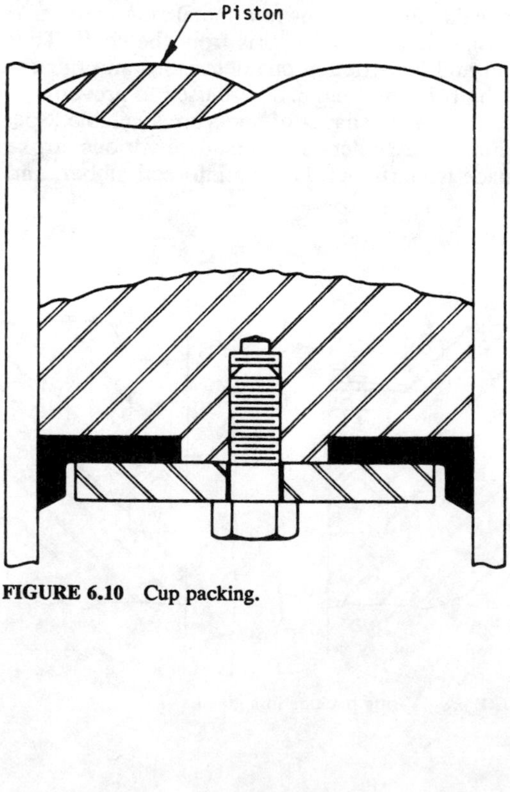

FIGURE 6.10 Cup packing.

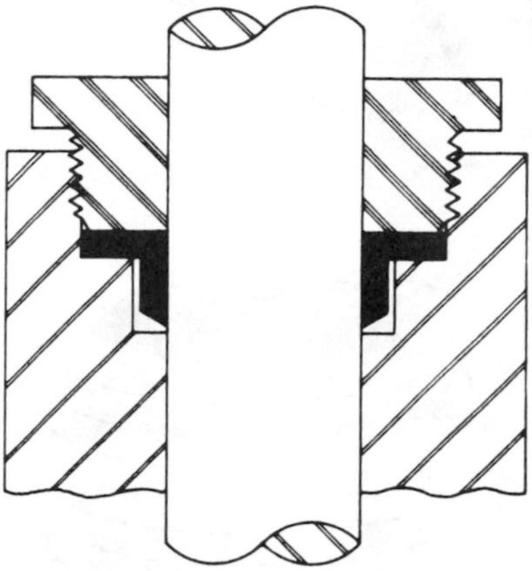

FIGURE 6.11 Flange packing.

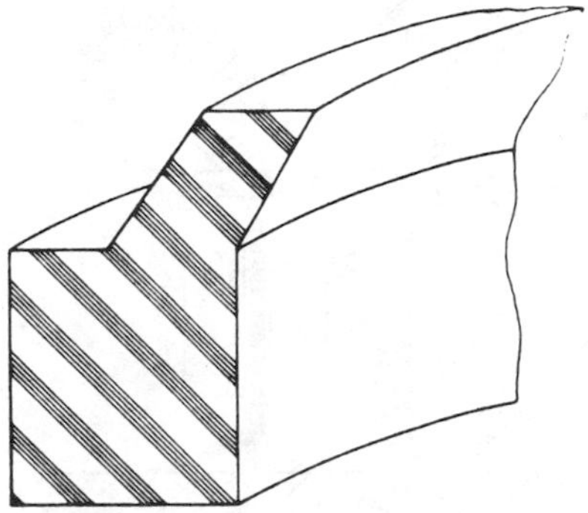

FIGURE 6.12 Rod wiper seal.

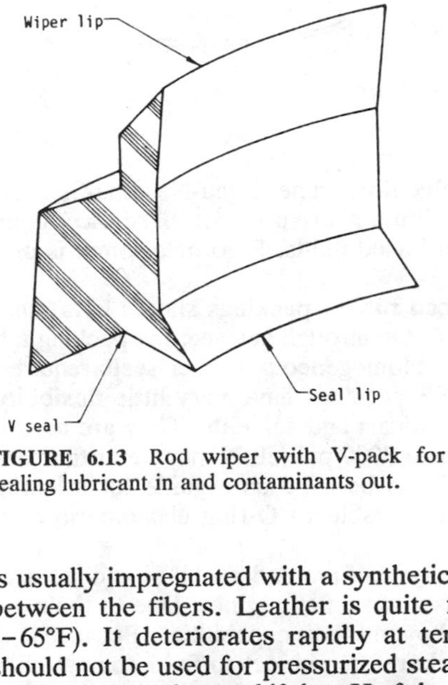

FIGURE 6.13 Rod wiper with V-pack for sealing lubricant in and contaminants out.

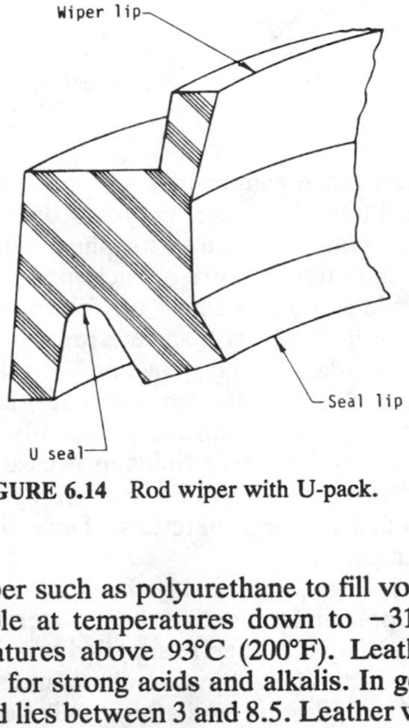

FIGURE 6.14 Rod wiper with U-pack.

is usually impregnated with a synthetic rubber such as polyurethane to fill voids between the fibers. Leather is quite flexible at temperatures down to −31°C (−65°F). It deteriorates rapidly at temperatures above 93°C (200°F). Leather should not be used for pressurized steam or for strong acids and alkalis. In general leather can be used if the pH of the liquid lies between 3 and 8.5. Leather will conform to rough surfaces up to 60 μin centerline average (CLA).

Packings are also made from synthetic rubber. The upper pressure limit of homogeneous rubber packings is 5000 psi (35 N/mm^2). This limit can be extended to 8000 psi (55 N/mm^2) if the material is reinforced with fabric. Cotton duck is commonly used to reinforce rubber when the temperature is less than 121°C (250°F). Asbestos is used for temperatures above 121°C (250°F). Nylon is used when strength and flexibility are required.

The synthetic rubber used depends on the type of fluid and the temperature.

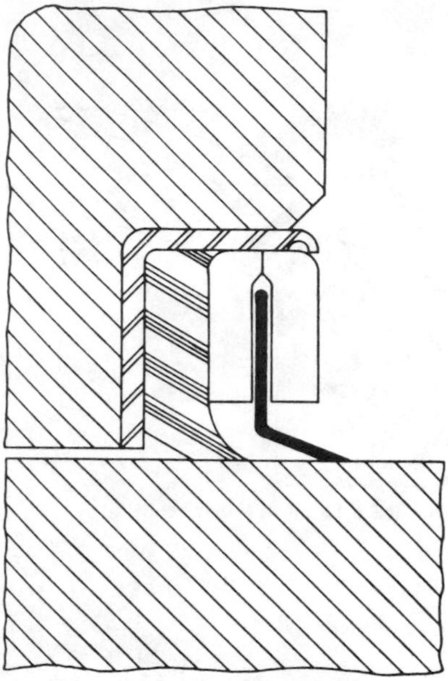

FIGURE 6.15 A wiper-scraper seal.

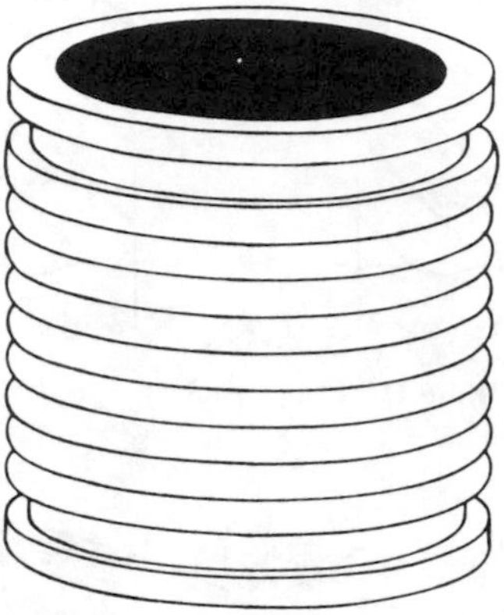

FIGURE 6.16 Boots for knuckles and joints.

The most common base polymers are polychloroprene, buna-N, buna-S, butyl, and fluoroelastomer. Polychloroprene and buna-N are used for oil service, buna-S for water, and butyl for phosphate ester-based fluids. Fluoroelastomer is used for high-temperature applications.

Metal surface finish for fabric reinforced rubber packings should be a maximum of 32 μin with a preferred value of 16. On a rough surface, the packing fabric abrades quickly, causing early failure. Homogeneous rubber seals require a surface finish between 8 and 16 μin. PTFE packings have very little flexibility, but they are resistant to practically all chemicals and solvents. They are used to seal very corrosive fluids at pressures up to 3000 psi (20 N/mm^2) and temperatures of 149°C (300°F) or less. Table 6.2 can be used as a guide for selecting molded packing materials. Table 6.3 is applicable to O-ring elastomeric compounds.

O-rings are probably the most common type of molded packings and are used in applications as diverse as water faucets and rocket engines. Due to their low cost, they are often misapplied or applied beyond their capabilities. Because they are made of elastomeric compounds and therefore subject to fluid compatibility and temperature limitations, it is very important to work closely with the manufacturer's technical specifications (see Tables 6.4 through 6.7). O-rings are sold in inch and metric sizes and are standardized by various engineering and military organizations. Military standards combine static and dynamic specifications whereas industrial standards have separate specifications for static and dynamic conditions. O-rings are sometimes used for rotary applications if the shaft rotation is slow or oscillatory, but one is best advised to secure the assistance of the manufacturer in this very critical type of application. There can be problems associated with twisting or spiral leakage. Some elastomers suffer from the

TABLE 6.2 Material Choices for Various Applications

Condition	Leather	Homogeneous rubber	Rubber fabric-reinforced
Oil	Good	Good	Good
Air	Good	Good	Good
Water	Good	Good	Good
Steam	Not recommended	Good	Good
Solvents	Not recommended	Good	Good
Acids	Not recommended	Good	Good
Alkalis	Not recommended	Good	Good
Fire-resistant fluids:			
Phosphate ester	Wax or polysulfide impregnation	Butyl base polymer	Butyl base polymer
Water-glycol	Not recommended	Buna-N base polymer	Buna-N base polymer
Water-oil emulsion	Wax, polyurethane, or polysulfide impregnation	Buna-N base polymer	Buna-N base polymer
Temperature range	−65 to +180°F	−65 to +400°F	−40 to +400°F*
Types of metal	Ferrous and nonferrous	Chrome-plated steel and nonferrous alloys with hard, smooth surfaces	Chrome-plated steel and nonferrous alloys with hard, smooth surfaces
Metal finish, rms (max)	60	16	32
Clearances	Medium	Very close	Close
Extrusions or cold flow	Good	Poor	Fair
Friction coefficient	Low	Medium and high	Medium
Resistance to abrasion	Good	Fair	Fair
Maximum pressure, lb/in^2	125,000	5000	8000
Concentricity	Medium	Very close	Close
Side loads	Fair	Poor	Fair
High shock loads	Good	Poor to fair	Fair

*Depending on specific formulation or combination of materials. (*Courtesy,* Machine Design)

TABLE 6.3 Temperature Limitations for Various Seal Elastomers

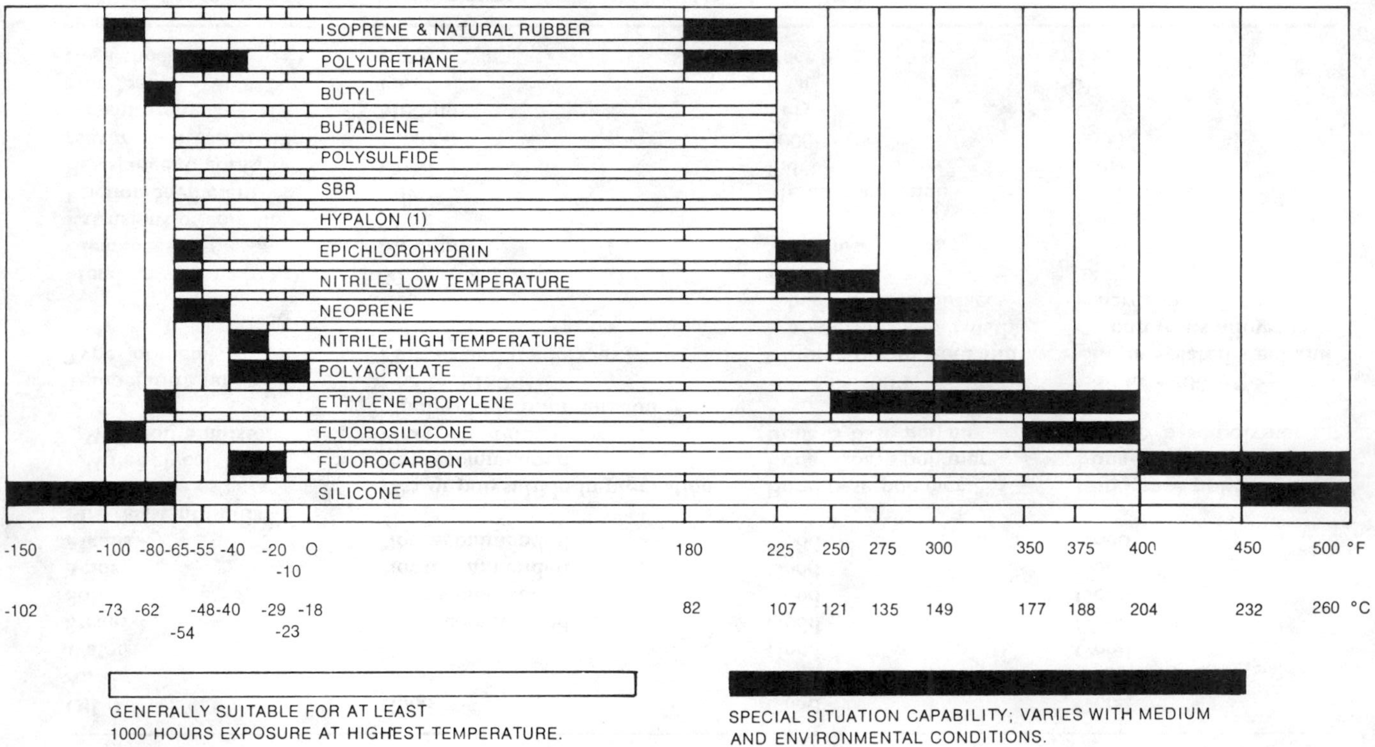

(1) TRADEMARK E. I. DuPONT
DE NEMOURS & CO., INC.

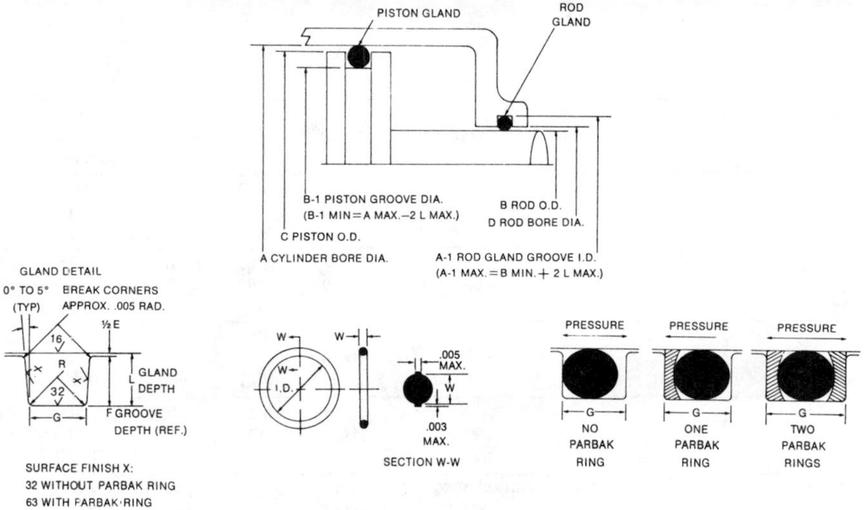

TABLE 6.4 Design Recommendations for Industrial Reciprocating Seals

O-ring size Parker 2-	W Cross section: Nominal	W Cross section: Actual	L gland depth	Squeeze: Actual	Squeeze: %	E* Diametral clearance	G Groove width: No parbak rings	G Groove width: One parbak ring	G Groove width: Two parbak rings	R Groove radius	Eccentricity max.†
006 through 012	1/16	0.070 ±0.003	0.055 to 0.057	0.010 to 0.018	15 to 25	0.002 to 0.005	0.093 to 0.098	0.138 to 0.143	0.205 to 0.210	0.005 to 0.015	0.002
104 through 116	3/32	0.103 ±0.003	0.088 to 0.090	0.010 to 0.018	10 to 17	0.002 to 0.005	0.140 to 0.145	0.171 to 0.176	0.238 to 0.243	0.005 to 0.015	0.002
201 through 222	1/8	0.139 ±0.004	0.121 to 0.123	0.012 to 0.022	9 to 16	0.003 to 0.006	0.187 to 0.192	0.208 to 0.213	0.275 to 0.280	0.010 to 0.025	0.003
309 through 349	3/16	0.210 ±0.005	0.185 to 0.188	0.017 to 0.030	8 to 14	0.003 to 0.006	0.281 to 0.286	0.311 to 0.316	0.410 to 0.415	0.020 to 0.035	0.004
425 through 460	1/4	0.275 ±0.006	0.237 to 0.240	0.029 to 0.044	11 to 16	0.004 to 0.007	0.375 to 0.380	0.408 to 0.413	0.538 to 0.543	0.020 to 0.035	0.005

*Clearance (extrusion gap) must be held to a minimum consistent with design requirements for temperature range variation.
†Total indicator reading between groove and adjacent bearing surface.

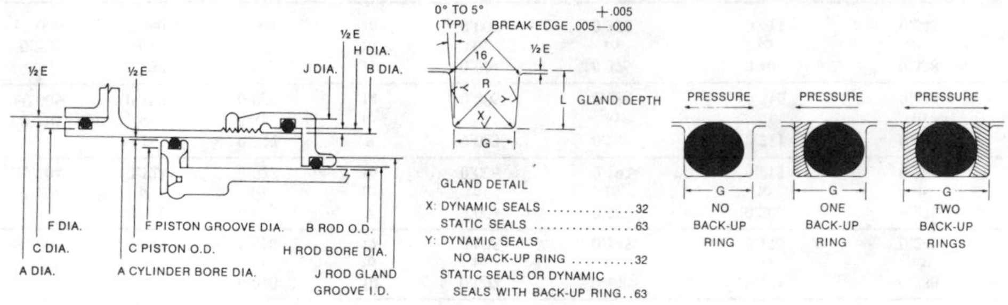

TABLE 6.5 Military O-Ring Design Recommendations

MS28775 O-ring dash size	W Cross section		L Gland depth	Squeeze		E* Diametral clear-ance max.	G +0.010 −0.000 Groove width			R Groove radius	Eccen-tricity max†
	Nominal	Actual		Actual	%		No back-up rings	One back- up ring	Two back- up rings		
001	3/64	0.040 ±0.003	0.031 to 0.032	0.005 to 0.012	13.5 to 27.9	0.004	0.063			0.005 to 0.015	0.002
002	3/64	0.050 ±0.003	0.040 to 0.041	0.006 to 0.013	12.8 to 24.5	0.004	0.073			0.005 to 0.015	0.002
003	1/16	0.060 ±0.003	0.048 to 0.049	0.008 to 0.015	14 to 23.8	0.004	0.083			0.005 to 0.015	0.002
004	1/16	0.070 ±0.003	0.057 to 0.058	0.009 to 0.016	13.4 to 21.9	0.004	0.094	0.149	0.207	0.005 to 0.015	0.002
005	1/16	0.070 ±0.003	0.0565 to 0.0575	0.0095 to 0.0165	14.2 to 22.6	0.004	0.094	0.149	0.207	0.005 to 0.015	0.002

006 through 012	1/16	0.070 ±0.003	0.056 to 0.057	0.010 to 0.017	13.4 to 23.3	0.004	0.094	0.149	0.207	0.005 to 0.015	0.002
013 through 028	1/16	0.070 ±0.003	0.056 to 0.058	0.009 to 0.017	13.4 to 23.3	0.005	0.094	0.149	0.207	0.005 to 0.015	0.002
110 through 116	3/32	0.103 ±0.003	0.089 to 0.091	0.009 to 0.017	9 to 16	0.005	0.141	0.183	0.245	0.005 to 0.015	0.002
117 through 149	3/32	0.103 ±0.003	0.089 to 0.091	0.009 to 0.017	9 to 16	0.005‡ to 0.007	0.141	0.183	0.245	0.005 to 0.015	0.002
210 through 222	1/8	0.139 ±0.004	0.121 to 0.123	0.0115 to 0.0215	8.5 to 15	0.006	0.188	0.235	0.304	0.010 to 0.025	0.003
223 through 247	1/8	0.139 ±0.004	0.121 to 0.123	0.0115 to 0.215	8.5 to 15.0	0.006‡ to 0.008	0.188	0.235	0.304	0.010 to 0.025	0.003
325 through 349	3/16	0.210 ±0.005	0.186 to 0.188	0.017 to 0.029	8.1 to 13.5	0.006‡ to 0.008	0.281	0.334	0.424	0.020 to 0.035	0.004
425 through 460	1/4	0.275 ±0.006	0.238 to 0.241	0.027 to 0.425	9.9 to 15.1	0.009‡ to 0.011	0.375	0.475	0.579	0.020 to 0.035	0.005

*Clearance gap must be held to a minimum consistent with design requirements for temperature range variation.
†Total indicator reading between groove and adjacent bearing surface.
‡Maximum diametral clearance varies with size.

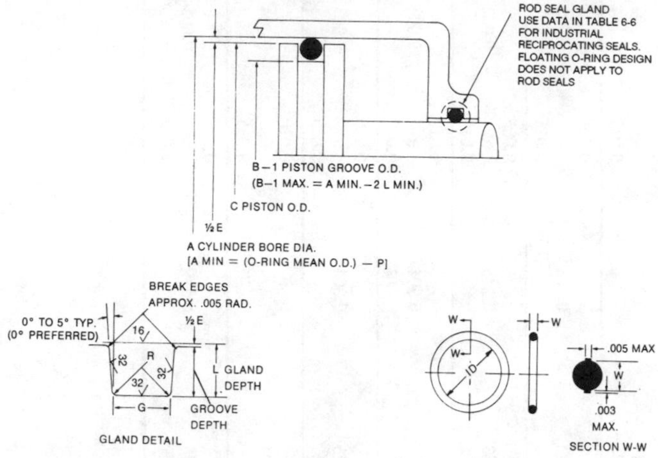

TABLE 6.6 Floating O-Ring Design Recommendations

O-ring size* Parker no. 2-	W Cross section Nominal	W Cross section Actual	P† Peripheral squeeze (variable)	L Gland depth	G Groove width	E Diametral clearance	Eccen-tricity max.‡	R Groove radius
006 through 012	1/16	0.070 ±0.003	0.035 to 0.042	0.072 to 0.076	0.075 to 0.079	0.002 to 0.010	0.002	0.005 to 0.015
104 through 116	3/32	0.103 ±0.003	0.038 to 0.062	0.105 to 0.109	0.111 to 0.115	0.002 to 0.010	0.002	0.005 to 0.015
201 through 222	1/8	0.139 ±0.004	0.061 to 0.082	0.143 to 0.147	0.151 to 0.155	0.003 to 0.011	0.003	0.010 to 0.025
309 through 349	3/16	0.210 ±0.005	0.084 to 0.124	0.214 to 0.218	0.229 to 0.233	0.003 to 0.011	0.004	0.020 to 0.035
425 through 460	1/4	0.275 ±0.006	0.140 to 0.175	0.282 to 0.286	0.301 to 0.305	0.004 to 0.012	0.005	0.020 to 0.035

*Only sizes listed are recommended for this design.
†Use to calculate A_{min} diameter.
‡Total indicator reading between groove and adjacent bearing surface.

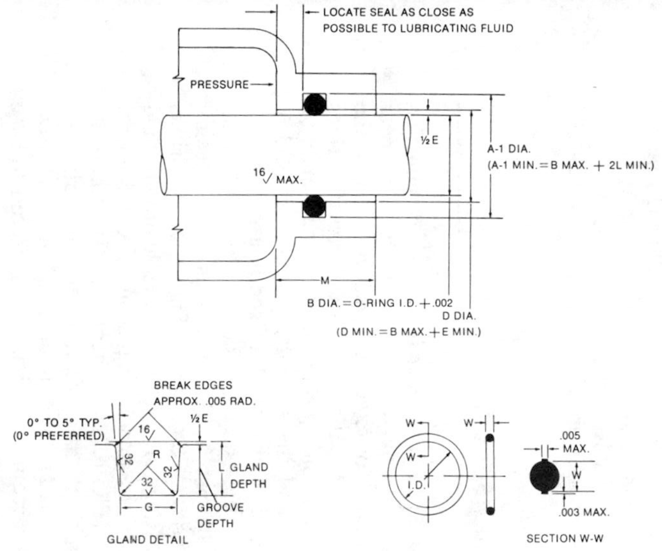

NOTE: DUE TO EFFECT OF CENTRIFUGAL FORCE, DO NOT LOCATE GROOVE IN SHAFT.

TABLE 6.7 Rotary O-Ring Design Recommendations

O-ring size Parker no. 2-	W Cross section		Maximum speed, FPM*	Squeeze, %	L Gland depth	G Groove width	E† Diametral clearance	Eccen-tricity, max.‡	M Bearing length, min.†	R Groove radius
	Nominal	Actual								
004 through 045	1/16	0.070 ±0.003	200—1500	0–11	0.065 to 0.067	0.075 to 0.079	0.012 to 0.016	0.002	0.700	0.005 to 0.015
102 through 163	3/32	0.103 ±0.003	200—600	1–8½	0.097 to 0.099	0.108 to 0.112	0.012 to 0.016	0.002	1.030	0.005 to 0.015
201 through 258	1/8	0.139 ±0.004	200—400	0–7	0.133 to 0.135	0.144 to 0.148	0.016 to 0.020	0.003	1.390	0.010 to 0.025

*Feet per minute = 0.26 × shaft diameter (inches) × rpm.

†If clearance (extrusion gap) must be reduced for higher pressures, bearing length *M* must be no less than the minimum figures given. Clearances given are based on the use of 80 shore durometer minimum O-ring for 800 psi max.

‡Total indicator reading between groove O.D. shaft, and adjacent bearing surface.

Gow-Joule effect whereby if the O-ring is slightly strained, it will contract under temperature. Shaft speed can heat the seal, causing contraction and possible failure of the application. Shaft material is very important and one should avoid stainless steel altogether (Fig. 6.17).

The O-ring is extremely versatile and is used extensively in reciprocating applications (Fig. 6.18). Standard groove dimensions must be smooth and free of sharp corners and burrs (5 to 16 μin, or 0.13 to 0.4 μm). Groove width must be sufficient to allow for oil swell, thermal expansion, pressure distortion, friction wear, and extrusion. The manufacturer will have limiting specifications, but because often a combination of variables will present a unique outcome, direct discussions and further testing might be necessary to verify the use of an O-ring in a given application. Back-up rings are sometimes used to help prevent extrusion (Fig. 6.19). Table 6.8 provides some material selection guidance for choosing O-rings. Other elastomeric seal ring cross sections are also available (Fig. 6.20).

FLOATING PACKINGS (SPLIT RING SEALS)

Expanding split piston ring seals are used to seal gases in internal combustion engines and fluids in pumps. Contracting split ring or rod seals are used in linear actuators whenever space, high temperature, or excessive pressure prohibit the use of packings. In all applications the application pressure forces the rings against the surfaces that require sealing. Expanding split ring seals must mate on the inner diameter of a cylindrical bore and the side of a piston (Fig. 6.21*a*). Contracting-type seals must contact the side surface of a fixed housing and the outer diameter of a reciprocating rod (Fig. 6.21*b*). The simplest and least-expensive split ring has a straight cut joint (Fig. 6.22). It is used as a low-pressure piston seal where joint leakage is not critical. Step joints (Fig. 6.22) of various types are used to reduce joint leakage. Rings are balanced when hydraulic fluid pressure is high. This is accomplished by machining small circumferential grooves in the wear surface of the ring (Fig. 6.22). System fluid flows into the groove and forms a very thin dam which acts as a pressure reducer. Balancing usually increase leakage. Multiring systems are used to seal high internal pressure. Table 6.9 gives the number of rings required for various pressure levels.

Metal rings are usually used in lubricated applications. If the application is not lubricated, PTFE rings impregnated with carbon-graphite, or molybdenum disulfide can be used. Table 6.10 lists the temperature limitations for materials used in ring applications. The limitations of various plating treatments appear in Table 6.11.

DIAPHRAGM SEALS

Diaphragms are membranes that are used to prevent movement of fluid or contamination from one chamber to another (Fig. 6.23). There are static diaphragms that merely separate fluids. These diaphragms are subject to very little displacement. Dynamic diaphragms are attached to the stationary and moving members and usually transmit force or pressure. Dynamic diaphragms can be flat or rolling. A flat diaphragm has no convolutions or has convolutions that are less than

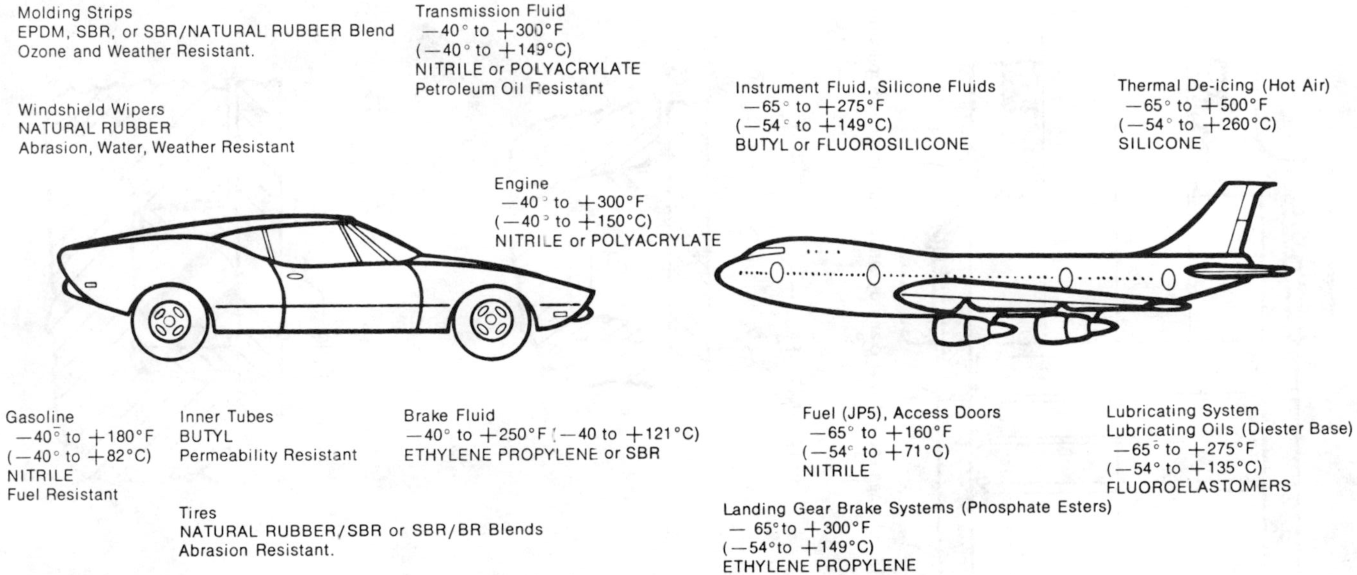
Molding Strips
EPDM, SBR, or SBR/NATURAL RUBBER Blend
Ozone and Weather Resistant.
Windshield Wipers
NATURAL RUBBER
Abrasion, Water, Weather Resistant
Transmission Fluid
−40° to +300°F
(−40° to +149°C)
NITRILE or POLYACRYLATE
Petroleum Oil Resistant
Engine
−40° to +300°F
(−40° to +150°C)
NITRILE or POLYACRYLATE
Instrument Fluid, Silicone Fluids
−65° to +275°F
(−54° to +149°C)
BUTYL or FLUOROSILICONE
Thermal De-icing (Hot Air)
−65° to +500°F
(−54° to +260°C)
SILICONE
Gasoline
−40° to +180°F
(−40° to +82°C)
NITRILE
Fuel Resistant
Inner Tubes
BUTYL
Permeability Resistant
Brake Fluid
−40° to +250°F (−40 to +121°C)
ETHYLENE PROPYLENE or SBR
Tires
NATURAL RUBBER/SBR or SBR/BR Blends
Abrasion Resistant.
Fuel (JP5), Access Doors
−65° to +160°F
(−54° to +71°C)
NITRILE
Lubricating System
Lubricating Oils (Diester Base)
−65° to +275°F
(−54° to +135°C)
FLUOROELASTOMERS
Landing Gear Brake Systems (Phosphate Esters)
− 65° to +300°F
(−54° to +149°C)
ETHYLENE PROPYLENE

FIGURE 6.17 Typical applications and types of seals used.

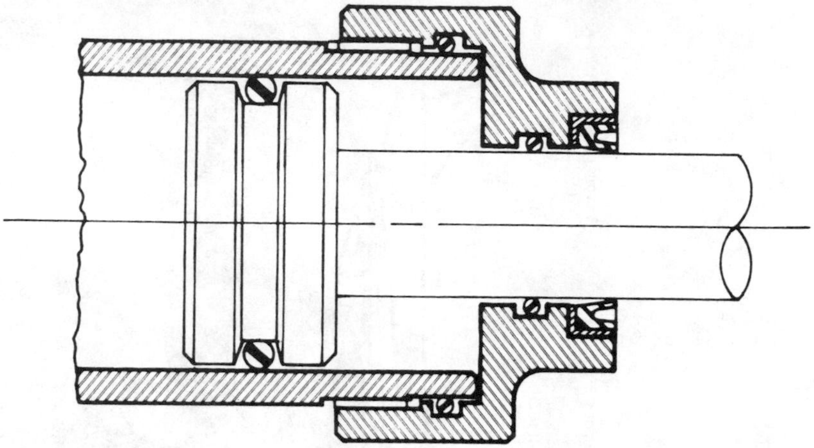

FIGURE 6.18 O-rings used as static and dynamic seals in same application.

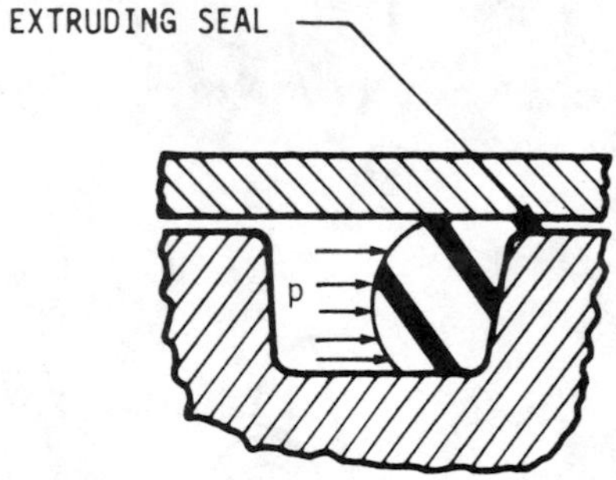

(a)

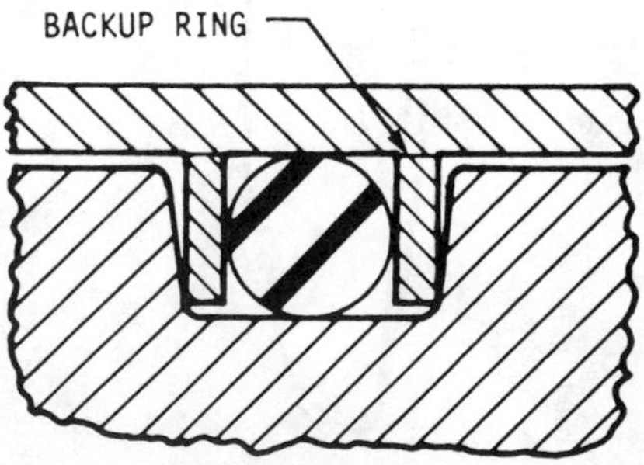

(b)

FIGURE 6.19 Back-up rings can prevent extrusion.

TABLE 6.8 Comparison of Elastomers in Various Fluids

	Properties																
Elastomer type (polymer)	Parker compound prefix letter	Abrasion resistance	Acid resistance	Chemical resistance	Cold resistance	Dynamic properties	Electrical properties	Flame resistance	Heat resistance	Impermeability	Oil resistance	Ozone resistance	Set resistance	Tear resistance	Tensile strength	Water/steam resistance	Weather resistance
Butadiene	D	E	FG	FG	G	F	G	P	F	F	P	P	G	GE	E	FG	F
Butyl	B	FG	G	E	G	F	G	P	GE	E	P	GE	FG	G	G	G	GE
Chlorinated polyethylene	K	G	F	FG	FP	G	G	GE	G		FG	E	F	FG	G	F	E
Chlorosulfonated polyethylene	H	G	G	E	FG	F	F	G	G	G	F	E	F	G	F	F	E
Epichlorohydrin	Y	G	FG	G	GE	G	F	FG	FG	GE	E	E	PF	G	G	F	E
Ethylene acrylic	A	F	F	FG	G	F	F	P	E	E	F	E	G	F	G	PF	E
Ethylene propylene	E	GE	G	E	GE	GE	G	P	E	G	P	E	GE	GE	GE	E	E
Fluorocarbon	V	G	E	E	FP	GE	F	E	E	G	E	E	GE	F	GE	FG	E
Fluorosilicone	L	P	FG	E	GE	P	E	G	E	P	G	E	GE	P	F	F	E
Isoprene	I	E	FG	FG	G	F	G	P	F	F	P	P	G	GE	E	FG	F
Natural rubber	R	E	FG	FG	G	E	G	P	F	F	P	P	G	GE	E	FG	F
Neoprene	C	G	FG	FG	FG	F	F	G	G	G	FG	GE	F	FG	G	F	E
Nitrile or buna N	N	G	F	FG	G	GE	F	P	G	G	E	P	GE	FG	GE	FG	F
Phosphonitrilic fluoroelastomer	F	F	P	G	E	F	FG	G	E	G	E	E	G	FP	F	F	E
Polyacrylate	A	G	P	P	P	F	F	P	E	E	E	E	F	FG	F	P	E
Polysulfide	T	P	P	G	G	F	F	P	P	E	E	E	P	P	F	F	E
Polyurethane	P	E	P	F	G	E	FG	P	F	G	G	E	F	GE	E	P	E
SBR or buna S	G	G	F	FG	G	G	G	P	FG	F	P	P	G	FG	GE	FG	F
Silicone	S	P	FG	GE	E	P	E	F	E	P	PG	E	GE	P	P	F	E

P = poor; F = fair; G = good; E = excellent.

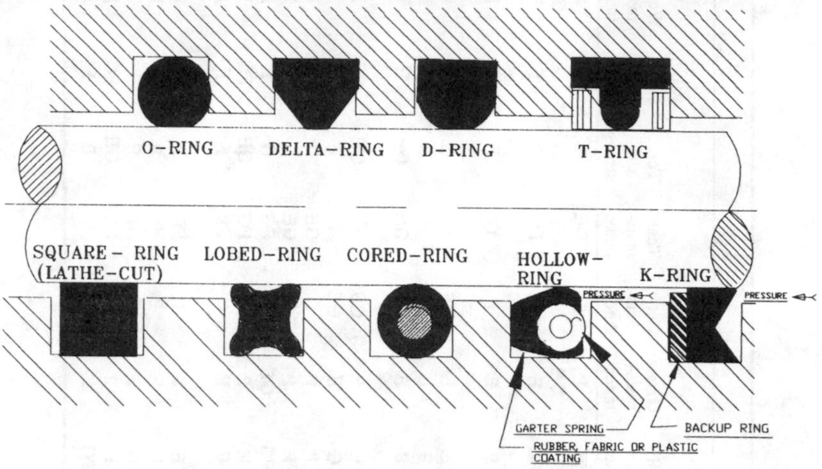

O-RING IS MOST COMMON FOR RECIPROCATING AND OSCILLATING MOTION. USE CAREFULLY IN ROTARY MOTION. REQUIRES A MINIMUM OF 6% SQUEEZE EXCEPT IN FLOATING APPLICATIONS. 1500 PSI TOPS UNLESS SUPPLIER TECHNICAL DATA PERMITS OTHERWISE.

T-RING USED EXTENSIVELY IN AIRCRAFT AND AEROSPACE APPLICATIONS. NOT SUSCEPTIBLE TO SPIRAL FAILURE. LOW SQUEEZE FORCE REDUCES FRICTION IN ROD AND PISTON APPLICATIONS. REQUIRES BACKUP RINGS. CAN RESIST HIGH PRESSURES. CAN BE USED IN CONVENTIONAL O-RING GROOVES.

LOBED-RING CAN BE USED IN STANDARD O-RING GROOVES. NOT SUBJECT TO SPIRAL FAILURE. MORE ADAPTABLE TO ROTARY APPLICATIONS THAN O-RING DUE TO LOWER SQUEEZE REQUIRED (1-3 PERCENT) BUT SHOULD STILL BE USED WITH CAUTION FOR ROTARY APPLICATIONS.

D-RING WORKS WELL IN ROD AND PISTON APPLICATIONS FOR RECIPROCATING MOTION. NOT SUBJECT TO SPIRAL FAILURE BUT SHOULD BE USED WITH BACK-UP RINGS. OPERATES WELL WITH LOW SQUEEZE.

SQUARE-RING OR LATHE-CUT RING MUST BE FABRICATED FROM A MOLDED BILLET AND NOT AN EXTRUSION TO HAVE SUPERIOR ENGINEERING PROPERTIES. NOT AS VERSATILE AS O-RING BUT USUALLY LESS EXPENSIVE. SUBJECT TO LARGE CONTACT WIDTH AND MORE FRICTION IN SLIDING APPLICATIONS. USEFUL IN LESS DEMANDING STATIC APPLICATIONS.

FIGURE 6.20 Some typical elastomer seal cross sections for dynamic applications.

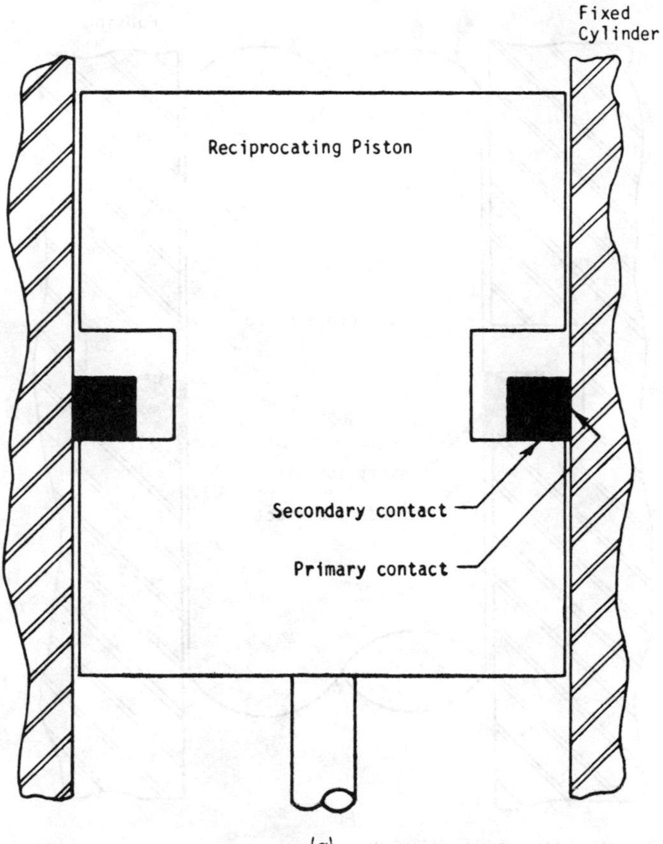

FIGURE 6.21a Expanding split ring seal.

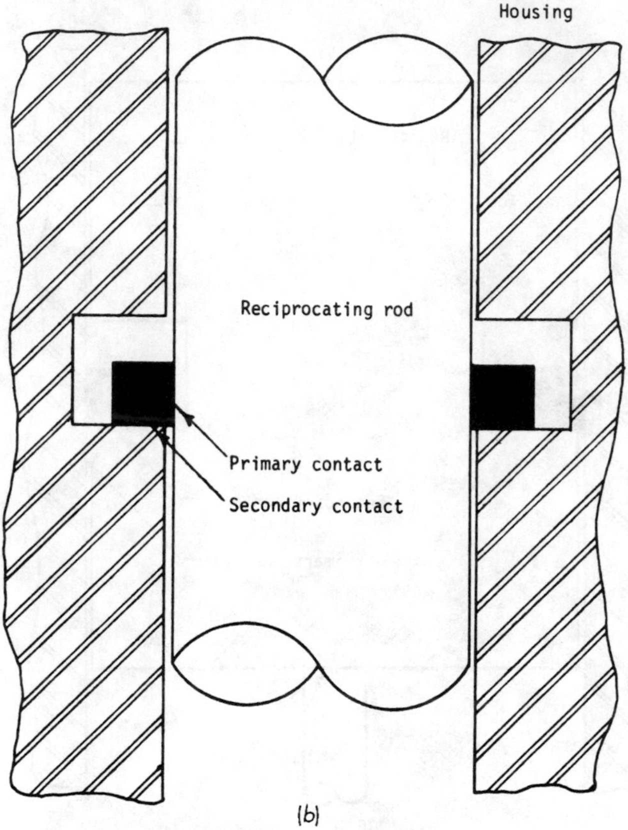

FIGURE 6.21*b* Contracting split ring seal.

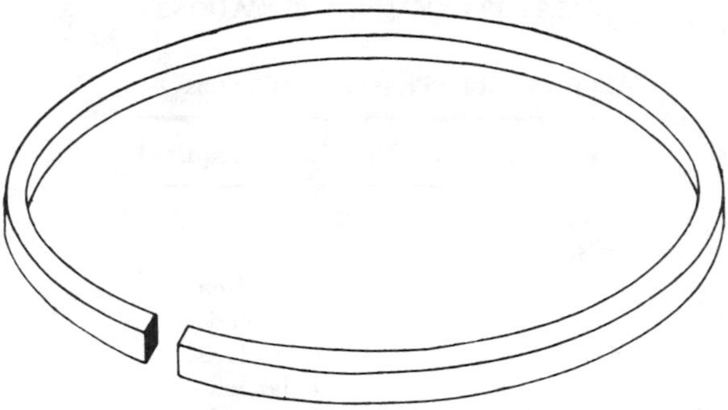

(a)

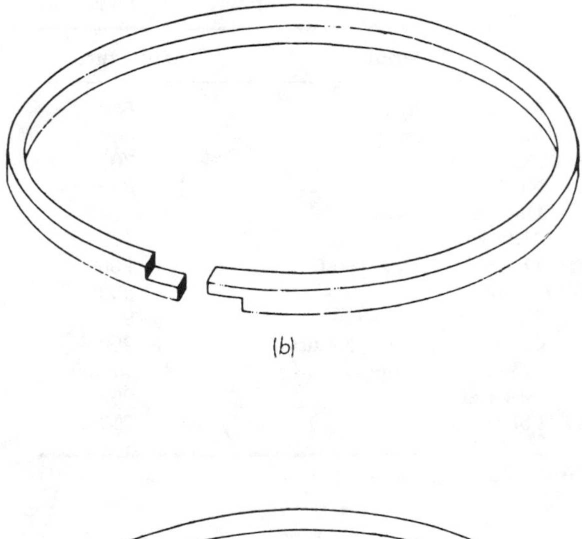

(b)

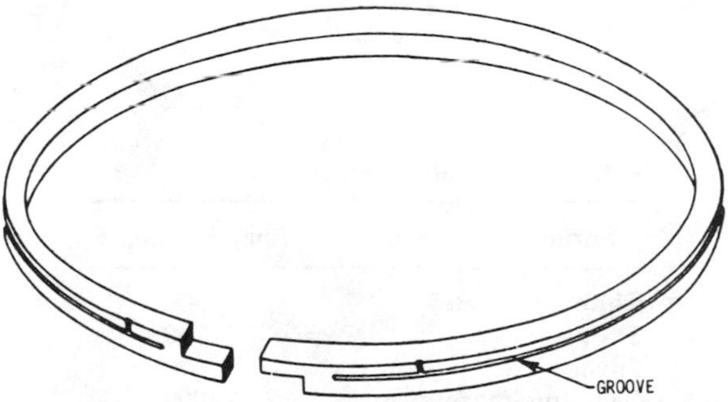

(c)

FIGURE 6.22 Several types of split rings. (*a*) Simple straight-cut split ring; (*b*) step-cut split ring; (*c*) balanced ring.

TABLE 6.9 Ring Pressure Limitations

Pressure, lb/in²	No. of rings required
Up to 300	2
300 to 900	3
900 to 1500	4 plain face 5 balanced
1500 to 3000	5 plain face 6 balanced
3000 and up	6 balanced

TABLE 6.10 Material Temperature Limitations

Material	Temperature, °F
Low-alloy gray irons	650
Malleable iron	700
Ductile iron	700
Ni-Resist	800
Ductile Ni-Resist	1000
410 stainless steel	900
17-4 pH stainless steel	900
Bronze	500
Tool steel, Tc 62-65	900
Carbon (high temperature)	950
K-30 (filled Teflons)	450 to 500
S-Monel	950
Polyimide	750

Courtesy: *Machine Design*.

TABLE 6.11 Plating Temperature

Surface treatment	Temperature, °F
Chromium plate	500
Tin plate	700
Silver plate	600
Cadmium-nickel plate	1000
Flame plate	1000–1600
Flame plate LC-1A	1600
Flame plate LA-2	1600

Courtesy: *Machine Design*.

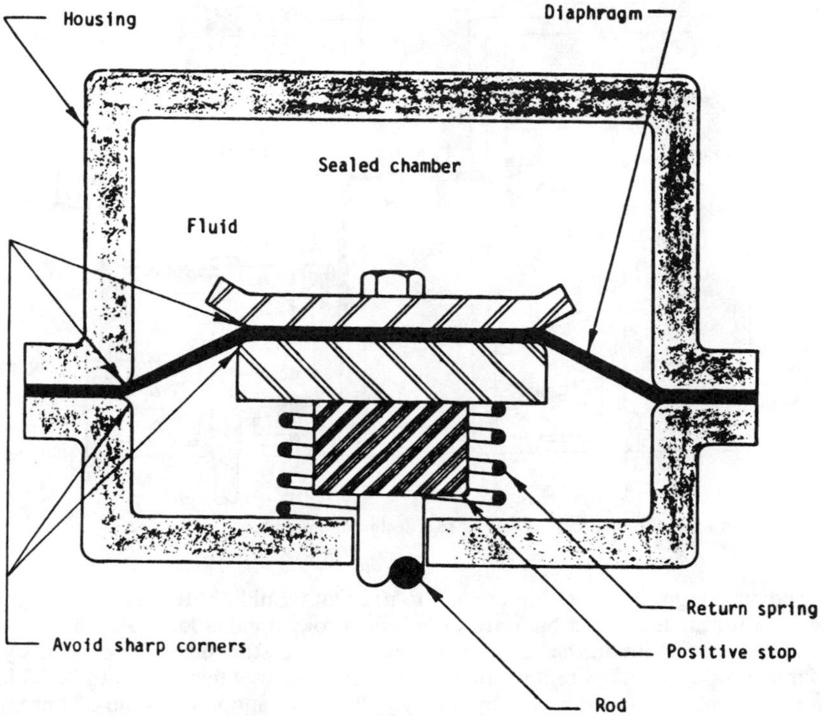

FIGURE 6.23 typical diaphragm application.

180°. Most heavy-duty flat diaphragms have molded convolutions to allow flexibility (Fig. 6.24). They are attached to the stationary housing at the edges. Plates are used to attach the diaphragm to the push rod. Springs are used to return the diaphragm to the neutral position; they ride in the central disk. Positive mechanical stops are used to prevent overstroking the diaphragm. Diaphragm motion should be less than 90 percent of the maximum possible stroke.

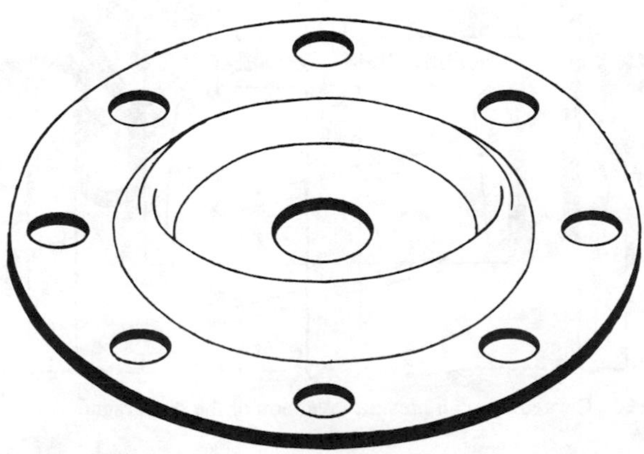

FIGURE 6.24 A flat convoluted diaphragm.

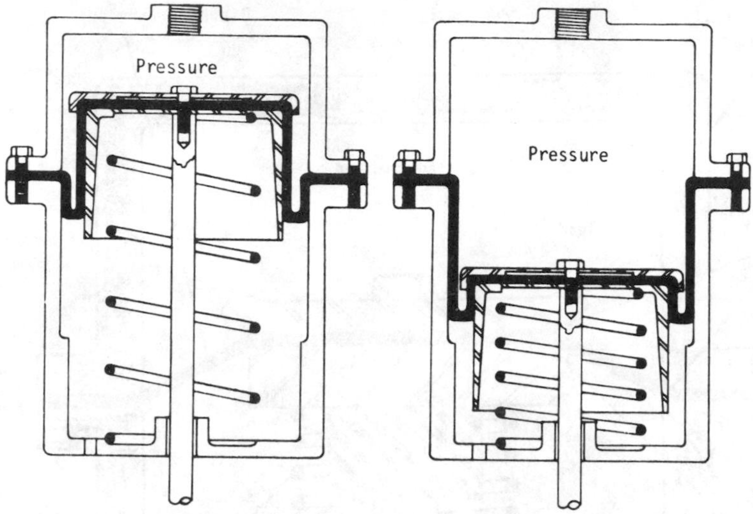

FIGURE 6.25 Rolling diaphragms offer a design alternative.

The diaphragm must be thin enough to prevent wrinkling. Rolling diaphragms are used for medium- and high-pressure long-stroke applications (Fig. 6.25). As the piston moves down, the diaphragm rolls off the piston sidewall onto the cylinder sidewall. The diaphragms are usually from 0.25 to 9 mm (0.010 to 0.035 in) thick. The pressure is supported by the cylinder head, and it holds the diaphragm against the piston and cylinder walls. Rolling diaphragms are usually molded in the shape of a top hat. The convolute is generated by inversion during assembly.

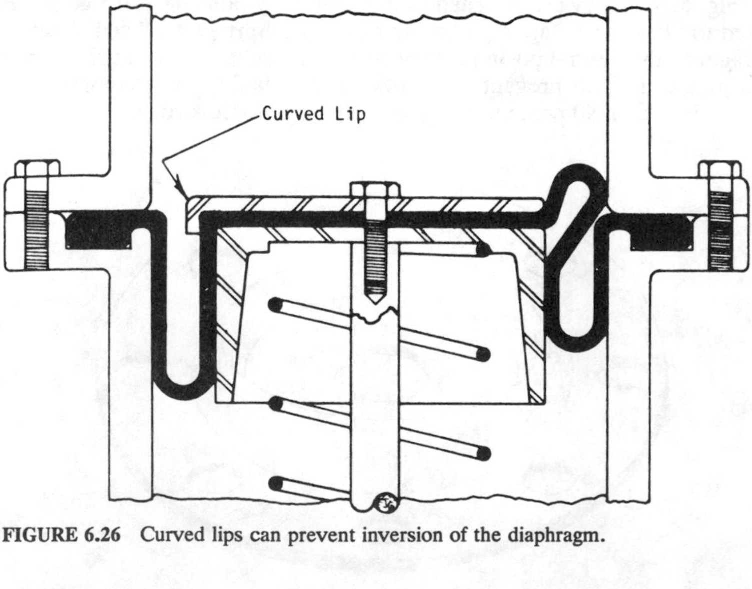

FIGURE 6.26 Curved lips can prevent inversion of the diaphragm.

TABLE 6.12 Some Common Sealing Elastomers and Their Typical Applications

Elastomeric type	Air permeability rating	Operating temperature limits, °C	Properties
Silicone	170 to 260	−80 to 260	General-purpose, low-temperature resistant; high permeability
Fluorosilicone	50	−60 to 230	Oil-resistant, high temperature
Nitrile	0.25 to 1.00	−40 to 120	Oil-resistant, low cost, attacked by ozone
Neoprene	1.40	−35 to 120	Weather-resistant; fair oil-resistance
Ethylene propylene	9.60	−40 to 150	Steam-, ozone-, acid-, and alkali-resistant
Fluorocarbon	0.32	−20 to 280	Oil-, fuel-, and chemical-resistant
Polyacrylate	1.50	−30 to 175	Hot-oil- and ozone-resistant
Epichlorohydrin	0.15 to 0.70	−40 to 150	Low permeability, oil-resistant

In some installations, the resiliency of the material might cause it to revert to its original position (Fig. 6.26). Sidewall scrubbing and high wear will result. This effect is prevented by using retained plates with curved lips. Rolling diaphragms cannot be used in applications where the low- and high-pressure chambers can be reversed. Pressure reversal will cause the diaphragm to wrinkle and scuff against the sidewall. High wear and premature failure can result.

Materials used to make diaphragms must have a high burst strength. This is the pressure that will rupture the material. Material modulus and tensile strength must be high. Blends of fabrics and elastomers are used in diaphragms. Cottons and nylons are used in flat diaphragms at temperatures less than 121°C (250°F). Nylon resists abrasion and fatigue but will creep under pressure and take a set. Davron can be used at temperatures up to 177°C (350°F), and Nomex is used for prolonged exposures at 260°C (500°F). All fabrics must be impregnated with elastomers to render the diaphragm impermeable. Elastomeric properties appear in Table 6.12.

CHAPTER 7
NONCONTACTING SEALS FOR ROTATING SHAFTS

INTRODUCTION

Some applications require seals with very low frictional losses. Other applications, such as sealing gases, may have little or no lubrication available at the seal location. Noncontacting seals may be the best choice for applications with these requirements. Typical noncontacting seal systems have a small clearance between the rotating shaft and the stationary housing. Direct contact of the seal surface and the rotating shaft does not occur, friction is minimized, and component wear is eliminated. Bushing seals are the simplest noncontacting seals, and the design principle allows leakage to occur. The amount of leakage can be controlled and reduced by decreasing the clearance between the bushing and the rotating shaft or by increasing the bushing length.

A labyrinth seal is a series of bushings joined together in such a way that the fluid must change directions as it flows from the sump to the atmosphere. The directional changes result in a longer path within a given space, and additional pressure drops occur wherever the fluid changes direction. Screw threads can be added to the rotating shaft or to the inside of the bushing. These screw threads act as an Archimedes pump that prevents dynamic leakage. Static leakage will occur when the shaft stops rotating. Some sophisticated noncontacting seals use a magnetic fluid placed in the clearance between the housing and the shaft. A magnet traps the fluid and prevents leakage. Shear of the magnetic fluid in the clearance will result in higher torque and power loss than seals without magnetic fluids.

BUSHINGS

The fixed bushing seal is the simplest noncontacting seal and consists of a sleeve which is rigidly attached to the housing (Fig. 7.1). Fixed bushings are primarily used to seal liquids. Leakage can be predicted and controlled by adjusting the radial clearance (h) between the rotating shaft and the housing and the bushing length (L). Incompressible fluid flow equations apply. The laminar leakage rate for fluids flowing in a perfectly aligned system can be predicted with Eq. (7.1). If the centerline of the shaft is not concentric with the centerline of the bushing, Eq.

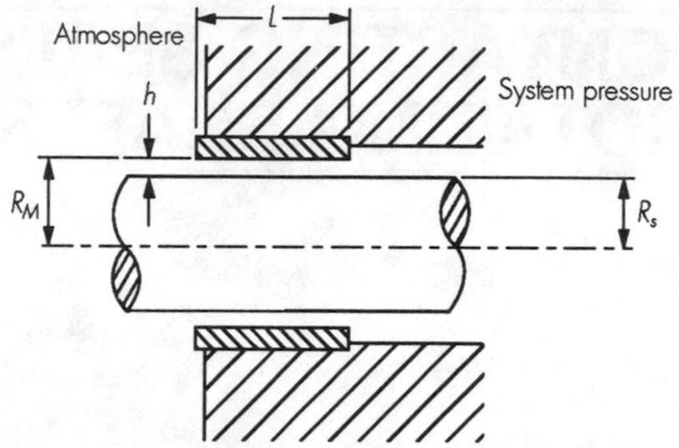

FIGURE 7.1 Fixed bushing.

(7.2) is used for laminar flow. The volumetric flow rate is directly proportional to the cube of the clearance and to the bushing length. The most effective way to minimize leakage is to minimize the clearance, but reducing the clearance can result in rubbing contact. Many times soft, low-friction metals such as babbitt metals are used for low-temperature applications. Bronze, carbon, and aluminum alloys are recommended for high-temperature service. PTFE, nylon, and other composite materials are often used.

As the bushing clearance increases, the flow rate increases and turbulence may result. Leakage rates can no longer be calculated with the laminar flow equations. Equation (7.3) must be used for turbulent flow for the concentric bushing and Eq. (7.4) is used for the eccentric case. The equations for turbulent flow must be used if the Reynolds number [Eq. (7.5)] is larger than 1000 for the concentric case and 550 for the fully eccentric case when $n = 1$.

$$Q_L = \frac{\pi R_M h^3 g \Delta\rho}{6\nu\rho L} \tag{7.1}$$

$$Q_{\mathrm{LE}} = Q_L\left(1 + \frac{3}{2}n^2\right) \tag{7.2}$$

$$Q_T = 2\pi R_M\left(\frac{1}{0.0665}\,\frac{h^3 g\Delta P}{\nu_{pL}^{1/4}}\right)^{4/7} \tag{7.3}$$

$$Q_{\mathrm{TE}} = 1.315Q_T \tag{7.4}$$

$$R_E = \frac{Vh}{\nu} = \frac{Q}{2\pi R_M \nu} \tag{7.5}$$

where Δp = pressure drop, g/cm^2
ρ = fluid density, g/cm^3
ν = kinematic viscosity, cm^2/s
g = gravitation constant, 980 cm/s^2
h = mean radial clearance, cm
L = bushing length, cm
n = eccentricity/radial clearance, e/h
Q = flow rate, cm^3/s
Q_L = laminar flow rate, cm^3/s
Q_{LE} = eccentric laminar flow rate, cm^3/s
Q_T = turbulent flow rate, cm^3/s
Q_{TE} = eccentric turbulent flow rate, cm^3/s
R_M = mean radius of annulus, cm
V = linear velocity, cm/s

Sample calculations follow.

Case 1: Consider a bushing with a mean radius of 0.6 cm with a length of 3.0 cm and a radial clearance of 0.02 cm. It is required to calculate the leakage rate of fresh water at 15°C (59°F) and 7 bar (7241 g/cm^2) for the perfectly concentric ($n = 0$) and fully eccentric case, $n = 1$. Start with laminar equations:

$$Q_L = \frac{\pi R_M h^3 g}{6\nu\rho}\,\frac{\Delta P}{L} \qquad Q_{LE} = Q_L\left(1 + \frac{3}{2}n^2\right)$$

The density is obtained from Table 7.1 and the kinematic viscosity from Fig. 7.2 ($\nu = 0.01$ cm^2/s: $\rho = 1.00$ g/cm^3):

$$Q_L = \frac{(3.1416)(0.6)(0.02)^3(980)(7241)}{6\,(0.01)(1.00)\,3.0} = 594.5\ cm^3/s \qquad Q_{LE} = 1486$$

Check to see if laminar assumption is correct:

$$R_E = \frac{Q_L}{2\pi R\nu} = \frac{594.5}{(6.2832)(0.6)\,0.01} = 15899 \text{ for } n = 0$$

$$R_E = \frac{Q_{LE}}{2\pi R\nu} = \frac{1486}{(6.2832)(0.6)\,0.01} = 39420 \text{ for } n = 1$$

TABLE 7.1 Specific Weight of Liquids

Liquid	°C	Specific weight, g/cm^3
Alcohol	20	0.800
Fresh water	15	1.000
Sea water	15	1.024
Lubricating oil	15	0.881–10.946
Glycerin	0	1.260
Fuel oil	15	0.897–1.026

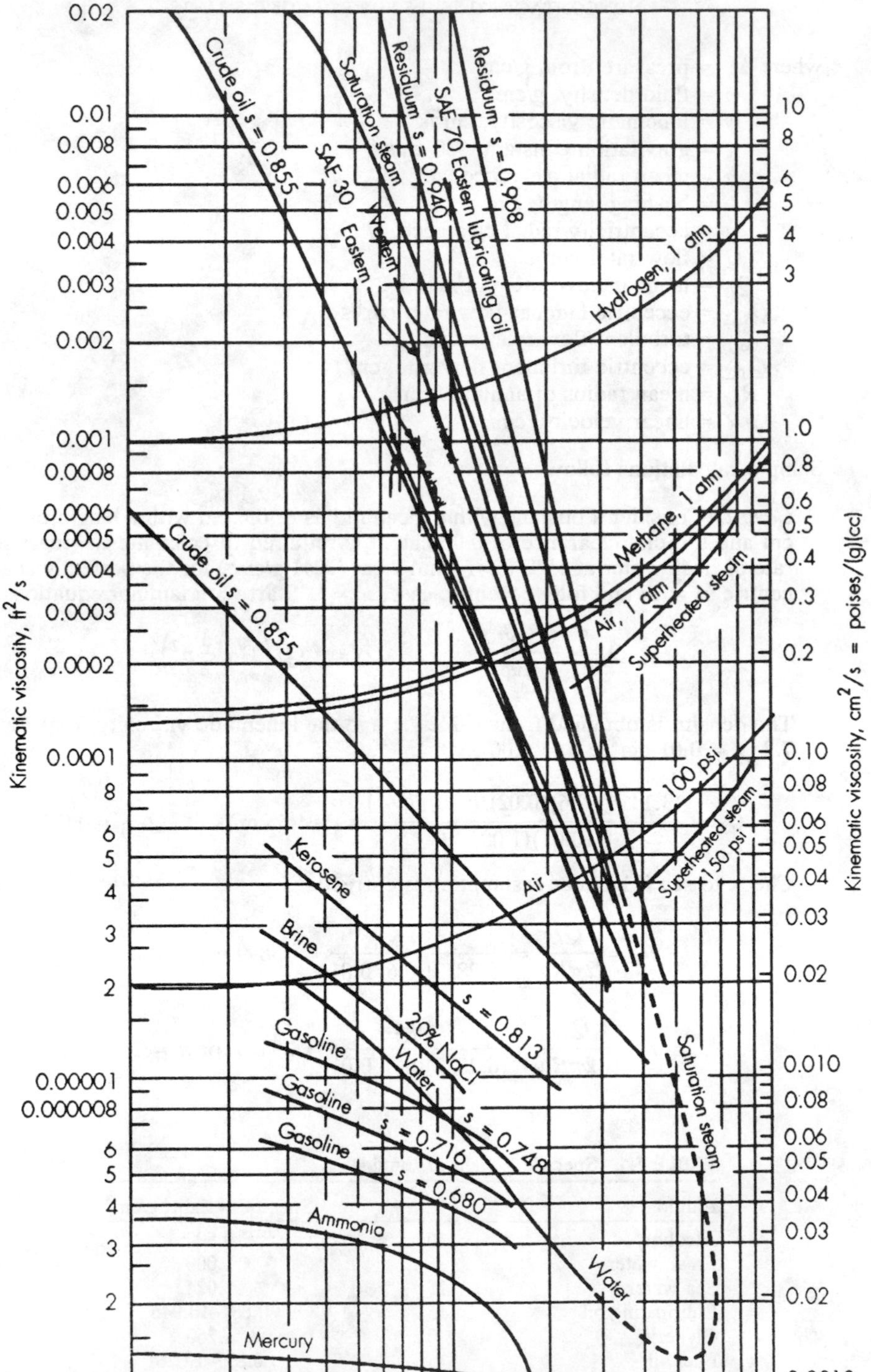

FIGURE 7.2 Kinematic viscosity of fluids.

The value of the Reynolds number for both the concentric and full eccentric case indicates the laminar assumption is invalid; therefore, the turbulent equations must be used:

$$Q_T = 2\pi(0.6)\left[\frac{1}{0.0665}\,\frac{(0.02)^3(980)(7241)}{(0.01)^{1/4}(1.00)3.0}\right]^{4/7} = 183.8\ \text{cm}^3/\text{s} \qquad R_E = 4875$$

$$Q_{\text{TE}} = 1.315 \qquad Q_T = 241.7 \qquad R_E = 6411$$

Case 2: What clearance must be used with the bushing described in case 1 to obtain a leakage rate of 1 cm^3/s for the concentric and fully eccentric case?

$$R_E = \frac{Q_L}{2\pi R\nu} = \frac{1}{(6.2832)(0.6)(0.01)} = 26.5 \text{ for } n = 0$$

$$R_E = \frac{Q_{\text{LE}}}{2\pi R\nu} = \frac{1}{(6.2832)(0.6)(0.01)} = 26.5 \text{ for } n = 1$$

The flow is in the laminar zone; therefore, the laminar equations can be used:

$$h_L = \left(\frac{Q_L 6\nu\rho L}{\pi R_M g\Delta P}\right)^{1/3} = 0.0024 \text{ cm for } n = 0$$

$$h_{\text{LE}} = \left(\frac{Q_{\text{LE}} 6\nu\rho L}{\pi R_M g\Delta P(1 + 3/2\,n^2)}\right)^{1/3} = 0.0018 \text{ cm for } n = 1$$

The bushing sleeve can be allowed to follow shaft radial eccentricities by letting it float. This will allow smaller clearances and thus reduce leakage. Floating bushings are often used to seal gases (Fig. 7.3). An axial spring provides an axial

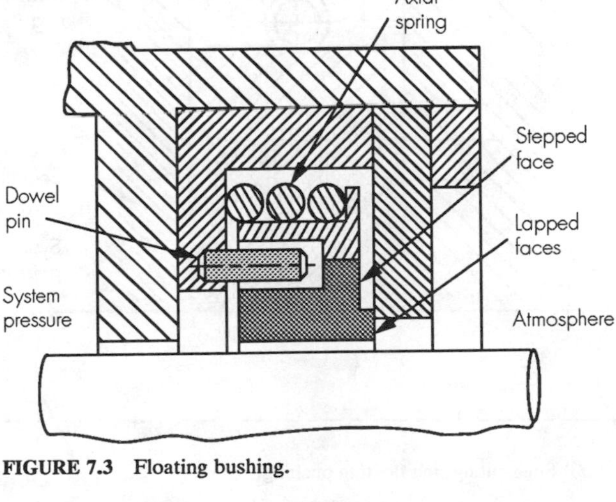

FIGURE 7.3 Floating bushing.

load against the housing face which prevents static leakage. If pressures are high or fluctuate, the bushing may be stepped to balance the seal. This step relieves the axial sealing force generated by fluid pressure. Excessive axial loads may cause premature wear and may cause the bushing to stick to the housing face and lose the ability to float. A dowel pin is sometimes used to prevent rotation.

In most cases, the axis of the shaft may not be parallel with the axis of the bushing. This cocked condition can result in undesirable contact even though the bushing floats radially. This condition can be relieved by using a series of short rings instead of one long bushing (Fig. 7.4). Each ring will adjust automatically to shaft motions over its own length. Each ring carries only a portion of the pressure drop; thus each one can move more freely than a single long bushing. The spring and static seal are required only with the final ring assembly of the series.

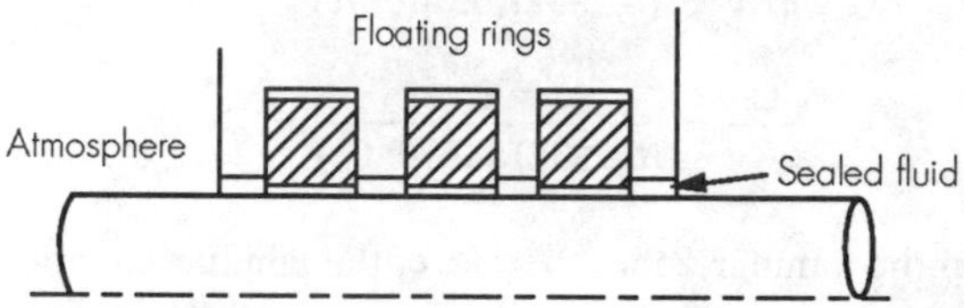

FIGURE 7.4 Multiple rings.

Another design that is often used is the segmented ring floating bushing (Fig. 7.5). The garter spring provides a load to push the seal ring against the shaft and the housing wall. This is accomplished because of the angled interface between the retaining and the sealing ring. The pressure against the shaft throttle leakage and the pressure against the housing wall prevents static leakage.

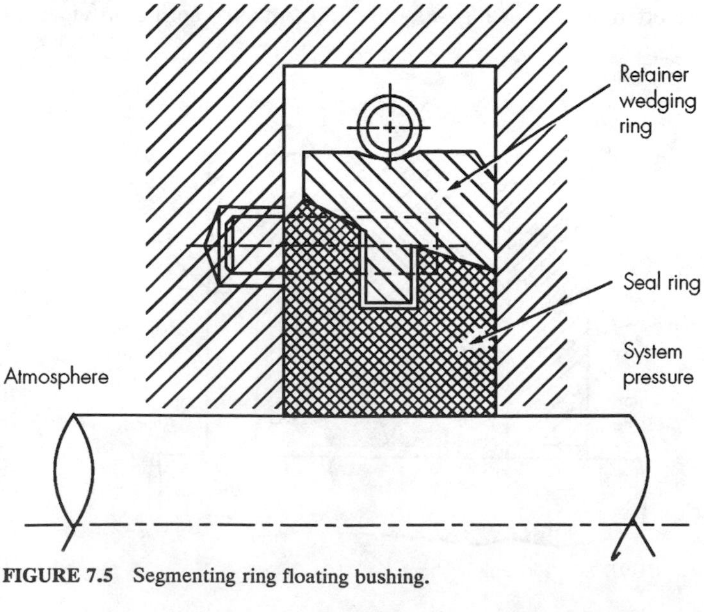

FIGURE 7.5 Segmenting ring floating bushing.

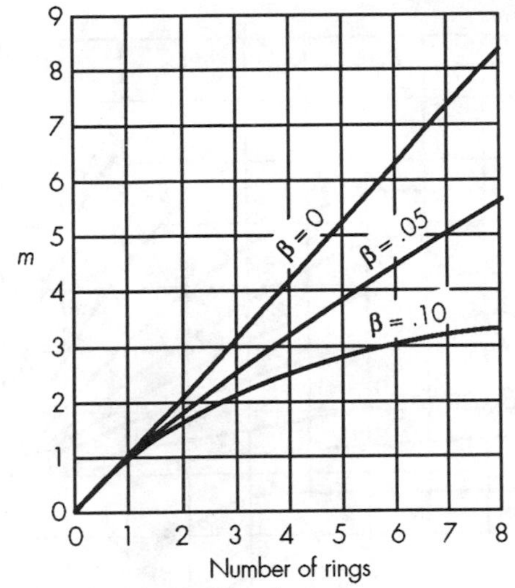

FIGURE 7.6 Multiring factor *m*.

If the differential pressure across the seal is low (0.5 bar), the recommended load between the seal ring and bushing inner diameter and the shaft is about 1 bar. The same load should be maintained between the faces that contact in the axial direction. The springs must be properly selected to ensure the recommended pressures are not exceeded.

Floating bushings are used for sealing gases. The density of gas changes with pressure. Fluid mechanics must be used to predict leakage rates. Equation (7.6) is used for laminar, isothermal, steady, and concentric geometry. If the shaft is eccentric, Eq. (7.7) is used for laminar flow. Calculations are more complicated if the flow is turbulent. The concentric case is represented by Eq. (7.8), and the fully eccentric case is estimated with Eq. (7.9).

The flow resistance coefficient α is a function of seal geometry, pressure ratio, and shaft speed. The following procedure is used to determine α. The velocity ratio β is calculated from Eq. (7.10) through (7.12). The multiring factor m is found in Fig. 7.6. The ring resistance factor R_r is determined from Eq. (7.13), and the pressure ratio γ is defined by Eq. (7.14). The values of R_r and β are used with Fig. 7.7 to determine thc flow resistance coefficient. The highest value of β given on the figures is 0.1. If the calculated value of β exceeds 0.1, the data presented for $\beta = 0.1$ should provide a satisfactory approximation for α.

$$G_{\text{LC}} = \frac{\pi R h^3 g}{12 \nu L}\left(\frac{P_i^2 - P_o^2}{P_s}\right)\frac{T_s}{T_i} \tag{7.6}$$

$$G_{\text{LEC}} = G_{\text{LC}}\left(1 + \frac{3}{2}n^2\right) \tag{7.7}$$

$$G_{\text{TC}} = 2\pi R h \alpha \sqrt{g P_i \rho_i} \tag{7.8}$$

$$G_{\text{TEC}} = 1.5\, G_{\text{TC}} \tag{7.9}$$

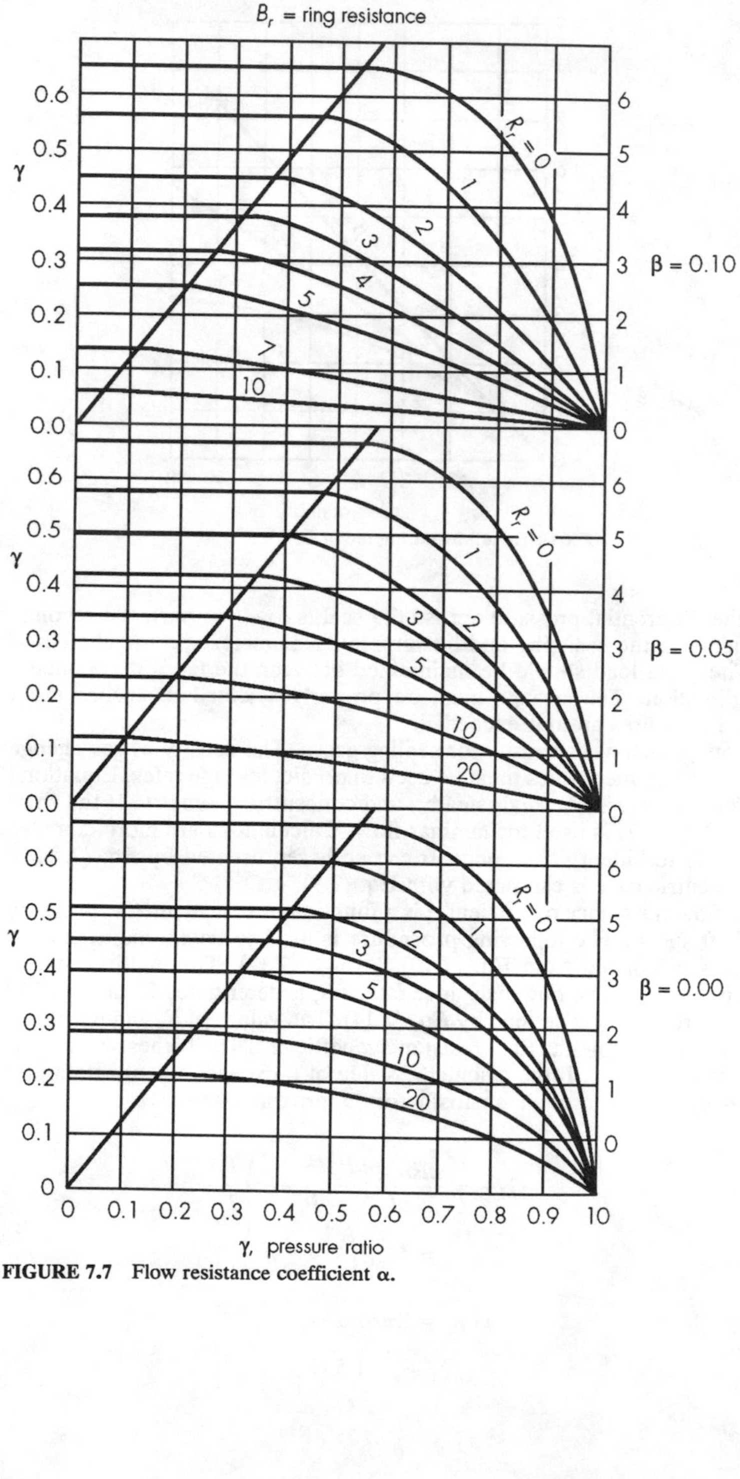

FIGURE 7.7 Flow resistance coefficient α.

$$V_c = \sqrt{\frac{2kgP_i}{(k+1)\rho_i}} \tag{7.10}$$

$$V_s = 2\pi RN \tag{7.11}$$

$$\beta = \frac{V_s}{V_c} \tag{7.12}$$

$$R_r = m\left(\frac{W}{h}\right) \tag{7.13}$$

$$\gamma = \frac{P_o}{P_i} \tag{7.14}$$

where α = flow resistance coefficient (see Figs. 7.6 and 7.7)
g = gravitation constant, 980 cm/s^2
h = clearance, cm
k = exponent of adiabatic expansion for gas (1.4 for air)
N = shaft rotational speed, rev/s
P_i = upstream pressure, g/cm^2
ρ_i = gas density at upstream condition, g/cm^3
P_o = downstream pressure, g/cm^2
P_s = pressure at std. temp., g/cm^2
G_{LC} = leakage rate of laminar compressible fluid, concentric, g/s
G_{LEC} = leakage rate of laminar compressible fluid, eccentric, g/s
G_{TEC} = leakage rate of turbulent compressible fluid, eccentric, g/s
G_{TC} = leakage rate of turbulent compressible fluid, concentric, g/s
R = shaft radius, cm
T_i = system temperature, K
T_s = standard temperature, K
V_c = critical fluid velocity, cm/s
V_s = linear shaft velocity, cm/s
W = ring width, cm

For proper performance with a minimum of leakage, the shaft surface must be round within 0.0125 mm (0.0005 in) for a shaft diameter of 125 mm (4.921 in) or less. If the shaft is larger than 125 mm (4.921 in), the roundness specification is 0.0001 mm per millimeter of shaft diameter (0.0025 in). The hardness of the shaft should be Rockwell C 55 minimum, and the surface finish should be less than 0.2 μm (7.87 μin) (rms). The shaft radial eccentricity should be less than 0.025 mm (0.001 in). Segmented ring seals can tolerate larger runout by increasing the garter spring load, but this practice will increase wear. If recommended shaft properties are maintained, shaft axial movement is unlimited. If axial cocking or axial runout (whip) occurs during the axial shaft motion, secondary leakage may result when the sealing ring is pulled away from the housing. Increasing the axial spring load will prevent this leakage, but excessive wear may result on the axial faces, and the radial response may be reduced.

The secondary seal interface lies between the housing and the side of the sealing ring. The flatness of the contacting faces must be held to four to six helium light bands to prevent secondary leakage. Material selection is very important and must be compatible with the system fluid (Table 7.2). The system fluid must be clean and should not tend to polymerize or crystallize. If the system fluid is water, it should be softened to remove minerals that may deposit in the bushing clearance. Combinations of materials are often used. Care must be exercised to ensure that differences in thermal expansion coefficients do not cause binding or slippage of components. In some cases, plasma coatings can be sprayed on the shaft for high-temperature applications.

TABLE 7.2 Bushing Material Selection Chart

Environment	Seal ring material	Shaft material
Gases (Air, H_2, O_2, He, N_2, CO_2)	Carbon graphite	Hardened tool steel, chrome carbide, stainless steels (300, 400), plating (chrome, ceramic, tungsten carbide)
Water	Bronze, stellite, carbon graphite, ceramic, 416 stainless steel	Chrome plated steel
Oil	Babbitt, bronze, aluminum, carbon graphite, filled PTFE	Hardened steel, nitrided steel, chrome plated steel

LABYRINTH SEALS

Labyrinth seals are an extension of the bushing seal concept. They are used to seal gases in turbine applications and also act as seals for rolling bearings, machine spindles, and other applications where some leakage can be tolerated.

The simplest labyrinth seal has a straight-through design and consists of circumferential strips of material that extend from the bore toward the shaft (Fig. 7.8*a*). These strips form orifices that throttle the flow and create high fluid velocities at the constriction. The fluid then expands into the chamber beyond the constriction, turbulence results, and a pressure drop occurs.

The efficiency of the labyrinth is increased by stepping (Fig. 7.8*b*) or staggering (Fig. 7.8*c*) the constrictions to increase the length and tortuosity of the path the fluid must travel from inside the system to atmosphere. The velocity carryover inherent in the straight-through design is minimized. The housing of the staggered labyrinth must be split to allow for assembly. This added cost and complexity limits the usage of this concept. Combinations of labyrinth designs are also used. The combination shown in Fig. 7.8*d* also uses a barrier flow of inert gas or air at a pressure higher than the process gas to prevent leakage of the process gas to the atmosphere.

The most efficient form of the labyrinth rings for sealing gases are the tapered teeth shown in Fig. 7.8*d.* The efficiency also increases as the clearance gap decreases. A line-to-line contact or a slight interference is often used. Design engineers must consider machining tolerances on components, eccentricity, differential thermal expansion, and dynamic deflection of components when selecting the optimum clearance.

Materials are selected that will wear or bed in during operation. The tooth-wearing process should maintain a clean and sharp tooth tip without pitting, mushrooming, or gouging. Soft brass teeth work well at low temperatures. Leaded nickel-tin bronze is satisfactory in steam up to 500°C (932°F). If air or oxygen is present, the maximum temperature is about 200 to 230°C (392 to

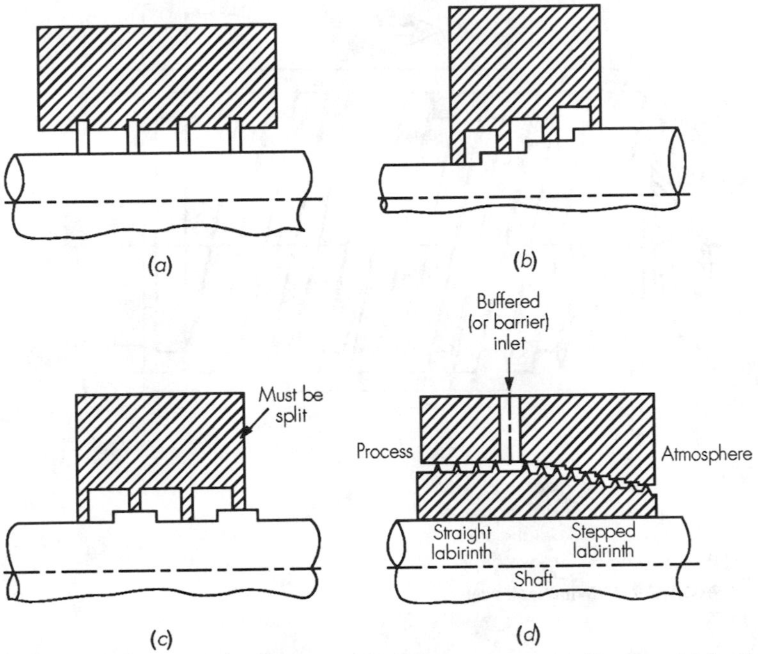

FIGURE 7.8 Labyrinth configurations. (*a*) Straight labyrinth; (*b*) stepped labyrinth; (*c*) staggered labyrinth; (*d*) combination labyrinth.

446°F). Aluminum bronzes, unanodized aluminum, and soft stainless steels can also be used.

The knife geometry and the number of knives are influenced by the space and allowable leakage. The knife tip will usually have a width of 0.12 to 0.35 mm (0.005 to 0.014 in) with a root width of 3.5 to 5.0 mm (0.138 to 0.197 in). The angle of the knife sides is typically 8 to 12°. In some cases, particularly with aircraft, an abradable, soft honeycomb material is used to throttle leakage.

Leakage of steam through a labyrinth can be estimated with Eq. (7.15). The coefficient K_T varies for different types of seal teeth. For simple straight-through labyrinths, K_T varies from 60 to 120. If the teeth are staggered, K_T varies from 30 to 65.

$$G = 0.554 K_T A \frac{\sqrt{(P_i g/V_i)\,[1 - (P_o/P_i)^2]}}{N_s - \ln(P_o/P_i)} \tag{7.15}$$

where A = area of clearance, cm^2
g = gravitational constant, cm/s^2
G = flow rate of steam, g/s
K_T = flow coefficient
N_s = number of stages in labyrinth
P_i = system pressure g/cm^2
P_o = exhaust pressure, g/cm^2
V_i = specific volume of steam at inlet, cm^3/g

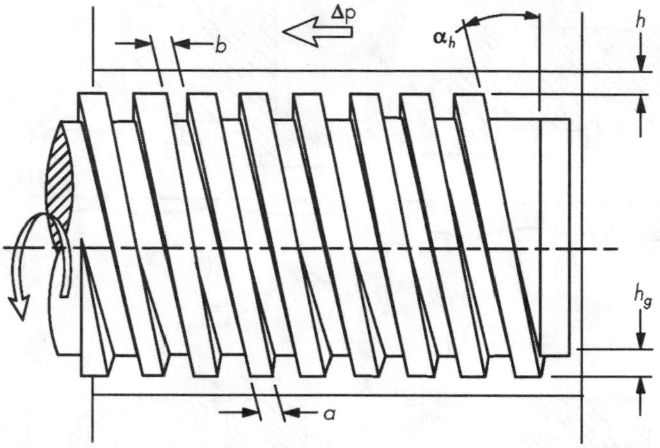

α_h = Helix angle
a = Axial land width
b = Axial groove width
h = Radial clearance
h_g = Groove depth

FIGURE 7.9 **Windback seal.**

WINDBACK SEALS

Windback seals are also known as visco seals, positive action seals, and hydrodynamic seals. They incorporate a screw thread on the shaft or housing that pushes or pumps the fluid from the outside toward the interior of the sump. They are very effective when sealing liquids and produce a pumping pressure which is proportional to shaft speed and liquid viscosity. The concept can be used only when the shaft rotates constantly in one direction. If the shaft is reversed, the liquid will be screwed out of the sump and gross leakage will occur. Many times a radial lip seal or O-ring will be employed to provide a static seal. Some designs use centrifugal force during rotation to prevent lip seals from contacting the shaft. When the shaft stops, the lips make contact and the static seal is formed.

Because the viscosity of gases is low compared to liquids, the windback seal is ineffective for sealing gases. A liquid-filled screw seal complete with a fluid feed, recovery, and recirculating system to contain the fluid can be used to seal gases. The screw thread feature has been successfully employed on bushing and radial lip seals.

The key design features of a windback seal appear in Fig. 7.9. Since there is no contact with this configuration, long reliable life occurs. Recommended value of the key design parameters are given in Table 7.3. The parameters A and B are defined by Eqs. (7.16) and (7.17).

$$A = \frac{b}{a + b} \tag{7.16}$$

$$B = \frac{h + h_g}{h_g} \tag{7.17}$$

where α_h = helix angle
a = axial land width
b = axial groove width
h = radial clearance
h_g = groove depth

TABLE 7.3 Recommended Design Parameters

	Laminar flow	Turbulent flow
α_h	15–20°	10–15°
A	0.5	0.62
B	3.6–4.1	4.1–6.5

MAGNETIC SEALS

Magnetic seals employ a magnetized fluid that is trapped by a magnetic field in the clearance gap of a labyrinth-type seal. The primary use is to prevent leakage of gases or the entrance of contaminants. The fluid consists of a carrier liquid that contains very small particles of a magnetic solid. The particles are coated, and brownian motion results in random collisions with the other magnetic particles and the molecules of the carrier liquid. These collisions keep the particles in colloidal suspension for an indefinite period of time.

Properties of magnetic liquids appear in Table 7.4. They have high electrical resistance to provide an essentially nonconductive media. The temperature of the system should be 110°C (230°F) or less to prevent evaporation of the fluid. The other chemical and mechanical properties of the magnetic fluid parallel that of the carrier liquid.

A typical magnetic seal design concept appears in Fig. 7.10. A circular magnet is placed around the shaft with a pole block on either side. A series of serrations are machined on the shaft directly opposite the faces of the pole block. A magnetic field across the serrations is generated and holds the magnetic fluid in place at the clearance gap between the serrations and the pole block faces. The magnetic fluid prevents gas leakage and contaminant entry with very low frictional losses during shaft rotation. The concept can be expanded by adding more stages. These additional stages are required to seal larger pressure drops. The seals typically use clearance gaps of 0.05 to 0.12 mm (0.002 to 0.005 in) and operate successfully at speeds up to 10,000 rpm.

TROUBLESHOOTING NONCONTACTING SEALS

The problem description, probable causes, and recommended corrective action when problems occur with noncontacting seals appear in Table 7.5.

TABLE 7.4 Properties of Magnetic Fluids

Carrier fluid	Saturation magnitization, gauss	Viscosity, cP, 27°C	Evaporation rate, g/s · cm², 340°C	Density, g/mL	Thermal conductivity, mW/m · K	Initial susceptibility	Pour point, °C	Permeability	Electrical resistivity, Ω · cm
Ester	450	450	1.39×10^{-6}	1.490	209	2.10	−28	1.0–1.4	1.60×10^{9}
Ester	450	450	1.07×10^{-5}	1.410	168	2.20	−44	1.0–1.4	0.27×10^{9}
Synthetic petroleum	300	120	3.69×10^{-6}	1.195	170	1.49	−54	1.0–1.3	0.94×10^{9}
Petroleum	200	200	2.17×10^{-6}	1.080	—	1.14	−51	1.0–1.2	1.32×10^{9}
Ester	300	75	5.95×10^{-6}	1.258	185	1.37	−54	1.0–1.3	11.30×10^{7}
Ester	450	100	1.27×10^{-5}	1.440	186	2.06	−63	1.0–1.4	9.30×10^{7}
Fluorocarbon	300	3500	2.24×10^{-6}	2.245	117	1.36	−27	1.0–1.2	13.20×10^{9}
Polyphenyl ether	450	4500	9.92×10^{-7}	1.665	—	2.13	−12	1.0–1.4	0.18×10^{9}
Synthetic petroleum	200	1000	6.94×10^{-6}	1.125	158	0.67	−38	1.0–1.1	9.90×10^{9}

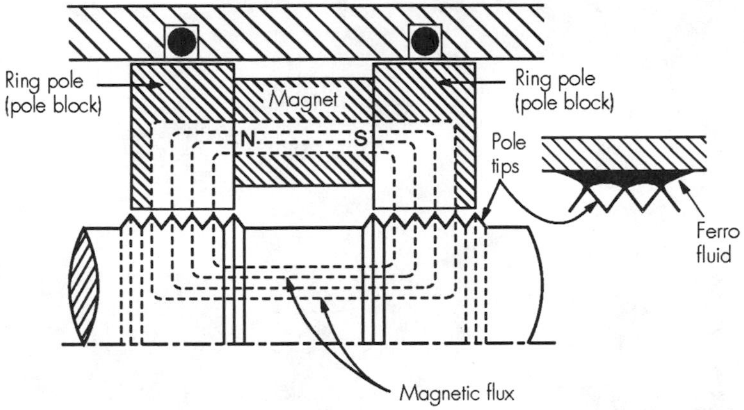

FIGURE 7.10 Magnetic seal principles.

TABLE 7.5 Troubleshooting Noncontacting Seals

Problem	Causes	Corrective action
Excessive leakage (all seals leak at all times)	Excessive radial clearance	Decrease radial clearance
	Excessive pressure	Decrease pressure or add vent
	Excessive lubricant fill	Decrease fill or increase cavity
	Excessive temperature	Decrease fill or increase cavity
	Excessive vibration	Increase lube cavity
	Excessive end play	Increase lube cavity
Excessive leakage (some seals leak at all times)	Excessive radial clearance	Improve quality control
	Radial contact	Decrease misalignment
	Bypass leakage	Fill scratches and voids
Excessive leakage (all seals leak after a period of time)	Loss of lubricant viscosity	Improve lubricant
	Excessive relubrication	Decrease relubrication
	Contaminant ingress	Add contaminant shield
Water ingress	Static leakage	Add static seal
	Partial vacuum	Improve or add vent
	No lube leakage	Increase lube fill
	Excessive wet environment	Add water shield
Dirt ingress	Excessive radial clearance	Decrease radial clearance
	Excessively dirty environment	Add dirt seal
	Partial vacuum	Improve or add vent
	No lube leakage	Increase lube fill

CHAPTER 8
FUNDAMENTALS OF MECHANICAL FACE SEALS

BASIC FUNCTIONS OF MECHANICAL FACE SEALS

Mechanical face seals are used to prevent leakage of gases and liquids in rotating shaft applications that exceed the capabilities of radial lip shaft seals and packings. A rotating face forms a seal with a mating face or ring. Successful operation depends on maintaining a thin (0.0006 to 0.006 μm, or 0.024 to 0.24 μin) lubricating film of fluid between the faces. Most mechanical seals are used as seals for pumps. They can withstand high operating pressures, temperatures, and shaft speeds and give longer life with less leakage than packings and radial lip seals. Face seals seal both statically and dynamically, can withstand large pressure changes, are compatible with many fluids, and will function in applications where shaft rotation changes direction.

Face seals can withstand some shaft to bore misalignment and shaft eccentricity with very little wear. Conventional designs and materials can be selected to function at pressures up to 200 atmospheres, at speeds up to 50,000 rpm, and with a temperature range from −200 to 650°C (−328 to 1200°F). Higher pressures can be accommodated by mounting several seals in tandem, which splits the pressure drop. The advantages and disadvantages of mechanical face seals appear in Table 8.1.

GENERAL DESIGN FEATURES

Mechanical Seal Components

There are a broad variety of mechanical seal designs to choose from, and they all employ two wearing rings or faces that form a barrier to prevent leakage. The seal head is usually spring loaded and is pressed against a mating ring or seat. Secondary seals between components prevent static leakage (Fig. 8.1*a*). Seals with rotating heads and stationary seats (Fig. 8.1*a*, 8.1*b*) have the simplest design configuration and are used for speeds up to a maximum of 5000 to 6000 rpm. When speeds exceed this maximum, stationary heads provide better dynamic balance and are used with rotating seats (Fig. 8.1*c*, 8.1*d*).

Seals can be mounted inside the equipment housing to provide protection from

TABLE 8.1

Advantages	Disadvantages
1. Handle all types of fluids (acids, salts, abrasive particles).	1. Requires more space than radial lip seals, etc.
2. Handle slightly misaligned/nonconcentric.	2. Cannot handle axial end play.
3. Handle bidirectional shaft rotation, large pressure, temperature, and speed excursions.	3. Sealing faces must be finished smooth (0.08 to 0.4 μm) and can be easily damaged.
4. Shaft condition (finish roughness, roundness, hardness and material) is not critical.	4. High initial cost.
5. Operation does not cause shaft wear.	
6. Long operating life.	
7. Positive sealing for food processing, hazardous chemicals, and radioactive fluids.	

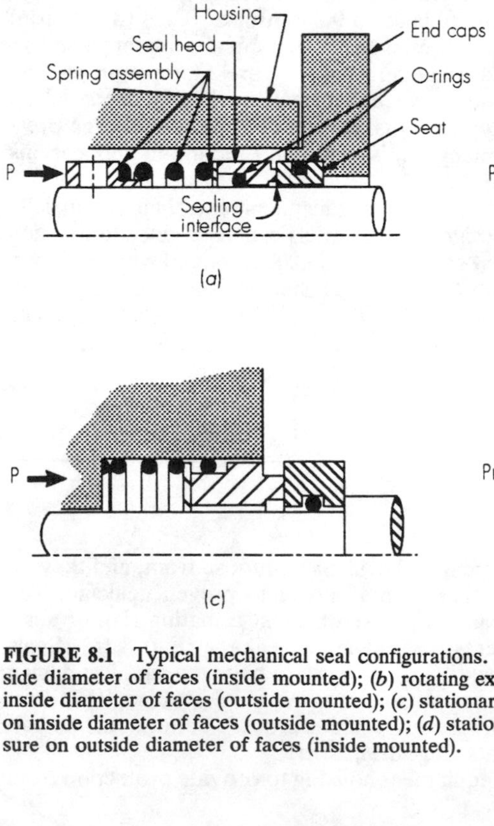

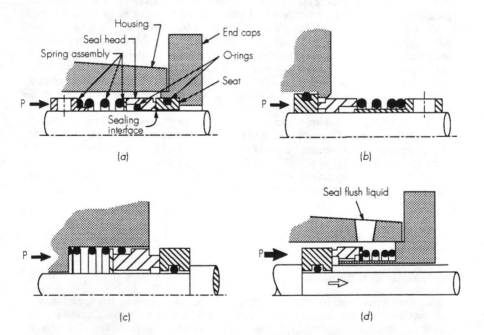

FIGURE 8.1 Typical mechanical seal configurations. (*a*) Rotating seal head—pressure on outside diameter of faces (inside mounted); (*b*) rotating externally mounted seal head—pressure on inside diameter of faces (outside mounted); (*c*) stationary internally mounted seal head—pressure on inside diameter of faces (outside mounted); (*d*) stationary externally mounted seal head—pressure on outside diameter of faces (inside mounted).

external contaminants, but they are more difficult to service. The mountings shown in Fig. 8.1*b* and 8.1*c* are more accessible but are exposed to the elements. Mechanical seals are described as inside mounted if the pressure acts on the outside diameter of the sealing faces. Any leakage that occurs will go from the O.D. of the faces to the I.D. The pressure tends to close the faces and results in a stable situation when pressure surges occur. An outside-mounted seal has the pressure acting at the I.D. of the faces, which means leakage would flow from the I.D. of the faces to the O.D. The pressure also acts to force the faces open. Inside-mounted seals can withstand pressures up to 10 times that of outside-mounted seals.

In some cases, a liquid from an external source (Fig. 8.1*d*) at a higher pressure than the sealed sump is used to flush the seal chamber. This provides lubrication and cooling for the seal. If the sealed fluid contains harmful abrasives, the nonabrasive flush fluid will prevent contaminants from harming the seal. If the sealed fluid is toxic, the flush media acts as a barrier. If leakage does occur, it will be the nontoxic flush material and not the more dangerous fluid in the sump.

The secondary seal also acts as a lock to prevent slippage of the seal head or the seat as the shaft rotates. In some cases, keyways or pins are employed to lock the elements in place to ensure slippage does not occur.

In summary, the components found in a mechanical seal include the following:

1. A stationary sealing face
2. A static secondary seal for the stationary face
3. A rotating sealing face
4. A static secondary seal for the rotating ring
5. A spring or bellows to press the sealing faces together
6. A method to prevent slippage of sealing faces (keyways, pins, or secondary seal friction)
7. System to flush seal area

Most metal components in a mechanical seal (including the spring) are made from corrosion-resistant material such as stainless steel, nickel-based alloys, chrome- or nickel-plated mild steel, and, in some cases, titanium.

Spring or Bellows Design

The two faces of a mechanical seal must have a positive load to ensure against leakage. In many cases, system pressure will provide that load. Springs of various designs are employed to ensure the faces are loaded if the pressure is low or cyclic or if the seals are balanced; Table 8.2 lists advantages and disadvantages. A single coil spring is the least expensive solution and the large coil size provides some protection against corrosion (Fig. 8.2*a*). The disadvantages include uneven face loading, and the spring requires a long axial space to provide the necessary loads. It will tend to unwind due to centrifugal force in one direction of rotation and tighten in the other direction. As a result, the large single coil spring is restricted to relatively low shaft speeds.

Multiple springs (Fig. 8.2*b*) provide an even face load, can be used at higher speeds, and require less axial space than the single coil spring. The disadvantages

TABLE 8.2 Advantages and Disadvantages of Spring Types

Spring type	Advantages	Disadvantages
Single coil	Corrosion, blockage resistant Low stress levels Low cost Greater axial tolerance	Uneven loading Requires more axial space Difficult to compress as size increases May unwind/tighten at high speeds
Multiple coils	Less axial space required Even face loads Resists high speeds	Less corrosion/blockage resistance High stress levels More cost
Wave spring/Belleville washer	Saves space	High spring rate High cost
Elastomeric bellows	Also provides secondary seal Relatively inexpensive	Cannot be used in all fluids Has temperature limitations
Corrugated/welded metal bellows	Provides secondary seal Corrosion resistant High temperature High controlled rate	Expensive Requires more space than coil springs

of multiple springs are cost, low corrosion tolerance due to smaller coils, a tendency to become blocked if contaminants are present, and high stress levels in the coils. Hardened Belleville washers or wave springs (Fig. 8.2*c*) can also be used to provide face loading. This solution has a high cost and a high spring rate but is used if the axial space is limited.

Elastomeric springs are sometimes used (Fig. 8.2*d*), but they can deteriorate when exposed to certain fluids and high temperature. They are usually inexpensive and can also function as secondary seals. Corrugated metal bellows (Fig. 8.2*e*) are resistant to most fluids and also act as a secondary seal. They are typically made of corrosion-resistant metal and are expensive. Welded metal bellows (Fig. 8.2*f*) provide a spring rate that is more controlled than the corrugated type, but they are very expensive. The bellows-type springs require more axial space than coil- or wave-type springs.

Secondary Seal

The most common and least expensive secondary seal design is the elastomeric O-ring (Fig. 8.3*a*). Other designs, such as the chevron (Fig. 8.3*b*), the U-cup (Fig. 8.3*c*), and the wedge (Fig. 8.3*d*), can be molded from elastomers and are pressure activated. As the seal faces wear, the internal pressure and the spring push the secondary seals along the shaft. These seals are commonly called *pusher seals.* The shaft must be in good condition with no corrosion, pitting, or solid deposits to inhibit seal movement. Shaft concerns can be eliminated if noncontacting bellows are employed. Care must be taken to choose the right material for the secondary seal since most elastomeric materials are sensitive to temperature extremes. Plastics, such as PTFE, are sometimes used. Properties of secondary seal materials and bellows appear in Table 8.3.

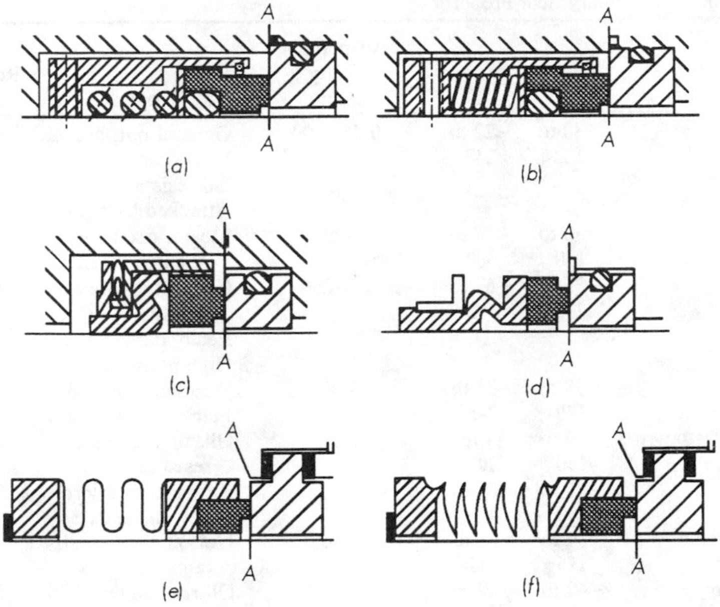

FIGURE 8.2 Various spring configurations (A–A = sealing faces). (*a*) Single coil spring; (*b*) multiple coil spring; (*c*) wave spring; (*d*) elastomeric bellows; (*e*) corrugated metal bellows; (*f*) welded metal bellows.

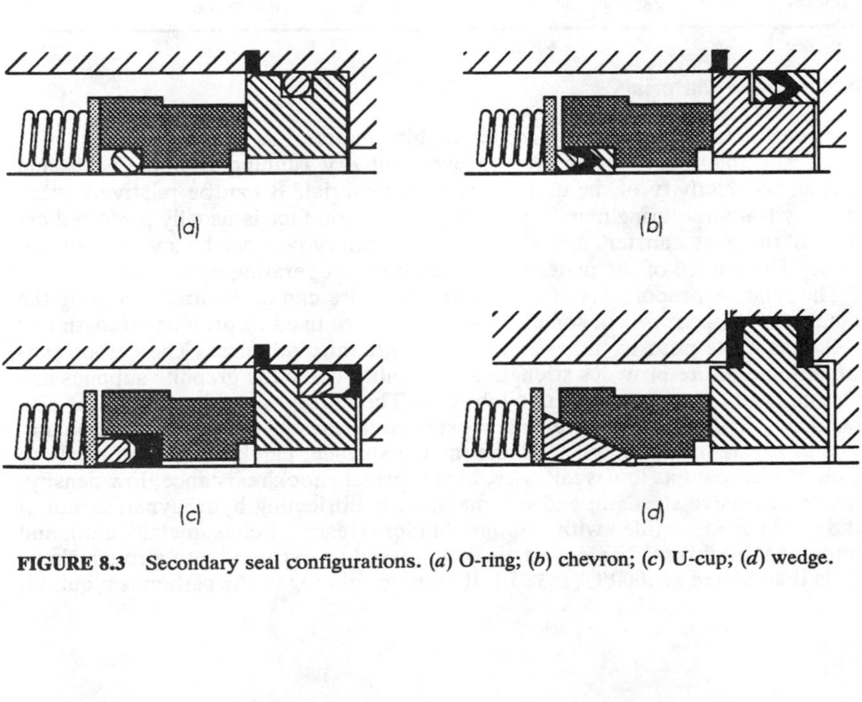

FIGURE 8.3 Secondary seal configurations. (*a*) O-ring; (*b*) chevron; (*c*) U-cup; (*d*) wedge.

TABLE 8.3 Secondary Seal Properties

Material	Oper. temp. limits °C	Oper. temp. limits °F	Air permeability rating	Properties	Relative cost
Nitrile	−30 to 120	−22 to 248	0.25–1.00	General purpose Low cost Oil resistant Attacked by ozone	1.0
Ethylene propylene	−50 to 150	−58 to 302	9.60	Steam, ozone acid, & alkali resistant	1.5
Silicone	−55 to 200	−67 to 392	170–260	Good at low temperature Easily damaged High permeability	6.25
Neoprene	−35 to 120	−31 to 248	1.40	Weather resistant Fair oil resistance	
Fluoroelastomer	−10 to 150	14 to 302	0.32	Oil, fuel, chemical resistance	22.5
PTFE	−55 to 230	−67 to 446		Resistance to virtually all fluids	22.5
Polyacrylate	−30 to 175	−22 to 347	1.50	Hot oil/ozone resistant	2.5
Epichlorohydrin	−40 to 150	−40 to 302	0.15–0.70	Oil resistant Low permeability	2.5
Metal bellows	−200 to 650	−328 to 1202		Positive seal, chemical resistance	Can be high constr. costs
High-temperature fluoroelastomer	−10 to 285	14 to 545	0.32	Excellent chemical resistance	1000

Sealing Face Materials

Most mechanical seals use a carbon-graphite material as one of the contacting faces. The main advantages are the excellent dry running characteristics and thermal conductivity of the carbon-graphite material. It can be relatively weak and may fracture during handling. A rotating carbon face is usually preferred because of the heat transfer capabilities. The stationary face can be a variety of materials. The choice of the material depends on the operating conditions.

The relative proportions of carbon and graphite can be controlled during the manufacturing process. In some cases, binders are used to provide strength and to eliminate the porosity inherent in carbon-graphite mixtures. The carbon portion of the mixture provides strength and stability while the graphite supplies low friction, oxidation resistance, and lubricity. The carbon-graphite material is also chemically inert; does not swell when exposed to various fluids; will not contaminate the system; will not weld to the mating surface; can be machined easily to close tolerances; has low wear rates, high thermal shock resistance, low density, high compressive strength; and will maintain a lubricating hydrodynamic film of fluid. The carbon is filled with graphite, binders (resin, pitches, metals, etc.), and antioxidant additives. The powder is compressed or extruded into a rough shape and is then kilned at 1000°C (1832°F). If more graphiting of the carbon is required, kilning is carried out at 2000°C (3632°F). The part is machined to the final shape and can then be impregnated with metal and resin materials. Properties of typical carbon faces appears in Table 8.4.

The materials used to impregnate the carbon faces must be compatible with the environment. Organic resins add strength and wear resistance. Inorganic salts assist in the formation of a lubricating film in vacuum and high-temperature applications and metals increase strength and thermal conductivity. The limitations of the various materials appear in Table 8.5.

Hydrofluoric acid is one of the most destructive fluids that a mechanical seal is exposed to. Carbon rings with a phenolic impregnant give the best results in this media. Blistering of the carbon face in hot oil is minimized by impregnating the carbon with metals (antimony, copper, or lead). Mechanical seals for food processing use a carbon-graphite face impregnated with resin.

The selection of the counterface material is also important. Cast iron, chrome-plated iron and steel, stainless steel, nickel irons, nickel alloys (stellite), and ceramic materials are often used. A selection chart of materials for mechanical seal faces in various media appears in Table 8.6.

The faces are usually machined out of the selected material and lapped to provide flatness to within three light bands of a helium lamp (0.87 μm, or 34.3 μin). Experience shows that this flatness will provide a positive seal with no leakage.

Seal Balance—Closing Force

The closing force of a mechanical seal is the sum of the spring force and the pressure differential across the seal. As the sealed pressure increases, the face load increases. High wear and loss of sealing ability can result. The load, because of internal fluid pressure, can be reduced by altering seal geometry so a portion of the sealing face is below the effective sealing diameter D_s.

The term *balance ratio* is employed to describe what fraction of the internal pressure is acting to close the seal faces. It is defined as the ratio of hydraulic loading area to the seal interface ratio. As the balance ratio decreases, the closing force, due to the internal pressure, also decreases [Eq. (8.1)].

The effective sealing diameter D_s is typically equal to the internal shaft diameter. If a bellows seal is used, the effective sealing diameter must be calculated

TABLE 8.4 Carbon and Graphite Material Properties

	Carbon	Carbon-graphite	Carbon resin impreg.	Carbon-graphite resin impreg.	Carbon-graphite metal-lized	Electro-graphite	Electro-graphite metal-lized
Density, g/cc	1.6	1.7	1.7	1.8	2.5	1.72	2.4
Hardness shore	90	65	100	85	70	50	65
Flex strength, psi	6000	6000	8500	10000	13000	4000	8000
Comp. strength, psi	23000	20000	30000	30000	40000	9000	24000
Modulus, psi × 10^6	2.0	2.5	3.0	3.5	4.5	1.2	2.0
Porosity, %	13.0	15.0	2.0	1.0	1.0	15.0	2.0
Relative thermal conductivity	9.0	13.0	9.0	13.0	20.0	60.0	70.0

TABLE 8.5 Resin Impregnants

Chemical medium	Phenolic	Modified phenolic	Furfural	Epoxy
		Acids		
Sulfuric, H_2SO_4	Satisfactory up to 70% at 70°C (158°F)	Satisfactory up to 80% at 100°C (212°F)	Satisfactory up to 80% at 170°C (338°F)	Satisfactory up to 80% at 100°C (212°F), 60% at 150°C (302°F), 50% at 200°C (392°F)
Hydrochloric, HCl	Satisfactory up to 100% at 300°C (572°F)	Satisfactory up to 100% at 300°C (572°F)	Satisfactory up to 100% at 170°C (338°F)	Satisfactory up to 100% at 250°C (482°F)
Nitric, HNO_3	Satisfactory up to 20% at 70°C (158°F)	Satisfactory up to 40% at 70°C (158°F)	Unsatisfactory	Satisfactory up to 20% at 100°C (212°F)
Organic	Satisfactory up to 100% at 300°C (572°F)	Satisfactory up to 100% at 300°C (572°F)	Satisfactory up to 100% at 170°C (338°F)	Satisfactory up to 100% at 250°C (482°F) except for acetic acid
		Alkalis		
Sodium hydroxide, NaOH	Satisfactory up to 20% at 70°C (158°F)	Satisfactory up to 40% at 100°C (212°F)	Satisfactory up to 90% at 170°C (338°F)	Satisfactory up to 25% at 200°C (392°F)
Potassium hydroxide, KOH				Satisfactory up to 50% at 150°C (302°F), 100% at 100°C (212°F)
		Organic solvents		
	Satisfactory except for those containing C-D or C-S bonds	Unsatisfactory	Inert to all commercial solvents	Inert to all commercial solvents
		Salts		
All concentrations	Satisfactory	Satisfactory	Satisfactory	Satisfactory
		Remarks		
			Max. temperature 170°C (338°F)	Max. temperature 250°C (482°F)

TABLE 8.6 Selection Chart for Face Materials in Various Media

Water	Salt solution	Sea water	Acids	Gasoline	Hydro-carbons
CG vs. {B, NIR, C, TC, SC} TC vs. SC	CG vs. {SS, C, MO} CB vs. PB	CB vs. {AB, SC} TC vs. SC B vs. LP	CG vs. {SS, C, CPTFF, HABC} C vs. {GPTFE, PTFE}	CG vs. {NIR, C, SF, SC, TC}	CG vs. {NIR, SF, TC, SC}

Rotating face materials:
B = bronze
CB = carbon-babbit
CG = carbon-graphite
TC = tungsten carbide
C = ceramic

Counterface materials:
AB = aluminum-bronze
B = bronze
C = ceramic
MO = monel
NIR = nickel-resist
TC = tungsten carbide
SC = silicon carbide
PB = phosphor-bronze
LP = laminated plastic
CPTFE = carbon-filled Teflon (nonoxidizing acids)
GPTFE = glass-filled Teflon (oxidizing acids)
PTFE = Teflon
SS = stainless steel
HABC = hasteloy, A, B, or C
SF = stellite hard-facing on stainless steel

with Eq. (8.2). The balance ratio for externally pressurized seals is given by Eq. (8.3); Eq. (8.4) applies if the seal is internally pressurized. The various balance conditions appear in Fig. 8.4. The imbalanced seal appears in 8.4*a* and has a balance ratio which is greater than 1 since the hydraulic area exceeds that of the face area.

In Fig. 8.4*b,* about one-third of the seal contacting face area is below the effective sealing diameter; therefore, the balance ratio is about 0.666. In Fig. 8.4*c* the outer diameter of the contacting faces is equal to the effective sealing diameter and the balance ratio is zero. Only the spring applies the closing force.

The outside-mounted case is shown in Fig. 8.4*d.* Unbalanced seals typically have balance ratios of 1.1 to 1.2 and are used for low-pressure applications (10 bar or less). Most commercially available balanced seals that are pressurized at the O.D., have a balance ratio between 0.65 to 0.85. This range provides reduced face loading while maintaining stability. Seals with balance ratios less than 0.65 can be hydraulically unstable if the pressure fluctuates.

$$F_{\mathrm{cl}} = F_s + F_h = F_s + A_f B_a \Delta_P + A_f P_o \tag{8.1}$$

$$D_{\mathrm{sb}} = \frac{(D_{\mathrm{bo}}^2 + D_{\mathrm{bi}}^2)^{1/2}}{2} \tag{8.2}$$

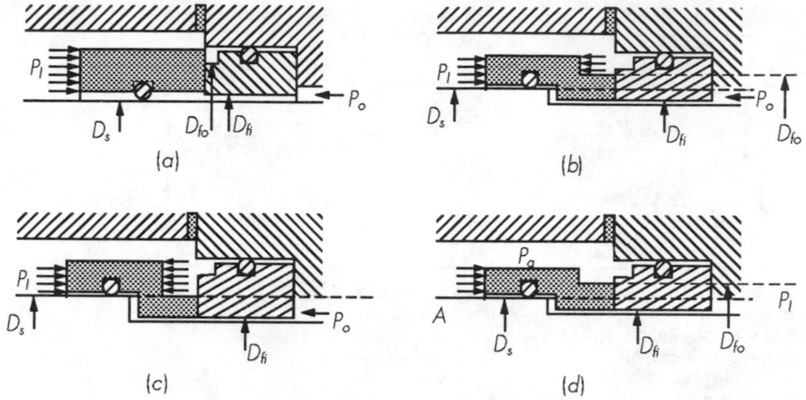

FIGURE 8.4 (*a*) Unbalanced seal $B_e > 1.0$; (*b*) balanced seal $B_e = 0.666$; (*c*) balanced seal $B_e = 0.0$; (*d*) balanced seal $B_i = 0.333$.

$$B_e = \frac{\text{hydraulic load area}}{\text{seal interface area}} = \frac{1/4\pi(D_{fo} - D_s^2)}{1/4\pi(D_{fo}^2 - D_{fi}^2)} = \frac{D_{fo}^2 - D_s^2}{D_{fo}^2 - D_{fi}^2} \tag{8.3}$$

$$B_i = \frac{\text{hydraulic load area}}{\text{seal interface area}} = \frac{1/4\pi(D_s^2 - D_{fi}^2)}{1/4\pi(D_{fo}^2 - D_{fi}^2)} = \frac{D_s^2 - D_{fi}^2}{D_{fo}^2 - D_{fi}^2} \tag{8.4}$$

where F_{cl} = closing force, N
F_s = spring force, N
F_h = hydraulic force, N
A_f = interface area, mm^2
B_a = balance ratio
B_e = balance ratio (externally pressurized)
B_i = balance ratio (internally pressurized)
ΔP = pressure drop $(P_i - P_a)$, N/mm^2
D_{fo} = outside interface diameter, mm
D_{fi} = inside interface diameter, mm
D_s = effective sealing diameter, mm
D_{sb} = effective sealing diameter for bellows, mm
D_{bo} = outside diameter of bellows, mm
D_{bi} = inside diameter of bellows, mm
P_i = upstream (system) pressure, N/mm^2
P_o = downstream (usually atmospheric) pressure, N/mm^2

For example, consider a seal of the type shown in Fig. 8.4 for a 75-mm shaft diameter D_s. The sealed pressure P_i is 1.137 MPa (165 psi), and the external pressure is atmospheric. The interface outer diameter D_{fo} is 80 mm, and the interface inner diameter D_{fi} is 73 mm. The spring force is 300 N. Calculate the total closing force for the case when the seal is (1) externally pressurized or (2) internally pressurized, using the following data:

$$P_i = 1.137 \text{ MPa} = 1.137 \text{ N/mm}^2$$

$$P_o = 0.1013 \text{ MPa} = 0.1013 \text{ N/mm}^2$$

$$\Delta P = P_i - P_o = 1.035$$

$$D_s = 75 \text{ mm}$$

$$D_{\text{fo}} = 80 \text{ mm}$$

$$D_{\text{fi}} = 73 \text{ mm}$$

$$F_s = 200 \text{ N}$$

$$A_f = 1/4\pi(D_{\text{fo}}^2 - D_{\text{fi}}^2) = 841.2 \text{ mm}^2$$

1. Externally pressurized:

$$B_e = \frac{D_{\text{fo}}^2 - D_s^2}{D_{\text{fo}}^2 - D_{\text{fi}}^2} = \frac{(80)^2 - (75)^2}{(80)^2 - (73)^2} = 0.724$$

$$F_h = (841.2)(0.724)(1.035) + (841.2)(0.1013) = 715.5 \text{ N}$$

$$F_{\text{cl}} = F_s + F_h = 300 + 715.5 = 1015.5 \text{ N}$$

2. Internally pressurized:

$$B_i = \frac{(75)^2 - (73)^2}{(80)^2 - (73)^2} = 0.2763$$

$$F_h = (841.2)(0.2763)(1.035) + 841.2(0.1013) = 325.8$$
$$F_{\text{cl}} = F_s + F_h = 300 + 325.8 = 625.8 \text{ N}$$

Multiple Seals

Some applications have sparse lubrication at the seal faces and other applications are sealing harsh chemicals that must not leak into the outside environment. Multiple seal arrangements (Fig. 8.5) are often used when these conditions exist.

The double mechanical seal (Fig. 8.5*a*) uses two mechanical seals mounted back to back to create a closed cavity. A cool fluid at a higher pressure than the sealed fluid is circulated in the cavity to provide lubrication to the seal faces, remove excess heat, and provide a barrier to prevent leakage of the sealed fluid.

Tandem mechanical seals (Fig. 8.5*b*) are oriented in the same direction and essentially provide a two-stage pressure drop. The inboard seal uses a flush which is typically the sealed fluid. The outboard seal uses a buffer fluid for cooling and lubrication at a pressure lower than the sealed fluid. When the inboard seal begins to leak, it will contaminate the buffer fluid used in the outboard seals. Sensors can be used to detect this contaminant, signaling the need for maintenance. Tandem seals are used to contain hazardous and toxic media and provide fail-safe performance for unattended applications.

Opposed mechanical seals (Fig. 8.5*c*) are used when changes in the fluid temperature may cause a phase change (vaporization or crystallization). The inboard seal is cooled by the sealed product. A barrier fluid between the inboard and outboard seal controls the temperature of both seals and protects the product fluid from the atmosphere. This will prevent the product fluid from vaporizing between the inboard seal faces at high temperatures or crystallizing at low temperatures.

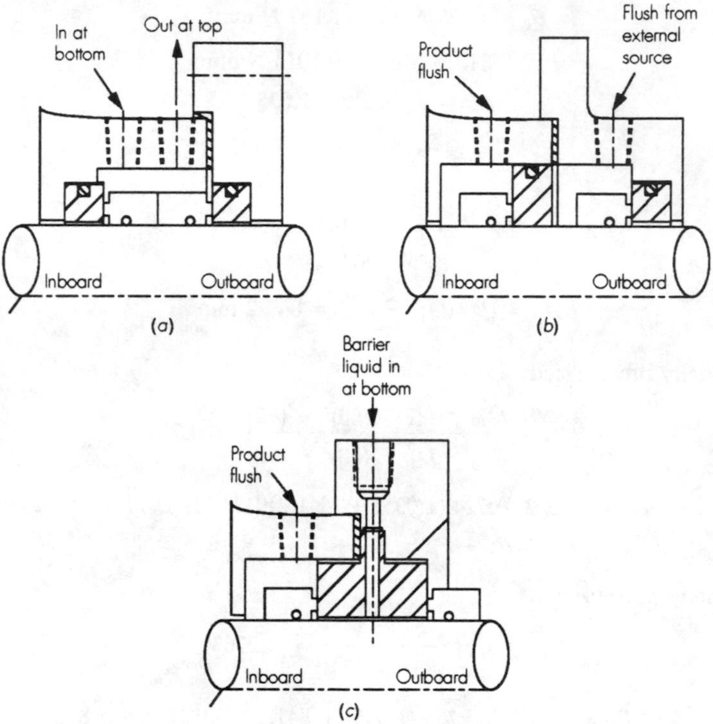

FIGURE 8.5 Multiple seals. (*a*) Double seal; (*b*) tandem seal; (*c*) opposed seal.

TABLE 8.7 Seal Environmental Considerations

Abrasives	Heat	Dry operation
Foreign material (dirt, casting, sand, etc.)	Oil degradation	Increased friction and wear
Corrosion (rust, scale)	Secondary seal degradation	Increased temperature
Sealed fluid (i.e., mud slurry)	Cracked faces	Increased oil degradation
Solids between sealing faces (sludge, crystals, etc.)	Heat-checked faces	Secondary seal failure
Wear debris from gears, etc.	Increased corrosion	Noise (stick/slip)
Outside abrasives	Vaporizes fluid film between faces	

TABLE 8.8 Control of Seal Environment

Flushing—Tapping discharge and piping fluid through filters and coolers into the seal area
1. Cool seal face.
2. Prevent vapor trap.
3. Prevent sludging.
4. Prevent abrasives from reaching seal faces.

Quenching—Flooding outside of seal cavity with water; control leakage with lip seal
1. Safeguard against leakage of toxic and corrosive liquids.
2. Cools external seal components.
3. Provide lubrication for vacuum conditions.
4. Prevents crystallization of certain fluids on outside of seal.

Jackets—Circulate fluids through a jacket around seal area
Heat or cool seal cavity area to prevent vapor lock or solids from forming.

Dead-end face lubrication
Stationary sealing ring ported to allow high-grade oils or grease into the system to lubricate faces.

Circulating face lubrication
Constant flow of oil lubricates and cools faces.

Grease lubrication
Dead-end cap with high temperature grease.

Auxiliary Equipment

A mechanical seal is often only one component in a complex system designed to protect the seal from harsh environmental conditions (Tables 8.7 and 8.8) while providing fail-safe sealing. The auxiliary equipment that is often used and the purpose of this equipment appears in Table 8.9.

TABLE 8.9 Auxiliary Equipment

Cyclone separators
Separates solids from fluids used to flush seal area

Pumps/pressurization units
Supply flow of fluid to seal faces to lubricate and control temperature

Heat exchangers
Control temperatures in seal face area

Air coolers
Used to cool seal areas

Rotameters and flow rate controllers
Used to control the flow of fluids that flush, quench, or lubricate the seal area

Pressure-sensitive switch
Will shut down system if pressure becomes too high or too low

Strainers/filters
Used to remove contaminants from fluids used to flush the seal area

Leakage detectors
Used to sound alarm or shut down unit if leakage does occur

MECHANICAL SEAL STANDARDS

Universal standards for mechanical seals do not exist since designs and applications vary widely. Many seals are custom made for specific customers and applications. As a result, some standards have been developed for specific applications. Water pump seal standards have been developed by the SAE and provide a general guide[1] as well as detailed dimensional standards, qualification test procedures, quality procedures, and recommendations for material selection.[2] A similar SAE document exists for passenger car air conditioning compressor face seals.[3] An attempt has been made to develop seal installation standards[4] and dimensional standards,[5] but many manufacturers and users have not adopted them. A code has been developed by Germany to identify single internal mechanical seals with rotating seal heads. This system employs three symbols to define size and type (Table 8.10) and five symbols to define the seal materials (Table 8.11).

TABLE 8.10

First symbol	Second symbol	Third symbol
U = unbalanced B = balanced	Nominal diameter, mm	Direction of rotation/ spring winding: R = right-hand, single spring L = left-hand, single spring S = independent of the direction of shaft rotation, single spring A = independent of the direction of shaft rotation, multiple springs

As an example, a seal with an identification of B75ABLVDD would have the following characteristics:

B	Balanced
75	75-mm nominal diameter
A	Multiple springs, independent of the direction of shaft rotation
B	Rotating ring material, hard carbon resin impregnated
L	Seat material, stellite
V	Secondary seal material, fluoroelastomer
D	Carbon steel spring
D	Carbon steel machined components

OPERATING LIMITATIONS

Sealed Fluid

Some fluids will corrode portions of sealing elements. It is essential that the nature of the fluid be understood and seal materials be chosen to reduce corrosion

TABLE 8.11 Material Code

First symbol—rotating ring material / Second symbol—seat material	Third symbol—secondary seal material	Fourth symbol—spring material / Fifth symbol—material for other parts*
Mechanical carbon	Elastomers	D = carbon steel
A = hard carbon impregnated	P = nitrile	E = Cr-steel
B = hard carbon resin impregnated	N = chloroprene	F = CrNi-steel
C = other carbon	B = butyl	G = CrNiMo-steel
Metal	E = EP	M = High nickel alloy
D = carbon steel	S = silicone	N = Bronze
E = Cr-steel	V = fluoroelastomer	T = Misc. metal
F = CrNi-steel	M = PTFE-encased	
G = CrNiMo-steel	X = misc. elastomers	
H = CrNi-steel, stellited	Nonelastomers	
K = CrNiMo-steel, stellited	T = PTFE	
L = stellite	A = asbestos impregnated	
M = high nickel alloy	F = IT seal	
N = bronze	Y = misc. nonelastomers	
P = cast iron	Special case	
R = alloyed cast iron	U = more than 1 secondary seal material used	
S = cast chrome		
T = misc. metals		
Metal carbide		
U = metal carbide		
Metal oxide (ceramic)		
V = Al-oxide		
W = Cr-oxide		
X = Misc. metal oxides		
Plastics		
Y = PTFE, strengthened		
Z = misc. plastics		

*Except end plate and shaft.

of seal elements. Specific gravity of the fluid is an important factor to consider when selecting a seal. Most seal manufacturers recommend a hydraulically balanced seal when the specific gravity is 0.63 or less.[6] The vapor pressure or boiling point of the sealed fluid must not be exceeded between the seal faces. If boiling occurs, the seal faces will open and tilt momentarily as gas builds up and then escapes. Cool fluid will enter, which allows the faces to close once again. Frictional heat will cause local temperatures to rise once again, and the cycle is repeated, which results in an unstable seal. The edges of the carbon ring will chip, and, in extreme cases, pitting of the carbon face will occur. Gross leakage will occur. To ensure safe operation, the seal cavity pressure should be 0.17 MPa (25 psi) below the fluid vapor pressure. The temperature within the seal cavity should be 28°C (50°F) below the fluid boiling point at the specific working pressure. Seals will function properly over a broad range of fluid viscosity. If the viscosity exceeds 20,000 Saybolt Seconds Universal (SSU), standard seals will give marginal performance since the friction generated by the thick fluid may cause the face temperature to exceed the fluid boiling point.

Balanced seals with a cooling quench fluid are recommended. Some fluids contain dissolved solids which may crystallize or polymerize between the seal faces. This can create catastrophic face wear, which will result in premature failure. A heated quench fluid may be required to prevent the plating of crystals between the seal faces. Fluids may also carry suspended solids which will wear the face materials. If a buffer fluid is not practical, filters or separators must be used to remove the solids from the fluid that enters the seal area. Any grit that enters the seal area may be trapped between the seal faces and create catastrophic damage within a short period of time.

Pressure Velocity (PV) Limitations

The wear rate of the face materials will dictate ultimate seal life and is dependent upon surface speed and interface loading. The PV factor is defined as the product of the pressure drop ΔP across the seal and the average rubbing velocity. The average rubbing velocity is determined with Eq. (8.5) and the PV factor is obtained with Eq. (8.6). In some cases, D_{ave} is estimated by using the shaft diameter D_s.

$$V = \frac{\pi D_{ave}}{1000}\left(\frac{N}{60}\right) = \frac{\pi}{1000}\,\frac{(D_{fo} + D_{fi})}{2}\,\frac{N}{60} \quad \text{(m/s)} \tag{8.5}$$

$$\text{PV} = \Delta P\,\frac{\pi}{1000}\,\frac{(D_{fo} + D_{fi})}{2}\,\frac{N}{60} \quad \text{(MPa m/s)} \tag{8.6}$$

As a general guide, seal capabilities can be classified as low if the PV value is 0.7 MPa m/s (about 20,000 psi ft/min), medium if the PV value is between 0.7 and 10 MPa m/s (about 285,000 psi ft/min), and high if the PV value ranges from 10 to 70 MPa m/s (about 2,000,000 psi ft/min). Balanced seals reduce the pressure acting on the seal faces; therefore, they have higher PV values than unbalanced seals. PV values for various face materials appear in Table 8.12.

Example. Determine the maximum rpm for an unbalanced seal with a mean face diameter of 25 mm (0.984 in) operating in an aqueous solution at 0.14 MPa (20 psi)

TABLE 8.12 PV Limits (MPa m/s) for Various Seal Face Material Combinations*

Face materials	Water and aqueous solutions: Unbalanced seals	Water and aqueous solutions: Balanced seals	Other fluids: Unbalanced seals	Other fluids: Balanced seals
Stainless steel carbon	0.55		3.0	
Lead bronze carbon	2.3		3.5	
Stellite/carbon	5	8.5	5.2	58
Chrome oxide/carbon	7	45		
Alumina ceramic/carbon	3.5	25	8.8	42
Tungsten carbide/tungsten carbide	4.5	50	7.0	42
Tungsten carbide/carbon	7	42	8.8	122

*Multiply PV value in MPa m/s by 28,530 to obtain psi ft/min. Multiply PV value in bar m/s by 2855 to obtain psi ft/min.

using ceramic-carbon face materials. How does the maximum rpm change if stainless steel-carbon faces are used instead?

From Table 8.12, the PV value for the unbalanced ceramic-carbon seal in an aqueous solution is 3.5 MPa m/s (99,860 psi ft/min). From Eq. (8.6):

$$N = \frac{60\ (1000)}{\Delta P \pi (D_{\text{ave}})}\ \text{PV} = \frac{60(1000)(3.5)}{(0.14)(3.1416)(25)} = 19098 \text{ rpm}$$

For stainless steel-carbon faces, the PV value is only 0.55:

$$N = \frac{60(1000)(0.55)}{0.14(3.1416)(25)} = 3000 \text{ rpm}$$

Obviously, the ceramic-carbon face combinations are preferred if the application is high speed.

Pressure-Temperature Operating Envelope

Most seals have a thin film of lubricant between the seal faces that provides lubrication. As temperatures increase, fluid viscosity decreases, the film may no longer be able to support the load between the faces, and dry running may occur. Elevated temperatures can also cause the fluid film to boil. A banging or popping sound may occur, and severe damage can result to the seal faces. It is thus essential to set temperature and pressure operating limits for seals that will provide a margin of safety to prevent the fluid film from boiling. The boiling temperature of fluids will increase as the pressure increases (Fig. 8.6). The temperature rise between the seal faces depends on many factors (heat transfer coefficients of the fluid, heat generation due to seal friction, thermal conductivity of the seal materials, seal dimensions, etc.) and can be calculated using typical heat transfer equations or measured on an actual seal. The upper temperature limit for the seal operating curve is obtained by subtracting the temperature rise at the seal faces from the fluid boiling point at the given pressure to ensure safety:

$$\text{Operating temperature limit} = \text{fluid boiling point at sealed pressure } P \text{ and shaft speed } N - \text{temperature rise at seal face}$$

This condition is illustrated graphically in Fig. 8.7. Points to define the operating envelope are developed for each pressure at a constant speed by estimating the temperature rise at the seal faces. The maximum pressure is defined by the PV value for the faces used in the seal and the speed of the application. The seal can usually be safely operated at any pressure and temperature within the operating envelope.

Effects of Speed and Seal Configuration on the Pressure-Temperature Operating Envelope

Increasing shaft rpm for a constant shaft diameter will lower the maximum pressure that a given combination of face materials can withstand. Higher speeds will

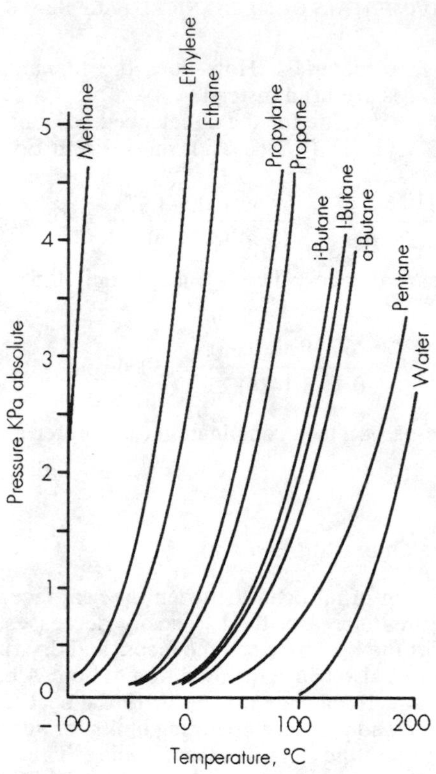

FIGURE 8.6 Vapor pressure curves for various fluids.

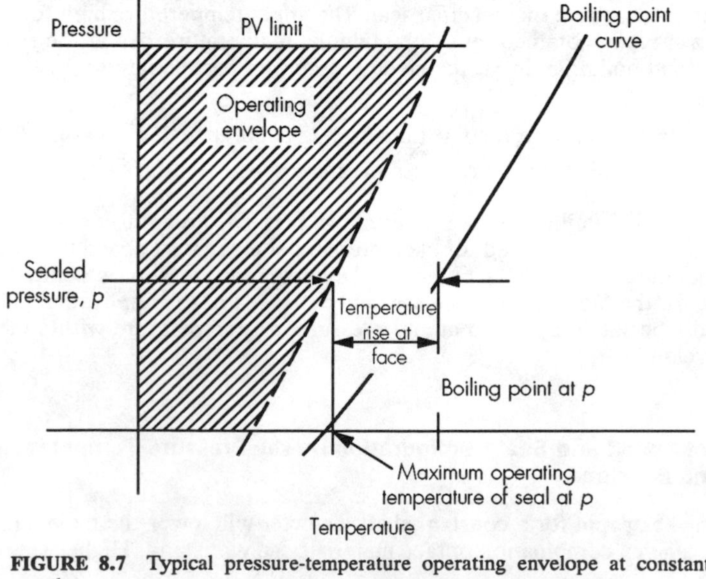

FIGURE 8.7 Typical pressure-temperature operating envelope at constant speed.

generate greater frictional torque, which will result in higher interface temperatures. The net effect is a reduction in the operating envelope as shaft speed increases (Fig. 8.8). The effect is essentially the same if the seal size is increased while holding shaft rpm constant. Increasing shaft size results in an increased linear velocity even though shaft rpm is held constant. The operating envelope will decrease as the shaft size increases (Fig. 8.9).

A balanced seal has a higher PV value than an unbalanced seal. The maximum pressure capability is thus higher for a balanced seal than for an unbalanced one at a constant speed. The forces between the faces are greater for an unbalanced seal than for a balanced one. The interface temperatures are thus greater. The operating envelope for an unbalanced seal is smaller than a balanced seal (Fig. 8.10).

Changing face materials will change the upper pressure limits and the interface temperature, which will result in various operating envelopes (Fig. 8.11). Each different fluid that is sealed will have a unique operating envelope. It is apparent that a large variety and quantity of operating envelopes can be developed for each seal size, shaft speed, and face materials. Seal engineers must always review the operating envelope for the specific fluid to be sealed to ensure the proper seal selection is made.

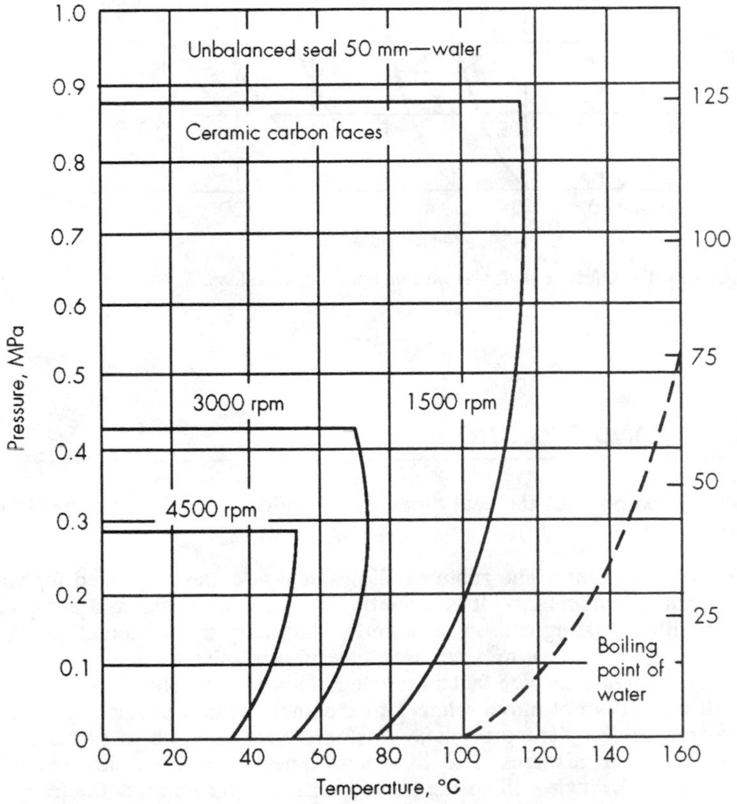

FIGURE 8.8 The effect of shaft speed on seal operating envelope.

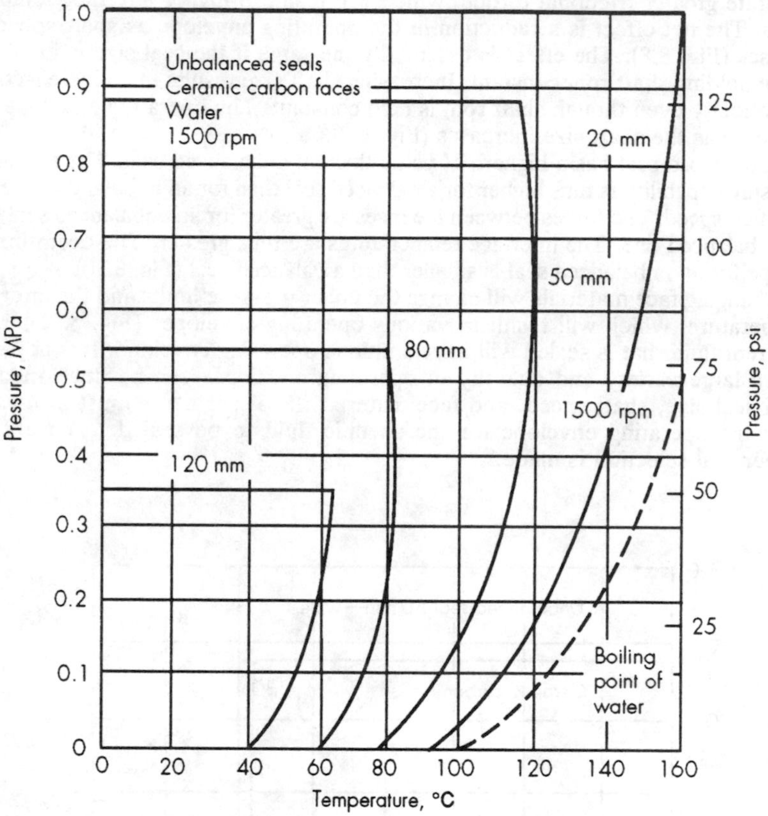

FIGURE 8.9 The effect of shaft size on seal operating envelope.

SEAL TYPES AND SELECTION

The following list outlines the seal types that should be used in various circumstances (see also Table 8.13):

Water and fuel pump: The rubber bellows seal is commonly used for water and fuel pump applications. It is a relatively inexpensive seal and is typically unitized with a rotating carbon head and a stationary metal face. To improve life and minimize abrasion, a ceramic face is often used. In extreme cases, tungsten or silicone carbide faces are used. The rubber bellows is pressed on the shaft and does not move relative to the shaft, which provides protection from corrosion. Ethylene propylene elastomers are used up to 140°C (284°F) with water-glycol mixtures, and fluoroelastomers are used with fuel up to 150°C (302°F). A single coil spring provides the proper load for the faces.

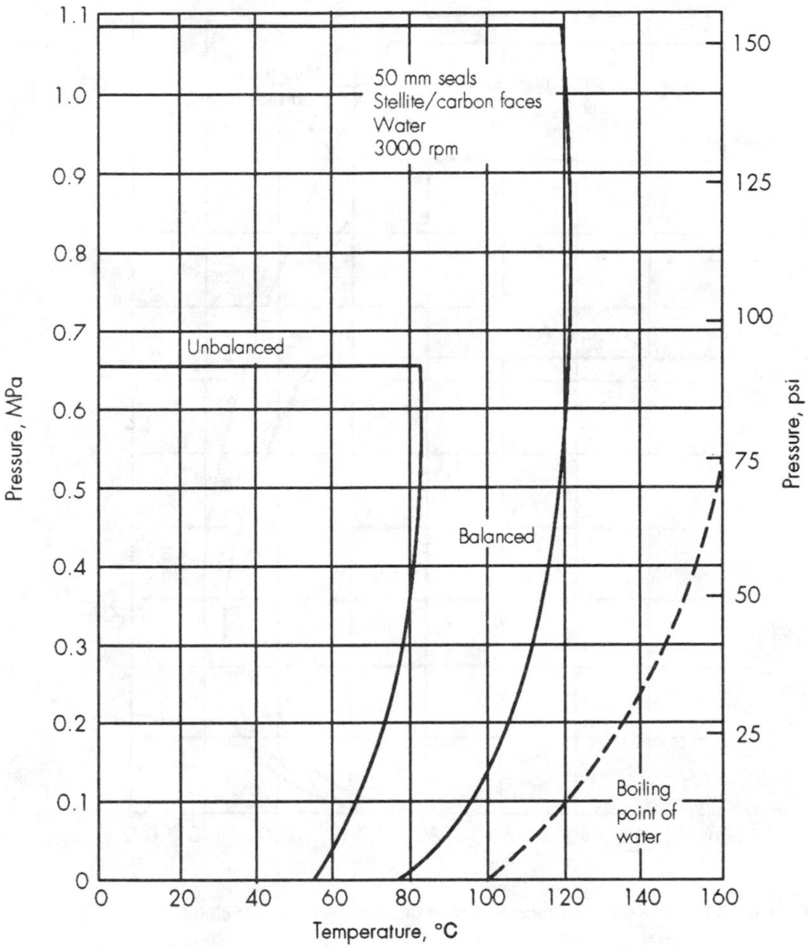

FIGURE 8.10 The effect of seal balance on seal operating envelope.

Slurry: The asymmetrically formed bellows has no sharp corners to trap contaminants contained in the slurry. It acts as both a spring and a seal. The bellows is made from corrosion-resistant materials. O-rings are used for the static secondary seals. Hard face materials are used to prevent face wear from abrasive materials contained within the slurry.

Hygienic Use: Seals exposed to food products must have components that are not harmful. These components must comply with stringent hygienic requirements that are prescribed by the federal government. Elastomers must withstand repeated sterilization with hot water or steam up to 120°C (248°F).

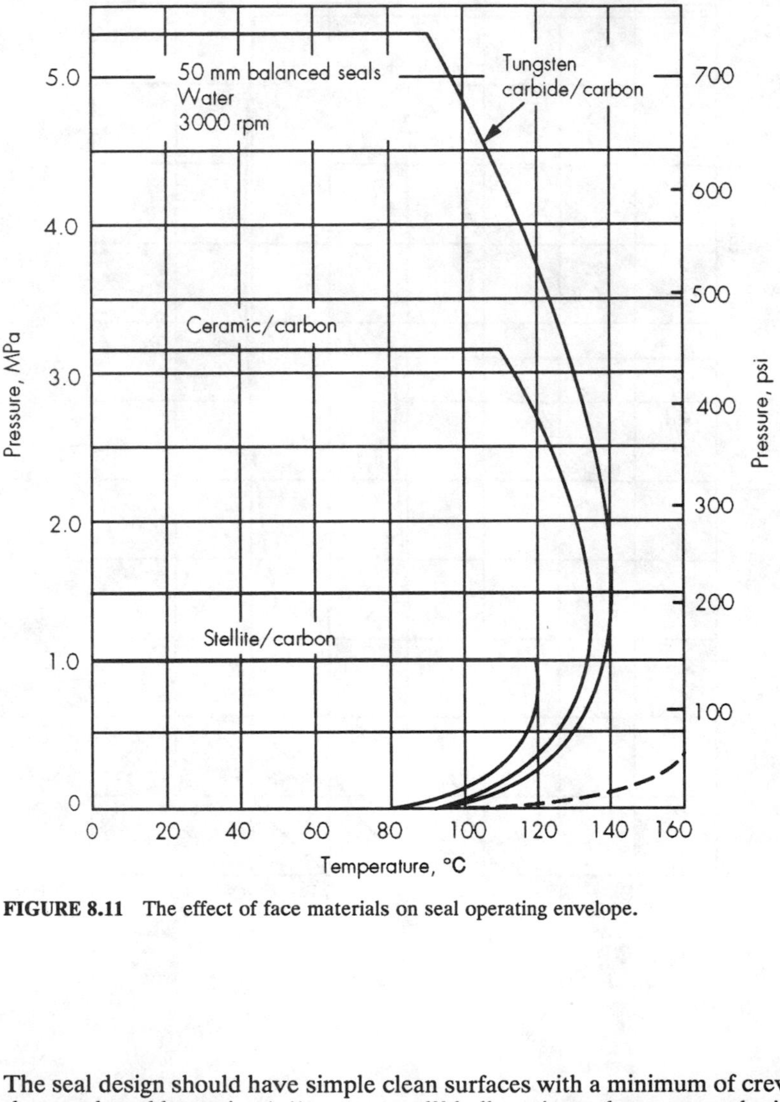

FIGURE 8.11 The effect of face materials on seal operating envelope.

The seal design should have simple clean surfaces with a minimum of crevices that can breed bacteria. A "wraparound" bellows is used to protect the internal contaminants form being exposed to the processed product. The asymmetrically formed metal bellows is sometimes used and can withstand temperatures up to 200°C (392°F). Multiple springs are used to provide a uniform load between the faces.

Mild Corrosive Fluids and Refrigeration Compressors: Conventional O-ring and elastomeric bellows seals sometimes are incompatible with mildly corrosive fluids and refrigerator coolants. PTFE wedge seals are used with multisprings to provide the required fluid compatibility. Asymmetrically formed bellows seals are also used.

TABLE 8.13 Seal Types and Applications

Application	Typical construction	Comments	Key seal requirements
Water, fuel		NBR elastomers up to100°C, FKM elastomers up to 150°C, carbon graphite vs. ceramic or silicone carbide-single spring.	LSE, HT, AR
Slurry	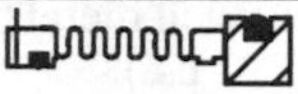	Asymmetric, formed metal bellows, hard face material	AR
Hygenic	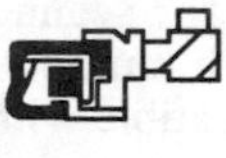	"Wrap-around" elastomeric bellows to withstand steril-ization up to 120•C, multiple springs. Also can use asym-metric formed metal bellows.	AR, CC, SF
Mild corrosive fluids and refrigeration compressors		PIFE wedge, multiple springs. Also use asymmetric formed metal bellows.	LSE, CR, AR, LL
Highly corrosive fluids		PIFE bellows, multiple springs, externally mounted. Also use asymmetric formed metal bellows.	LSE, CR, AR, LL, PL
Hot Hydrocarbon		Single spring wedge seal is used to prevent clogging. Multispring wedge and welded metal bellows designs are also used. Wedge material must resist high temperature.	LSE, HT, AR, LL
Cool Hydrocarbon	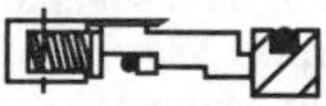	Many seal types can be used. Carbon face should be blister resistant, "O"-ring multiple spring designs are most popular.	HP, LL
Pipeline Pumps	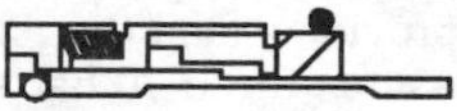	Sleeve mounted to handle high speeds (6000 rpm) and pressure (83 bar).	HP, AR, HSS, LL
Cryogenic		Welded metal bellows.	LT, LL
Boiler Feed	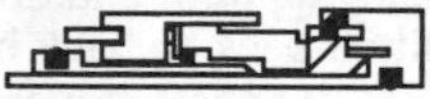	Welded springs, multiple springs are also used.	HT, HP, HSS, LL, PL
Gas Compressors	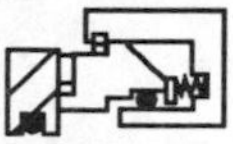	Spiral grooves on rotating face push gas inward to retard leakage.	HT, HP, HSS, LL
Mixer/ Agitators		Double seals mounted vertical to seal vertical shafts.	LSE, CR, HT AR

LSE = low shaft erosion; CR = corrosion resistant; HT = high temperature; LT = low temperature; HP = high pressure; AR = abrasion resistant; CC = clean contours; HSS = high shaft speed; LL = low leakage; PL = predictable life; SF = must resist sterilization.

Highly corrosive use: A PTFE bellows is used as the seal in conjunction with multiple stainless steel springs. All other parts of the seal in contact with the fluid are also made of PTFE. The seals are usually externally mounted and have visual wear indicators that signal when the seal must be changed. This will ensure that the highly corrosive fluids do not escape to the environment.

Hot hydrocarbon: A wedge seal with multiple springs is used to seal hot hydrocarbons. The wedge is typically made of a high-temperature graphite wedge if high pressures are encountered. If clogging can occur, a single coil spring is substituted for the multiple springs. The welded metal bellows is used for temperatures up to 300°C (572°F) and 20 bar (290 psi).

Cool hydrocarbon: Many seal types can be used for sealing cool hydrocarbons. The most popular seal type is the O-ring with multiple springs. The O-ring and other elastomeric materials must be compatible with the application fluid and temperature. The carbon used for the face material must be blister-resistant grade. The face and seat components must be designed to be flat and must be installed to minimize distortion.

Pipeline pumps: Pipeline pumps can be exposed to a variety of fluids such as crude oil, which is often mixed with water, solids, and entrained gas. Pipeline pumps also must handle salt water and other corrosive fluids. Materials selected must be compatible with the fluids and temperatures and must resist corrosion. The seals are often sleeve mounted and are exposed to pressures of 83 bar (1200 psi) at speeds up to 6000 rpm.

Cryogenic use: Low temperatures can cause elastomers to become brittle and crack. Thermal contraction can also cause secondary seal leakage. These problems are avoided by eliminating elastomers. A metal wedge coupled with a welded metal bellows is often used.

Boiler feed: Demineralized water is a poor lubricant, and the face materials must be selected to survive with sparse lubrication. The seals are often sleeve mounted since shaft speeds may approach 6000 rpm. Wave springs are sometimes used instead of multiple springs to conserve space.

Gas compressors: Gas compressors have no lubrication and are often run dry. Wider seal faces equipped with a spiral groove or a Rayleigh pad are used to push the gas toward the sealed system to provide a dam (Fig. 8.12). The seal faces actually ride on a thin gas film. Pressures up to 70 bar (1000 psi) and speeds up to 150 m/s (30000 ft/min) can be sealed with spiral groove seals.

Mixer agitators: Many mixers and agitators have vertical shafts that enter the top of the machine. Seals are mounted vertically as well, and double seals are normally used. Conventional seals may be fitted back to back but concentric seals are easier to install, can accommodate greater shaft movement, and provide higher reliability.

FAILURE ANALYSIS

Seal leakage may result for a variety of reasons and may come from several paths. In general, most premature leakage problems result from improper selection of seal and materials, improper use of the seal, and improper installation. Possible leakage paths from a typically balanced seal appear in Fig. 8.13. A more detailed troubleshooting list is given in Table 8.14.

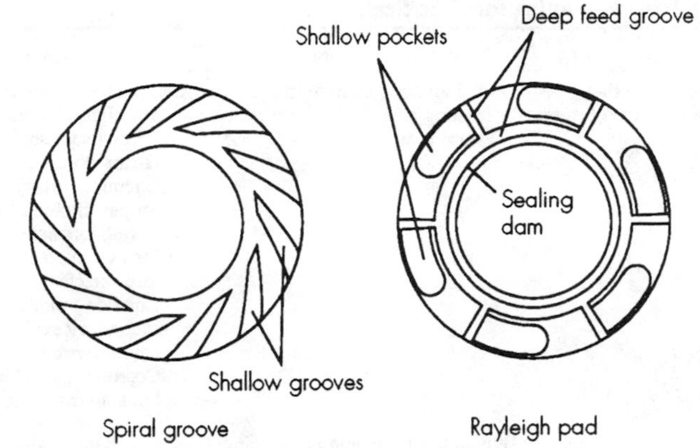

FIGURE 8.12 Hydrodynamic seals for gas applications.

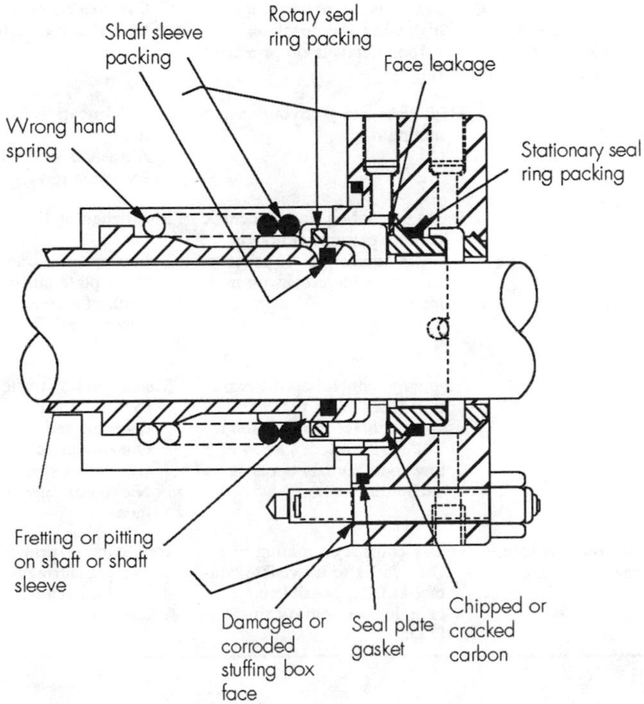

FIGURE 8.13 Possible leakage paths and causes.

TABLE 8.14 Troubleshooting for Face Seals

Symptoms	Appearance and observations	Causes
Seal leaks steadily whether shaft is stationary or rotating	Full 360° contact patterns on mating ring, little or no measurable wear	Secondary seal leakage caused by: Nicked, scratched, or porous seal surfaces O-ring compression set Chemical attack due to improper choice of elastomer O-ring extrusion due to excessive pressure, temperature, or elastomeric swell Corroded nickel or pitted secondary seal countersurface Corroded, cracked, or porous housing material
	Noise from flashing or popping; high wear or thermal distress on mating ring; high wear and carbon deposits on primary ring; possible edge chipping on primary ring; thermal distress over 1/3 of mating ring, located 180° from inlet of seal flush; high wear and possible carbon deposits on primary ring; thermal distress at 2–6 locations on mating ring; high wear and possible carbon deposits on primary ring	Sealed liquid vaporizing at seal interface, caused by: Low suction or stuffing box pressure Improper running clearance between shaft and primary ring Insufficient cooling Improper bushing clearance Circumferential flush groove in gland plate missing or blocked Overloaded seal PV value too high
	High wear and grooving on mating ring	Poor lubrication from sealed fluid Abrasives in fluid PV value too high
	Damaged mating ring; eccentric contact pattern, although width equals that of primary ring; possible cracks on mating ring.	Misaligned mating ring caused by: Improper clearance between gland plate and stuffing box Lack of concentricity between shaft O.D. and stuffing box I.D.
	Two large contact spots, pattern fades away between spot; contact through about 270°, pattern fades away at low spot; contact spots at each bolt location	Mechanical distortion caused by: Overtorqued bolts Out-of-square clamping parts Out-of-flat stuffing box faces Nicked or burred gland surface
Steady leakage at low pressure, little or no leakage at high pressure	Heavy contact on mating ring O.D. fades to no visible contact at I.D.; possible edge chipping on primary ring O.D.	Deflection of primary ring from overpressurization; seal faces not flat because of improper lapping

TABLE 8.14 Troubleshooting for Face Seals (*Continued*)

Symptoms	Appearance and observations	Causes
Steady leakage when shaft is rotating, little or no leakage when shaft is stationary	Heavy contact pattern on mating ring I.D.; fades to no visible contact at O.D.; possible edge chipping on primary ring I.D.	Thermal distortion of seal faces; seal faces not flat because of improper lapping
Steady leakage when shaft is rotating, no leakage when shaft is stationary	Contact pattern on mating ring slightly larger than primary ring width; possible high spot opposite drive pin hole.	Out-of-square mating caused by: Nicked or burred gland surfaces Improper drive pin extension Misaligned shaft Piping strain on pump casing causing distortion Bearing failure Shaft whirl

MECHANICAL SEAL THEORY

Complexity of Mechanical Seal Theory

The performance of mechanical end face seals is dependent on geometric, hydrodynamic, and thermal effects. The geometric factors include face deformation due to hydrostatic and centrifugal forces generated between the faces as well as shaft eccentricity, misalignment, and vibration. Other geometric factors include the amount of contact force between the faces, which results in face wear.

Hydrodynamic effects that must be considered are fluid properties such as viscosity, density, lubricity, and surface tension. The fluid pressure and its distribution between the seal faces is also important. Face friction produces heat that can deform the face geometry and vaporize the sealed fluid if the heat is not properly removed. Vaporization can destroy lubrication, which will eventually destroy the seal. All of these factors will interact and make a theoretical study of seal performance very complex. Seal performance factors include frictional torque, power consumption, face wear, and rate of leakage.

Net Closing Force

The total closing force on the seal face defined by Eq. (8.1) must be supported by the hydrostatic pressure due to the sealed pressure (F_{hs}), the hydrodynamic film support (F_{hd}), and mechanical support (F_m) that occurs by direct contact between face asperities [Eq. (8.7)]. Note that B_a is equal to B_e if the seal is externally pressurized [Eq. (8.3)] and B_i if the seal is internally pressurized [Eq. (8.4)].

The sealed fluid creates a pressure distribution from the high side to the low side. This distribution depends on face geometry and how parallel these faces are. The hydrostatic force F_{hs} can be calculated if the pressure distribution $P[r]$ is known [Eq. (8.8)]. Usually the pressure distribution is unknown, and a factor B_{hs} is defined that can be used for analysis [Eq. (8.9)].

If the faces are parallel, a linear pressure distribution results, and B_{hs} is equal to 0.5. If the faces converge, the film will converge, and B_{hs} is greater than 0.5. When the film diverges, B_{hs} is less than 0.5 (Fig. 8.14).

The net closing force is defined by Eq. (8.10) and is the difference between the total closing force F_{cl} and the hydrostatic force F_{hs}. It is interesting to note that the spring force will be the only closing force if the faces are parallel (B_{hs} = 0.5), and the balance factor B_a is chosen to be 0.5.

$$F_{cl} = F_s + A_f(B_a\Delta P + P_o) = F_{hs} + F_{hd} + F_m \tag{8.7}$$

$$F_{hs} = 2\pi \int_{D_{fi_2}}^{D_{fo_2}} rP(r)\,dr \tag{8.8}$$

$$F_{hs} = A_f(B_{hs}\Delta P + P_o) \tag{8.9}$$

$$F_{net} = F_{cl} - F_{hs} = F_{hd} + F_m = F_s + A_f\Delta P(B_a - B_{hs}) \tag{8.10}$$

Estimating Possible Face Separation

Face separation will occur when the hydrostatic pressure between the seal faces builds up and equals or exceeds the net closing force Eq. (8.11). This can only occur if the value of B_{hs} is larger than the seal balance number B_a. The maximum

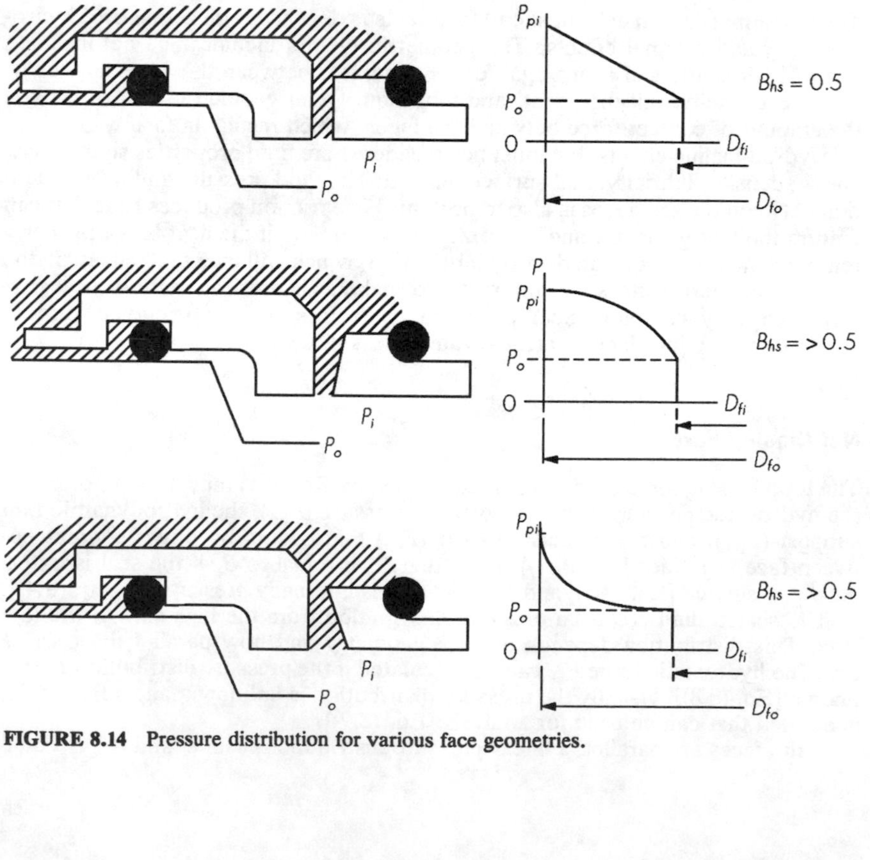

FIGURE 8.14 Pressure distribution for various face geometries.

value of B_{hs} is unity and the critical ΔP can be calculated using Eq. (8.12). Any value of ΔP larger than $(\Delta P)_{crit}$ could result in face separation if maximum face divergence exists. This is prevented by increasing the spring load, reducing the seal face area, or changing the balance factor.

The critical pressure distribution factor can be estimated for a given ΔP with Eq. (8.13). Any value of B_{hs} less than this critical value will not cause face separation. If B_{hs} is larger than the critical value $(B_{hs})_{crit}$, the seal face area must be reduced, the spring load increased, or the balance number B_a increased.

$$F_{net} = 0 = F_s - A_f \Delta P (B_{hs} - B_a) \tag{8.11}$$

$$(\Delta P)_{crit} = \frac{F_s}{A_f(1 - B_a)} \tag{8.12}$$

$$(B_{hs})_{crit} = \frac{F_s}{A_f \Delta P} + B_a \tag{8.13}$$

For example, for the seal data found in the section "Seal Balance—Closing Force" and an externally pressurized seal:

$$(\Delta P)_{crit} = \frac{F_s}{A_f(1 - B_a)} = \frac{300}{841.2(1 - 0.724)} = 1.292 \text{ N/mm}^2$$

$$(B_{hs})_{crit} = \frac{F_s}{A_f \Delta P} + B_a = \frac{300}{(841.2)(1.035)} + 0.724 = 1.068$$

$(\Delta P)_{crit}$ is greater than ΔP and $(B_{hs})_{crit}$ is larger than unity: therefore, this seal configuration will not blow with the designed pressure drop of 1.035 N/mm^2.

For an internally pressurized seal it is:

$$(\Delta P)_{crit} = \frac{F_s}{A_f(1 - B_a)} = \frac{300}{841.2(1 - 0.276)} = 0.493 \text{ N/mm}^2$$

$$(B_{hs})_{crit} = \frac{F_s}{A_f \Delta P} + B_a = \frac{300}{841.2(1.035)} + 0.276 = 0.620$$

This seal design presents a problem for the desired ΔP of 1.035 N/m^2. It can only be used for applications with a ΔP of 0.493 N/m^2 or less. The spring force F_s must be increased, the seal face area A_f reduced, and/or the seal balance value B_a increased to successfully seal the desired pressure of 1.035 N/m^2.

Centrifugal Effects

When face rotation begins, a fluid film forms if the seal is properly designed. Centrifugal force will develop a pressure gradient that will oppose the hydrostatic pressure gradient of an externally pressurized seal. If the seal is internally pressurized, the pressure gradients will be additive and leakage will increase. Many seals are externally pressurized so that the centrifugal forces will oppose flow and reduce leakage.

In large externally pressurized seals at high speeds, the centrifugal forces may be so great that the film between the faces is starved, lubrication is reduced, and the seal will fail due to high heat generation and excessive wear. It can be shown that the pressure drop required to ensure a full film penetration of the rotating seal face to provide adequate lubrication is given by Eq. (8.14):[7–10]

$$(\Delta P)_c = \rho \left(\frac{2\pi N}{60}\right)^2 \left(\frac{D_{fo}^2 - D_{fi}^2}{8000}\right) \tag{8.14}$$

For example, consider the externally pressurized seal in the section "Seal Balance—Closing Force" rotating at 10,000 rpm sealing a fluid with a density of 0.8 g/cm^3. The critical pressure drop required to ensure full penetration of the rotating face is

$$(\Delta P)_c = 0.8 \left(\frac{2\pi 10000}{60}\right)^2 \frac{80^2 - 75^2}{8000} = 0.085 \text{ MPa} = 0.085 \text{ N/mm}^2$$

Since the system pressure drop is 1.035 N/mm^2, the effects of centrifugal force for this application is negligible.

Seal Friction, Torque, and Power

Is is often difficult to predict seal friction, torque, and power using theoretical approaches. Empirical methods are often used with tables of friction coefficients that have been carefully measured for mechanical seals in various fluids (Table 8.15). These friction coefficients apply to the total closing force applied to the seal faces and will average the friction due to lubricant shear as well as rubbing friction that occurs when face asperities collide. The friction force is given by Eq. (8.15), the frictional torque by Eq. (8.16), and the power generated by Eq. (8.17). It should be noted that this power is generated by the forces on the seal faces only. Frictional heat can also be generated by fluid shear with other rotating seal components within the housing.

$$F_f = \mu(F_s + \Delta P A_f B_a) \quad \text{N} \tag{8.15}$$

TABLE 8.15 Typical Values of Coefficients of Friction for Mechanical Seals

Pressure, MPa	Water & aqueous solutions	Lt. hydrocarbons visc., <2 cSt	Light oils visc., 2–50 cSt	Heavy oils visc., >50 cSt
0.000	0.1	0.1	0.14	0.18
0.138	0.1	0.075	0.11	0.14
0.276	0.1	0.055	0.08	0.10
0.413	0.1	0.03	0.055	0.065
0.55	0.1	0.025	0.041	0.05
0.69	0.1	0.02	0.03	0.04
0.69+	0.1–0.4	0.02	0.03	0.04

$$\text{Torque} = F_f \frac{(D_{fo} + D_{fi})}{4000} \quad \text{N} \cdot \text{m} \tag{8.16}$$

$$P_{ow} = \text{torque} \frac{2\pi N}{60} = \frac{\text{N} \cdot \text{m}}{s} \quad \text{watts} \tag{8.17}$$

For example, consider the seal data described in the section "Seal Balance—Closing Force" for sealing light oils (2–50 cSt) at 3000 rpm. The seal is externally pressurized (ΔP of 1.035 N/mm^2). From Table 8.14, the coefficient of friction is estimated to be 0.03:

$$F_f = 0.03\,[300 + (1.035)(841.2)0.724]$$

$$F_f = 27.91 \text{ N}$$

$$\text{Torque} = F_f \frac{(D_{fo} + D_{fi})}{4000} = 27.91 \left(\frac{80 + 75}{4000}\right)$$

$$= 1.082 \text{ N} \cdot \text{m}$$

$$P_{ow} = (1.082) \frac{2\pi(3000)}{60}$$

$$= 340 \text{ watts}$$

This power is generated as heat and must be transported from the seal faces to prevent damage and premature leakage. In some cases, conduction is adequate. Many times external coolants must be applied to remove the heat.

REFERENCES

1. SAE J1234, "Guide to the Application and Use of Engine Coolant Pump Face Seals," June 1982.
2. SAE J790, "Engine Coolant Pump Seals," June 1990.
3. SAE J1954, "Guide to the Application and Use of Passenger Car Air-Conditioning Compressor Face Seals," May 1990.
4. ISO 3069, "Seal Installation Standards."
5. DIN 24960, "Seal Dimensional Sizes."
6. Bulletin S-207-19, Crane Packing Corp., 1972.
7. Sneck, H. J., "Reversed Flow in Face Seals," ASME Preprint 68-Lub-9, November 1968.
8. Findlay, J. A., "Inward Pumping in Mechanical Face Seals," ASME Preprint 68-Lub-2, November 1968.
9. Sneck, H. J., "The Effects of Geometry and Inertia on Face Seal Performance-Laminar Flow," *ASME Journal of Lubrication Technology,* April 1968.
10. Ishiwata and H. Hiraboyashi, "Friction and Sealing Characteristics of Mechanical Seals," *BHRA International Conference on Fluid Sealing,* April 1961.

CHAPTER 9
RADIAL LIP SEALS

BASIC FUNCTIONS

Radial lip seals are used to prevent fluids, normally lubricants, from leaking around rotating shafts and their housings. They are also used to prevent dust, dirt, and foreign contaminants from entering the lubricant chamber. Lip seals are also used to separate fluids and confine pressure. They are used throughout industry in a variety of applications with widely varying operating conditions. These conditions can vary from high-speed shaft rotation (6000 to 12,000 rpm) with light oil mist and no pressure in a clean environment to a completely flooded low-speed application in a muddy environment. External temperatures can range from −51°C (−60°F) in the arctic to 51°C (120°F) in the tropics. Sump temperatures can reach 149°C (300°F) or higher. The advantages of lip seals include low cost, small space requirements, easy installation, and an ability to seal a wide variety of applications. Sealing depends on maintaining adequate interference between the elastomeric lip member and the moving shaft throughout the life of the seal (Fig. 9.1). A garter spring is sometimes used to help maintain the sealing force over the life of the seal. An interference fit must also be maintained between the seal outside diameter and the bore inside diameter.

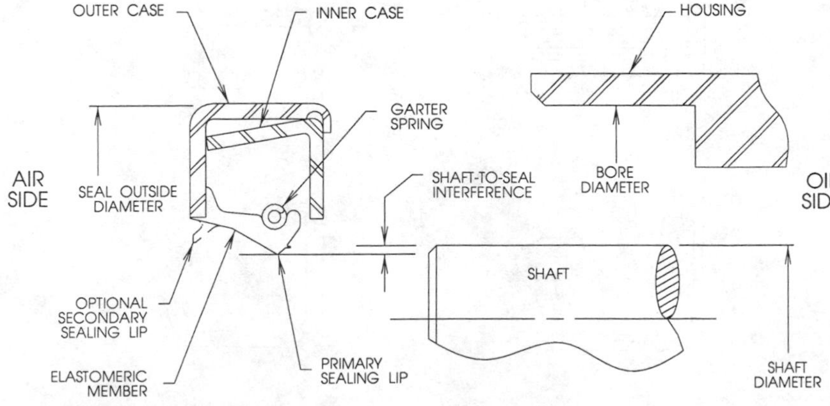

FIGURE 9.1 Typical radial lip seal.

GENERAL RADIAL LIP SEAL DESIGN

There are many seal designs and lip configurations that are used to meet various application conditions. The basic operating principles for these different configurations are the same. There are several general design factors that should be adhered to when designing or selecting a seal for a specific application.

Seal Design Evolution

The earliest seals were leather straps at the end of a wheel axle to hold grease or animal fat in place. The industrial revolution in the 1800s created demands for more sophisticated lubrication and sealing systems. Packings and rope were used successfully as seals, even though some leakage occurred. Higher shaft speeds and operating temperatures demanded improvements. About 1927, leather oil seals were produced with sealing surfaces that could follow shaft eccentricities (Fig. 9.2*a*). These assembled seals were compact, self-contained units. They could be press-fitted easily into a housing bore.

A major improvement in leather seal design took place about 1934. Leather was clinched in an outer case with a spring added to provide uniform loads (Fig. 9.2*b*). Both finger springs and garter springs were used. The garter spring proved to be a more effective way to get reliable sealing.

As lubrication problems became more complex, oil replaced grease as a lubricant. Operating temperatures increased. Seal design and materials were modified to prevent leakage. Leather treatments were introduced to prevent light oil seepage through leather fibers.

The development of the world's first oil-resistant rubbers led to synthetic rubber oil seals in 1935.[1] Early synthetic rubber seal designs were actually assembled leather seal designs, with rubber simply replacing leather. A lathe was used to trim the sealing edge.

By 1945, bonding cements were developed to bond rubber directly to the

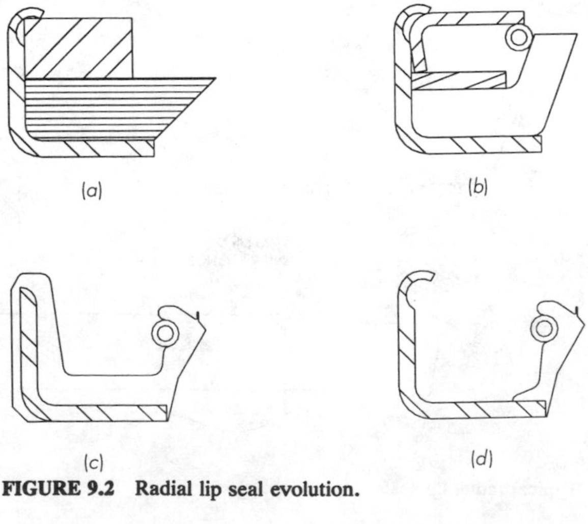

FIGURE 9.2 Radial lip seal evolution.

metal case. Seals had large bonding areas to ensure good adhesion (Fig. 9.2*c*). Additional improvements in seal bonding systems have been made since the late 1940s. Today, seals are designed with cemented joints that are actually stronger than the rubber itself.

Assembled seals are no longer recommended because of their high cost, possible internal leakage, and lack of dimensional control. Modern seal designs have a molded sealing lip that eliminates trimming. These seals provide excellent dimensional control.[2] Molded lip seals are the recommended designs for today's applications (Fig. 9.2*d*).

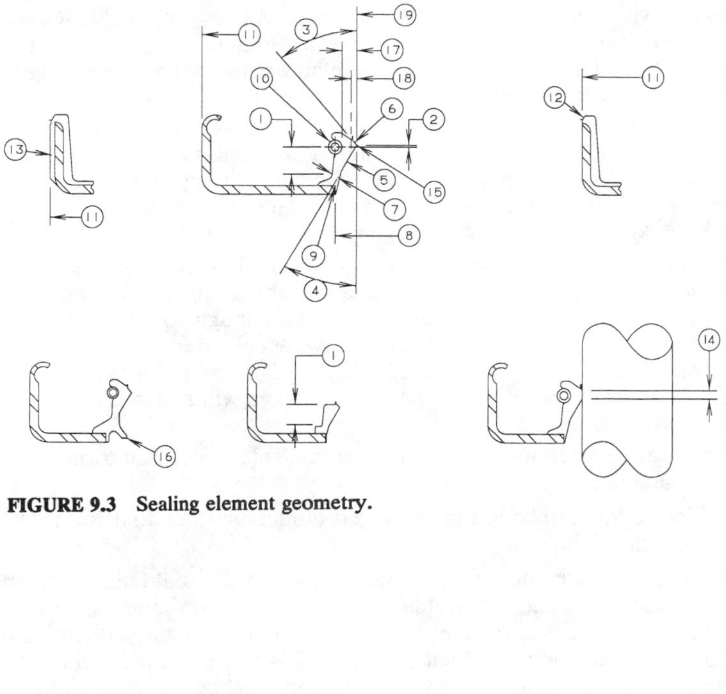

FIGURE 9.3 Sealing element geometry.

Sealing Element Geometry

There are many interrelated variables in seal design that affect the ability of a sealing element to function. The lip geometry (Fig. 9.3) plays an important role in obtaining the proper contact pattern and loading at the sealing interface. Methods of defining the radial lip seal cross section for quality checks are found in Ref. 3. Terminology of radial lip seals appears in Ref. 4 and is as follows:

1. Lip, or beam, length is the axial distance between lip contact point and the base of the flex thickness. A short beam length will tend to make a short, "stubby" element, thereby increasing the radial load and increasing the wear of the element. A long beam length will increase the flexibility of the element

and increase followability. Beam lengths are normally 2.5 to 5 mm (0.100 to 0.200 in), depending on the application and seal size.

2. The R value, or spring axial position, is the axial distance between the lip and shaft contact point and the centerline of the spring. For best performance, the R value should be between 0.25 and 0.75 mm (0.010 and 0.030 in) and should be positive (i.e., the contact point of the element should be toward the "oil side," and the centerline of the spring should be toward the "air side"). If the R value becomes negative, the spring exerts a higher radial load on the lip, reduces the oil film between the lip and the shaft, and changes the pressure distribution between the lip and the shaft. Instant leakage often results. However, it should be noted that installation of the seal on the shaft deflects the lip, causing the R value to shift toward the positive side. Therefore, some designs with a zero R value in the free state will have a positive R value when installed. If the R value is greater than 0.75 mm (0.030 in) in the positive direction, the spring causes the sealing element to collapse, resulting in accelerated wear and shortened life.
3. The oil side, or scraper, angle is the surface angle between the shaft surface and sealing lip surface on the oil side. This angle is larger than the air side angle and should be great enough to allow a scraping action. If it is too low, the desired pressure gradient across the interface will not be generated and leakage will occur. This angle is generally between 40 and 70°.
4. The barrel angle is the angle between the shaft surface and sealing element surface on the air side. This surface angle should be small to allow a meniscus of the lubricant film to form and large enough to allow the element to follow shaft runout. If the angle is too low, accelerated wear will occur on the barrel, which shortens seal life. If it is too high, the proper pressure gradient will not exist at the interface, and seal leakage will result. This angle should lie between 20 and 35°.
5. The outside lip surface is the truncated conical surface that forms the air side of the seal lip.
6. The inside lip surface is the truncated conical surface that forms the oil or sealing side of the seal lip.
7. The flex section is the area that deforms when the seal is installed and governs the seal load and followability. The flex section should be designed with sufficient thickness to reduce the tendency to deform in the direction of shaft rotation and to eliminate "stick-slip."[5] This will control interface temperature and heat generation. The flex thickness must be thin enough to allow the element to follow shaft dynamics. The flex thickness is normally 0.63 to 1.5 mm (0.024 to 0.060 in) and depends on the application and shaft size.
8. The case I.D. is the inside, or smallest, diameter of the case on the seal.
9. The heel section is that portion of the lip seal which is attached to the case. It is bound by the flex section and the outside face.
10. The garter spring is a helically coiled wire with its ends connected to form a ring. It is normally used in tension for maintaining a radial sealing force between the sealing element of a radial lip seal and the shaft.
11. The seal O.D. is the outer diameter of the seal.
12. The nose gasket is an elastomeric element which is attached to the face of the seal case O.D. It forms a gasket when compressed on the bottom of the housing bore.

13. The rubber O.D. is an elastomeric element bonded to the seal case O.D. It forms a gasket when pressed into a bore.
14. The contact width is the lip contact area when the seal is placed on the shaft. It is measured in the axial direction.
15. The contact line is formed between the outside and inside lip surfaces. In a cross-sectional view, this is represented as a point.
16. The radial dirt lip is a short lip which is located at the outside face of the seal to prevent ingress of contaminants.
17. Head thickness is the radial distance between the contact point and the spring groove. The head thickness maintains stability of the element. An element that is too thin in this area will be flimsy and will tend to lift off of the shaft. In applications where the element is subjected to vibration or stick slip, an increase in the head thickness will increase the stability of the design. Normal head thickness range is between 1.27 and 2.5 mm (0.050 and 0.100 in) and increases with seal size.
18. Interference is the diametrical difference between the sealing lip and the shaft. The diameter of the sealing element at the contact point should be less than the shaft diameter. Excessive interference will cause a high radial load at the contact point and will increase wear. Insufficient interference will not allow the element to follow the eccentricities of the shaft and will result in leakage.
19. Lip I.D. is the diameter of the primary sealing lip.

Hydrodynamic Sealing

Sealing effectiveness can be improved in many applications with the addition of "hydrodynamic" geometry to the oil seal surface. Raised helical or parabolic ribs, triangular pads, or sinuous wavy lip elements can, under dynamic conditions, redirect (or pump) excess oil away from the seal interface.

The hydrodynamic molded elements form a footprint or pattern on the shaft and enhance the pumping ability of the seal when there is relative motion of the shaft (Fig. 9.4). When the shaft stops, the seal must rely on the static dam formed by the seal lip geometry to form an effective seal. Brink and Horve[6] introduced a plain lip seal design with a wavy sealing edge (Fig. 9.5) that pumped more than conventional plain lip designs. It was found that increasing the number of waves and the amplitude of the wave increased the pump rate (Fig. 9.6).

Figure 9.6 also demonstrates that the helix pattern provides the greatest pumping ability. Helixes can be used only if the shaft rotates in one direction. If the shaft is reversed, the unidirectional ribs will screw oil out of the pump, causing gross leakage. The other bidirectional seal types will pump oil back into the sump in both directions of shaft rotation.

Seal Case Materials and Configurations

Normally, inner and outer cases are fabricated from 1005 to 1020 cold rolled carbon steel. The steel is zinc phosphated to provide a good surface for the rubber bonding. However, iron phosphating may also be used where a smoother surface is desired. The phosphating also provides a protective coating to minimize cor-

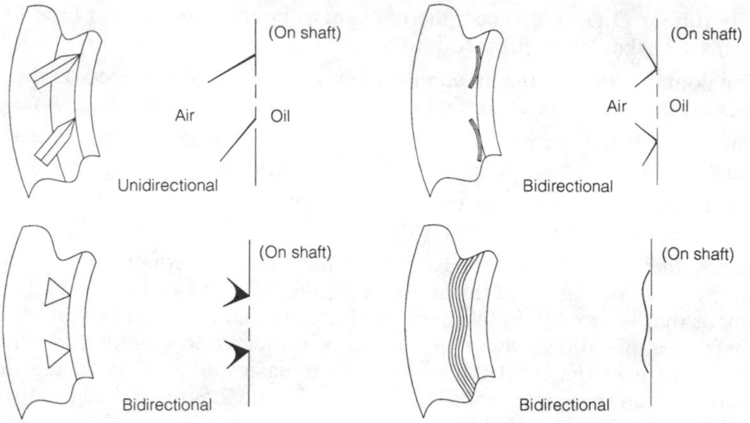

FIGURE 9.4 Hydrodynamic features. (*a*) Helix pattern; (*b*) parabolic ribs; (*c*) triangular pads; (*d*) sine wave.

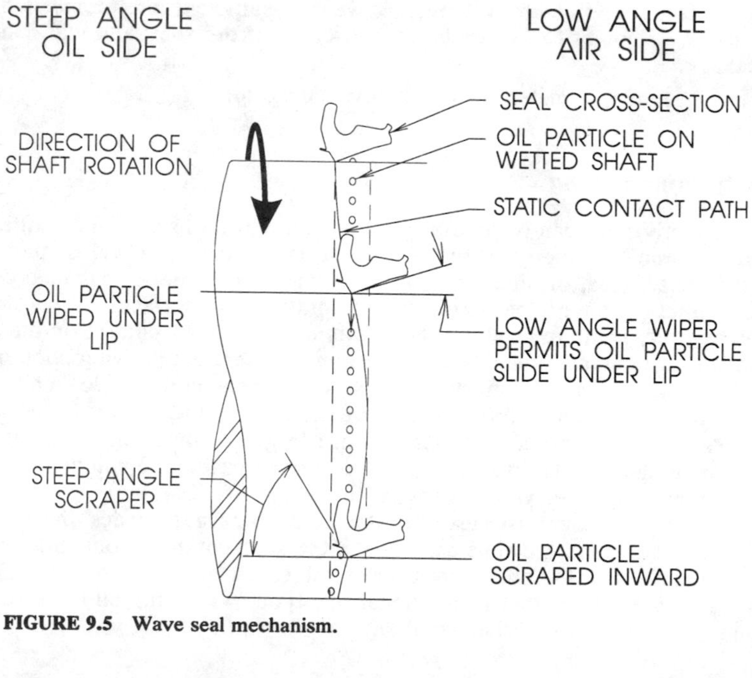

FIGURE 9.5 Wave seal mechanism.

rosion during handling and storage. In some cases, the steel cases may be oiled and roll wrapped in a protective paper as added protection against corrosion.

In special cases, zinc or cadmium plating with a clear chromate or dichromate dip may be applied for corrosion protection. Cases can also be fabricated in stainless steel, brass, or aluminum to minimize corrosion. These are premium materials which warrant a cost penalty. Stainless steel is very expensive and difficult

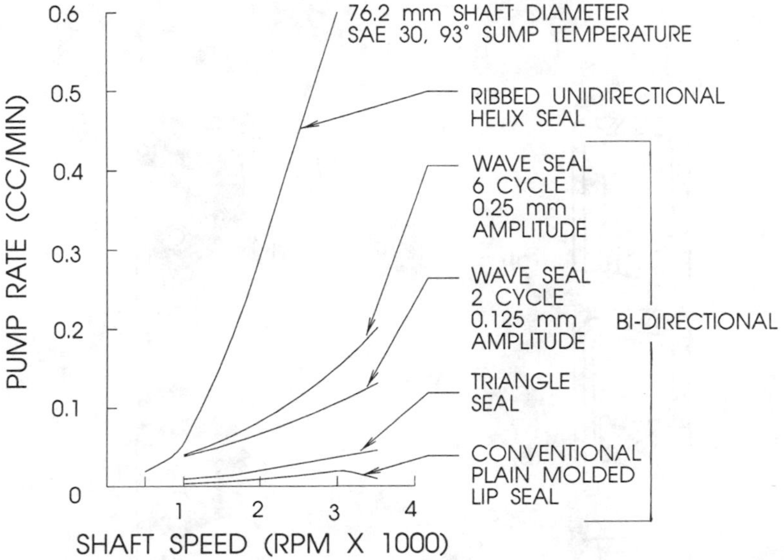

FIGURE 9.6 Pump rates for various seal types.

to draw. Aluminum and brass are normally used to minimize the differential thermal expansion between the case and nonferrous housing bores such as aluminum, titanium, magnesium, or bronze. Aluminum has the added advantage of being lightweight and is used in many aircraft applications.

The use of plastic cases is limited because of their high thermal expansion rate, low strength at higher temperatures, and low modulus, which requires a high interference to provide adequate bore retention. However, plastics may be used as backup rings in high-pressure seal designs.

Seal cases are available in a variety of shapes to suit application requirements (Fig. 9.7); they are described in the following list. As case configurations become more complicated, costs increase. Care must be taken to use the simplest case design for the application.

1. *Standard L outer case:* The L case stamping is the simplest and least-expensive configuration; it is often used in catalog, or "stock," seals. A curl or chamfer is provided at the open end to aid installation into the bore.
2. *Inner case:* An inner case can be incorporated within the outer case to provide additional strength to the case during installation and to protect the elastomeric sealing lip. The inner case can also be used as a spring retainer on designs where a spring groove is not present, such as on assembled seals. The inner case protects the seal lip when the shaft is blindly inserted from the oil side or when bulk packaging and handling could damage the lips.
3. *Shotgun case:* The shotgun case has a heavy outer flange to strengthen the case, position the seal, and minimize cocking in the bore. The flange retains followable gaskets which are sometimes placed on the seal O.D. and are scraped back during insertion of the seal. It provides a pry-out flange for removal of seals from the bore.
4. *Reverse channel:* The reverse channel case has a U-shaped section to accom-

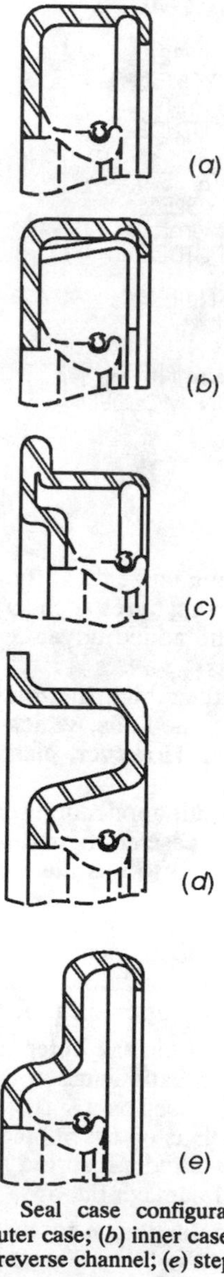

FIGURE 9.7 Seal case configurations. (*a*) Standard L outer case; (*b*) inner case; (*c*) shotgun case; (*d*) reverse channel; (*e*) stepped case.

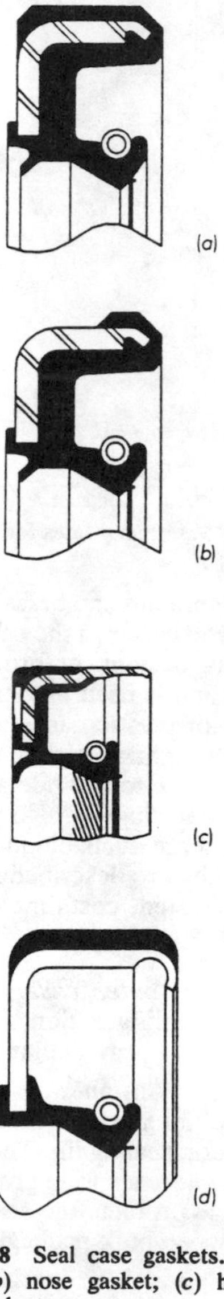

FIGURE 9.8 Seal case gaskets. (*a*) Rubber covered; (*b*) nose gasket; (*c*) head gasket; (*d*) O.D. sealants.

modate large clearances between the shaft and bore and to provide additional strength. In some cases, a pry-out flange is added.

5. *Stepped case:* The stepped case can accommodate large clearances between the shaft and bore. It also is used to properly position the sealing lip on the shaft if geometric obstacles exist.

Seal Case Gaskets

A secondary leakage path may exist between the bore and seal case interface. Rough bore finishes, axial scratches, corrosion of the bore or case, high-expansion bore materials such as aluminum, large bore tolerances, distorted seal cases, and low-viscosity fluids can result in bore leakage. Seal gaskets or bore sealants (Fig. 9.8) are recommended to ensure that a reliable seal is obtained at the bore interface. The types of gaskets are as follows:

1. *Rubber-covered case:* Rubber-covered cases offer the maximum protection against bore leakage due to deficiencies in the bore surface, and they prevent corrosive action on the metal case. They are used when bore finish recommendations are not followed or to compensate for differential thermal expansion when the bore and seal case material differ or in pressure applications where high bore retention forces are required.
2. *Nose gasket case:* Nose gaskets are used in low-pressure applications that do not require high-retention forces in the bore or in applications with narrow radial clearances. They also offer protection against corrosive action from the media being sealed. This design is difficult to manufacture and the rubber-covered case is preferred.
3. *Heel gasket case (half rubber/half metal):* Heel gaskets are an alternate method of bore sealing that are similar to the nose gasket but offer corrosion protection against the outside media. A metal to metal pressfit assures bore retention and the rubber gasket prevents bore leakage.[7]
4. *O.D. sealants:* O.D. sealants can be applied to the case O.D. in order to prevent leakage in rough or scratched bores. Some of these sealants are factory applied as a uniform coating on the case and act as a lubricant to reduce the installation forces while acting as an adhesive after curing in the application. When applying sealants during the installation of seals, care must be exercised to avoid getting the sealant on the shaft or sealing lip. Specifications for O.D. sealants are documented in RMA Technical Bulletin OS-14.[8]

RADIAL LIP SEAL TYPES

Seals without springs (Fig. 9.9*a*) are usually the simplest and most inexpensive solution when retaining highly viscous materials like grease at shaft speeds less than 10 m/s (1968 ft/min). An effective contaminant excluder is created when the seals are installed with the primary lip facing the atmosphere. Rubber can be molded all along the inside stamping face to minimize corrosion (Fig. 9.9*b*). A secondary lip can also be employed to exclude contaminants (Fig. 9.9*c*). Typical applications include conveyor rollers, vehicle

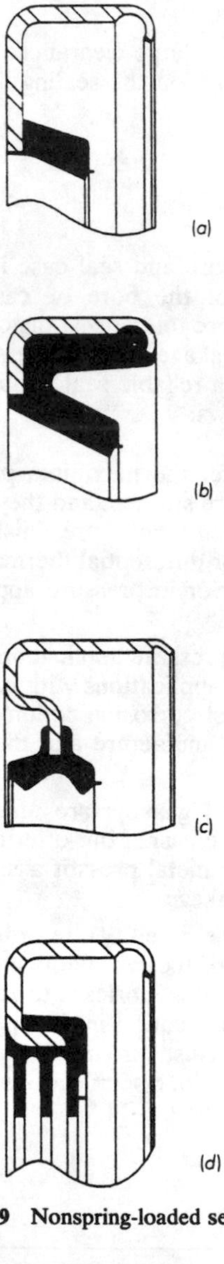

FIGURE 9.9 Nonspring-loaded seals.

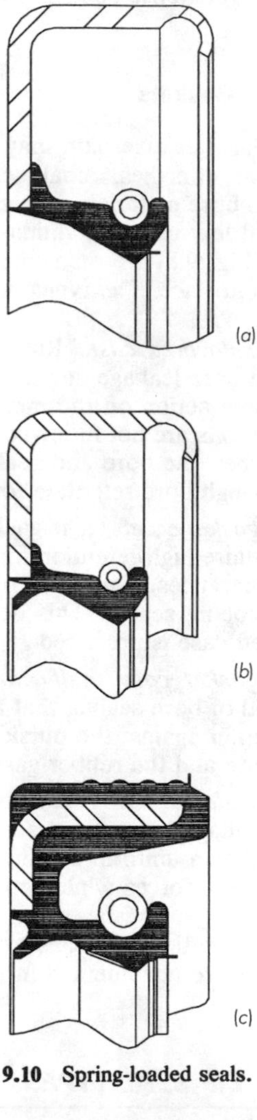

FIGURE 9.10 Spring-loaded seals.

wheels, and other grease lubricated devices. Heavy-duty multilip seals with a grease pack between the lips (Fig. 9.9*d*) are used to retain grease when external contamination is severe, such as in disk harrows. The shaft speed is usually limited to 2.5 m/s (492 ft/min).

Spring-loaded seals are used to retain low-viscosity lubricants such as oils at speeds up to 25 m/s (4920 ft/min). The single lip seal is the most economical general-purpose oil seal (Fig. 9.10*a*). Typical applications include engines, drive axles, transmission pumps, electric motors, and speed reducers. If light-duty dirt exclusion is required, a dual lip seal is used (Fig. 9.10*b*). These seals are commonly used in automotive, farm, and industrial applications where they are protected from light external contaminants during operation. Rubber O.D. seals (Fig. 9.10*c*) are typically more expensive than metal O.D. seals and are used only if application conditions warrant this design.

Many applications have unique requirements and special custom designs are used. A seal design (Fig. 9.11*a*) with a nylon backup ring is used for operations at intermittent pressures of 8000 to 10,000 kPa (1159 to 1449 lb/in^2). For applications where shaft dynamic runout is excessive (1.5 mm, or 0.059 in), seals with convoluted lips are recommended (Fig. 9.11*b*). Some seals are made with lip materials that cannot be bonded to the case. The lip material is tightly clinched between metal cases to prevent lip wafer slippage (Fig. 9.11*c*). An internal gasket must be supplied to eliminate internal leakage. The assembled seal is usually more expensive than a design that has the elastomeric material bonded to the case. Special seal designs with the lip at the O.D. rather than the shaft are sometimes used for wheel seal applications (Fig. 9.11*d*). In this case, the O.D. rotates and the spindle is stationary.

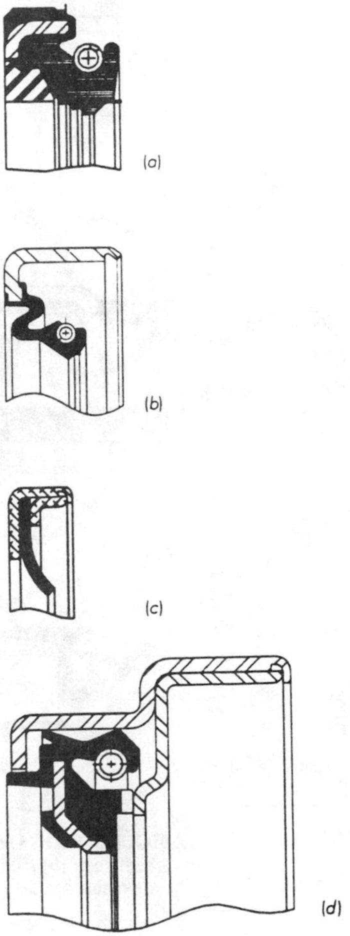

FIGURE 9.11 Special seal designs.

Seals can be supplied with a unitized wear surface to relieve the end user of the responsibility of providing a ground surface on the shaft (Fig. 9.12). The unitized grease seal shown in Fig. 9.12*a* is used for automotive wheel seal applications. A truck wheel seal[9] appears in Fig. 9.12*b*, and a unitized seal for diesel crankshaft applications[10] is a composite of PTFE and an elastomeric material (Fig. 9.12*c*).

In some cases, the oil seal is combined with other components of the application to reduce customer assembly costs and problems. A power steering rack seal can be combined with a dirt excluder seal and an oil filter element (Fig. 9.13) to provide the end user with one component instead of three.[11]

Another example is shown on Fig. 9.14. The lip seal is combined with a thermoplastic retainer and a gasket to seal an automotive engine rear crankshaft application.[12] This package combines three components into one.

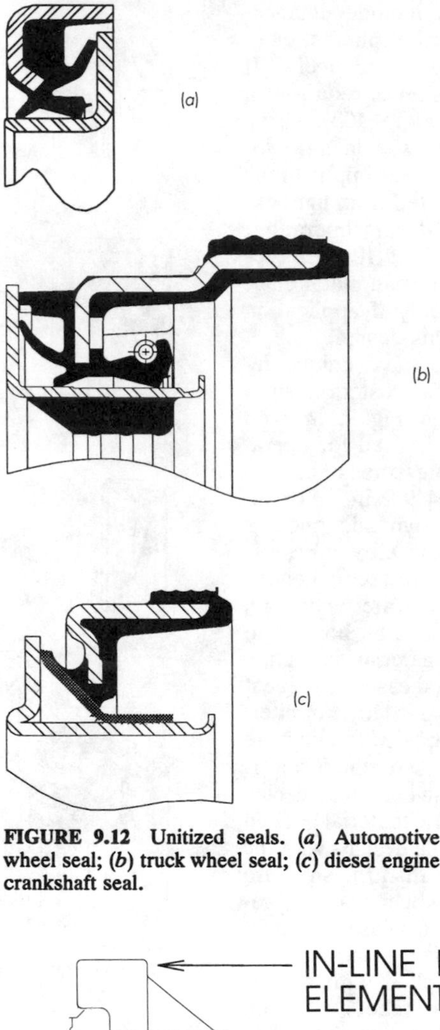

FIGURE 9.12 Unitized seals. (*a*) Automotive wheel seal; (*b*) truck wheel seal; (*c*) diesel engine crankshaft seal.

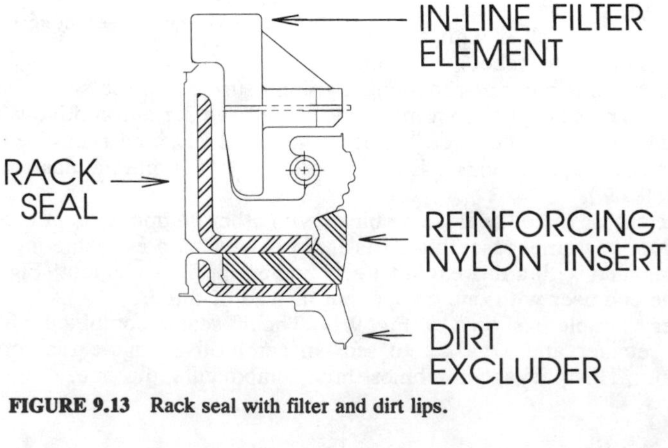

FIGURE 9.13 Rack seal with filter and dirt lips.

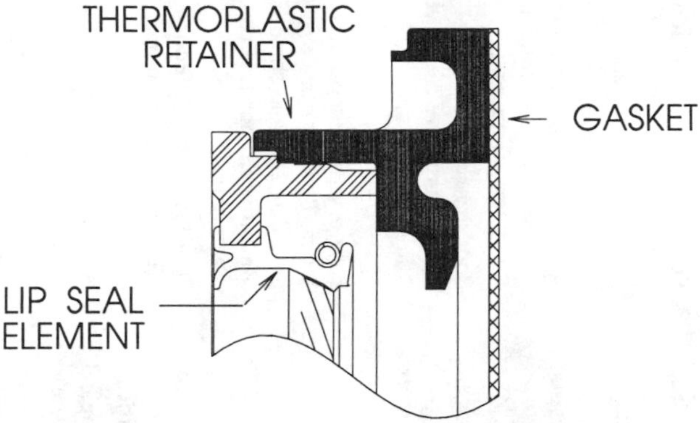

FIGURE 9.14 Automotive engine rear crankshaft seal application.

POLYMER

A number of oil- and grease-resistant polymers are available for lip seal applications. These polymers are mixed with other materials that give proprietary compounds to meet application requirements. Proper material selection is essential for good field performance. Table 9.1 provides a summary of polymers available. The temperature ranges refer to normal sump operating temperatures. Acceptable upper- and lower-temperature limits will vary depending upon seal design and application requirements. Available polymers are discussed in the following list:

Nitrile (NBR): Nitrile is a co-polymer of butadiene and acrylonitrile. The practical operating temperature range is −40 to 110°C (−40 to 230°F). The upper-temperature performance can be improved by increasing the percent of acrylonitrile. Unfortunately, the low-temperature performance is reduced. Conversely, reducing the percent of acrylonitrile improves the low-temperature properties, but the high-temperature limit is reduced. Nitriles are used extensively in mild operating conditions. They offer low cost and good oil resistance, wear resistance, and low temperature properties. They lack high-temperature resistance and will fail due to hardening and cracking when operated continuously at elevated temperatures.

Carboxylated nitriles are more expensive than standard nitriles. They provide better high-temperature wear resistance. Their oil resistance is identical to standard nitriles, but low-temperature flexibility is slightly less.

An advanced nitrile polymer (HNBR) has been developed that extends the upper temperature range to 135°C (275°F). Its use in oil seals has been limited to date because of its high cost (15 times that of conventional NBR) and manufacturing process difficulties.

Polyacrylic (ACM): Polyacrylic polymers offer improvements over nitrile in high-temperature applications but cost more. These compounds function well at 135°C (275°F) in engine or transmission fluids. They resist EP additives in gear oils and can replace nitrile in these applications, where nitrile hardening and cracking is a problem. Polyacrylic is not recommended in applications exposed to acids, bases, water,[13] and other polar solvents such as esters and ke-

TABLE 9.1 Polymer Selection Chart

Material	Approx. relative cost (polymer)	Approx. relative cost (seals)	Sump oil temperature °C	Sump oil temperature °F	Advantages	Disadvantages
Nitrile (NBR)	100	100	−40 to 110	−40 to 230	Low cost, low swell. Good wear and oil resistance at moderate temperatures.	Poor resistance to EP additives. Poor high-temperature resistance.
Polyacrylic (ACM)	250	115	−40 to 135	−40 to 275	Good oil resistance. Low swell. Generally resistant to EP additives.	Fair wear properties. Poor dry-running. Poor water resistance. May crack at low temperatures.
Ethylene acrylic (AEM)	250	115	−40 to 150	−40 to 302	Good temperature range. Good moisture resistance. Fair abrasion resistance.	Poor dry-run characteristics. May have high swell in some fluids.
Silicone (VMQ)	625	130	−70 to 150	−94 to 302	Very broad temperature range.	Poor dry-running properties. Poor resistance to oxidized oil and some EP additives. Poor tear characteristics.
Fluorosilicone (FVMQ)	2250	200	−70 to 150	−94 to 302	Excellent oil and chemical resistance. Good low-temperature properties.	Poor tear strength. Poor dry-running. Poor wear resistance.
Fluorocarbon (FKM)	2250	200	−40 to 150	−40 to 302	Excellent oil and chemical resistance. Good wear properties. Low swell.	Becomes stiff at low temperatures. Poor followability at low temperatures.
Polytetrafluoroethylene (PTFE)	2250	300	−80 to 200	−112 to 392	Excellent oil and chemical resistance. Excellent temperature range.	Easily damaged. Becomes stiff at low temperature. High cost. Poor followability at low temperature.

tones. Many polyacrylates become brittle and can fracture at low temperatures (−15°C, 5°F). Special polymers can extend low-temperature performance to −40°C (−40°F) if shaft runout is low. Extreme care must be used when selecting a polyacrylic compound if low-temperature properties are important. ACM materials have poor dry running properties.

Ethylene acrylic (AEM): AEM is a co-polymer of ethylene and methyl acrylate. Its properties are similar to polyacrylic, and it has good moisture resistance with good low-temperature properties. The temperature range is −40 to 150°C (−40 to 302°F).

Silicone (VMQ): This material is recommended for a temperature range of −70 to 150°C (−94 to 302°F) in engine and transmission oils. The material will decompose and cause failure when exposed to EP additives in gear oils. It is not recommended for use in gasoline, chlorinated oils, or aromatic solvents. The dry running characteristics are poor, and it is recommended that silicone seals be hot soaked in oil before use in sparsely lubricated conditions. The material can be torn easily and care must be used during handling and installation.

Fluorosilicone (FVMQ): FVMQ has the temperature range of silicone and excellent chemical resistance. It is quite expensive and unfortunately also has the poor tear strength and wear characteristics of silicone.

Fluorocarbon (FKM): Fluorocarbon has excellent chemical and high-temperature resistance (150°C, 302°F) and is expensive. The low-temperature range is often listed as −40°C (−40°F), but care must be exercised with its use. The material typically begins to stiffen around 0°C (32°F) and may not follow shaft eccentricities. Leakage may occur even though the material does not fracture. This material is typically recommended for high-speed, high-temperature applications. It has fair dry running characteristics.

Polytetrafluoroethylene (PTFE): PTFE is inert to virtually all fluids and has a temperature range from −80 to 200°C (−112 to 392°F). It is a plastic rather than an elastomer and cannot follow shaft motion as well as other materials. Bonding of PTFE to other materials is difficult and expensive. Seal construction is often of the costly assembled design. The material is easily damaged and care must be exercised during handling and assembly.

Miscellaneous polymers: Urethanes are tough materials with good wear resistance. They are used to exclude contaminants in very dirty applications such as off-highway earth-moving equipment. Epichlorohydrin lies between nitrile and polyacrylate in cost and physical properties. Butyl and ethylene propylene diene monomers (EPDMs) are used to seal polar solvents and non-petroleum-based brake fluids. These materials represent a small fraction of the seal population because of high cost and manufacturing process problems.

Lubricant Compatibility

Lubricants vary in the base oil stocks and the additives used to obtain specific characteristics that satisfy application requirements. These variations have significant effects upon the elastomers used for oil seals. It is recommended that elastomer samples be immersed and aged in application oils at temperatures that approximate application conditions. Attempts to shorten test time by using elevated temperatures can lead to erroneous conclusions since oils deteriorate at

TABLE 9.2 Elastomer Compatibility in Various Fluids

Medium	Nitrile	Poly-acrylate	Ethylene acrylic	Silicone	Fluoro-silicone	Fluoro-carbon	PTFE
Engine oil	Good	Good	Good	Good	Good	Good	Good
ATF-A	Fair	Good	Good	Good	Good	Good	Good
Grease	Good	Fair	Fair	Fair	Fair	Good	Good
EP lube	Fair-poor	Good	Fair	Poor	Fair	Good	Good
SAE 90	Good	Good	Good	Good	Good	Good	Good
Fuel oil	Good	Fair	Fair	Poor	Good	Good	Good
Kerosene	Good	Fair	Fair	Poor	Good	Good	Good
Gasoline	Good	Fair	Fair	Poor	Good	Good	Good
Petroleum-based hydraulic oil	Good	Good	Good	Good	Good	Good	Good
Brake fluid	Poor	Poor	Poor	Poor	Fair	Fair	Good
Skydrol 500	Poor	Poor	Poor	Good	Poor	Poor	Good
MIL-L-7808	Fair	Poor	Poor	Good	Good	Good	Good
MIL-L-23699	Fair	Poor	Poor	Good	Good	Good	Good
MIL-L-6082-A	Good	Good	Good	Good	Good	Good	Good
MIL-L-5606	Good	Good	Good	Poor	Good	Good	Good
MIL-L-2105	Fair	Good	Good	Poor	Fair	Fair	Good
MIL-G-10924	Good	Good	Good	Poor	Good	Good	Good
Butane	Good	Good	Good	Fair	Good	Good	Good
Ketone	Poor	Poor	Poor	Poor	Poor	Poor	Good
Ammonium gas cold	Good	Poor	Poor	Fair	Poor	Poor	Good
Fresh water	Good	Poor	Poor	Good	Good	Good	Good
Salt water	Good	Poor	Poor	Good	Good	Good	Good

high temperatures. Water content and air ingestion will also alter test results. The immersion test should duplicate the application conditions as closely as possible.

The physical properties of the elastomer samples are measured at time intervals (70, 108 h, etc.) and the change (or lack of) defines compatibility. Typical property measurements include hardness, tensile strength and elongation at break, and modulus values at 50 and 100 percent elongation. Compatibility of various elastomers to different fluids appears in Table 9.2.

RECOMMENDED PRACTICES AND SPECIFICATIONS

Nominal Dimensions

Many of the recommended practices and specifications in this and the following paragraphs have been accepted as standards by various organizations. Details can be found in Refs. 14 through 21. The recommended nominal dimensions for the seal envelope appear in Tables 9.3 and 9.4. Equipment designers should provide the proper space to ensure standard seals will be available.

TABLE 9.3 ISO 6294/1 Standards (dimensions in mm)

D_s	D_b	W	D_s	D_b	W	D_s	D_b	W	D_s	D_b	W
6	16	7	25	52	7	45	65	8	120	150	12
6	22	7	28	40	7	50	68	8	130	160	12
7	22	7	28	47	7	50	72	8	140	170	15
8	22	7	28	52	7	55	72	8	150	180	15
8	24	7	30	42	7	55	80	8	160	190	15
9	22	7	30	47	7	60	80	8	170	200	15
10	22	7	30	52	7	60	85	8	180	210	15
10	25	7	32	45	8	65	85	10	190	220	15
12	24	7	32	47	8	65	90	10	200	230	15
12	25	7.	32	52	8	70	90	10	220	250	15
12	30	7	35	50	8	70	95	10	240	270	15
15	26	7	35	52	8	75	95	10	260	300	20
15	30	7	35	55	8	75	100	10	280	320	20
15	35	7	38	55	8	80	100	10	300	340	20
16	30	7	38	58	8	80	110	10	320	360	20
18	30	7	38	62	8	85	110	12	340	380	20
18	35	7	40	55	8	85	120	12	360	400	20
20	35	7	40	62	8	90	120	12	380	420	20
20	40	7	42	55	8	95	120	12	400	440	20
22	35	7	42	62	8	100	125	12			
22	40	7	45	62	8	110	140	12			
22	47	7									
25	40	7									
25	47	7									

D_s = shaft diameter
D_b = bore diameter
W = seal width

Shaft Recommendations and Specifications

The elastomeric seal lip and the shaft surface must mate together to form the primary sealing system. Shaft seals will perform satisfactorily on mild steel, cast iron, or malleable iron shafts. The portion of the shaft that the seal contacts should be hardened to Rockwell C-30 minimum; to avoid handling damage, nicks, and scratches, Rockwell C-45 is recommended. Brass, bronze, aluminum alloys, zinc, magnesium, and other shaft materials should not be used. If these materials are used, a wear sleeve or ring of mild steel should be pressed over the shaft. The wear sleeve should be changed whenever the seal is replaced. Chrome or nickel plating is commonly used to prevent shaft wear in dirty corrosive environments. It can also provide a hard sealing surface over a softer material.

The shaft surface finish is critical to seal function. The specification is 0.25 to 0.50 μm (10 to 20 μin) RA with $0 \pm 0.05°$ (± 3 min.) lead angle. This finish will wear away the rubber skin from the sealing tip and create asperities in the wear track. The asperities support a lubricating film that protects the lip from further wear. If the shaft surface is initially too smooth, the seal lip may not break-in properly and leakage will occur. If the shaft surface is initially too rough, gross lip wear may occur before the protective lubricating film develops and leakage will result.

The molded skin at the sealing tip is broken very quickly as the shaft rotates (1 h or less) and a wear track begins to develop. If there are no large quantities of external or internal contaminants present, the width of the wear track will stabilize within 100 h (Fig. 9.15). The width of a typical stabilized wear track will vary

TABLE 9.4 SAE J946 Standard (dimensions in inches)

D_s	D_b	W	D_s	D_b	W	D_s	D_b	W	D_s	D_b	W
0.500	0.999	*A*	1.625	2.502	*B*	2.750	3.500	*C*	3.875	4.876	*D*
0.500	1.124	*A*	1.625	2.623	*B*	2.750	3.623	*C*	3.875	4.999	*D*
0.500	1.250	*A*	1.750	2.374	*B*	2.750	3.751	*C*	3.875	5.125	*D*
0.625	1.124	*A*	1.750	2.502	*B*	2.750	3.875	*C*	3.875	5.251	*D*
0.625	1.250	*A*	1.750	2.623	*B*	2.875	3.623	*C*	4.000	4.999	*D*
0.625	1.375	*A*	1.750	2.750	*B*	2.875	3.751	*C*	4.000	5.125	*D*
0.625	1.499	*A*	1.875	2.623	*B*	2.875	3.875	*C*	4.000	5.251	*D*
0.750	1.250	*A*	1.875	2.750	*B*	2.875	4.003	*C*	4.000	5.375	*D*
0.750	1.375	*A*	1.875	2.875	*B*	3.000	3.751	*C*	4.250	5.251	*D*
0.750	1.499	*A*	1.875	3.000	*B*	3.000	3.875	*C*	4.250	5.375	*D*
0.750	1.624	*A*	1.875	3.125	*B*	3.000	4.003	*C*	4.250	5.501	*D*
0.875	1.375	*A*	2.000	2.623	*B*	3.000	4.125	*C*	4.250	5.625	*D*
0.875	1.499	*A*	2.000	2.750	*B*	3.125	4.125	*D*	4.500	5.501	*D*
0.875	1.624	*A*	2.000	2.875	*B*	3.125	4.249	*D*	4.500	5.625	*D*
0.875	1.752	*A*	2.000	3.000	*B*	3.125	4.376	*D*	4.500	5.751	*D*
1.000	1.499	*A*	2.000	3.125	*B*	3.125	4.500	*D*	4.750	5.751	*E*
1.000	1.624	*A*	2.125	2.750	*C*	3.250	4.249	*D*	4.750	6.000	*E*
1.000	1.752	*A*	2.125	2.875	*C*	3.250	4.376	*D*	5.000	6.000	*E*
1.000	1.874	*A*	2.125	3.000	*C*	3.250	4.500	*D*	5.000	6.250	*E*
1.125	1.624	*A*	2.125	3.125	*C*	3.250	4.626	*D*	5.000	6.375	*E*
1.125	1.752	*A*	2.125	3.251	*C*	3.375	4.249	*D*	5.250	6.250	*E*
1.125	1.874	*A*	2.250	3.000	*C*	3.375	4.376	*D*	5.250	6.375	*E*
1.125	2.000	*A*	2.250	3.125	*C*	3.375	4.500	*D*	5.250	6.500	*E*
1.250	1.752	*A*	2.250	3.251	*C*	3.375	4.626	*D*	5.250	6.625	*E*
1.250	1.874	*A*	2.250	3.371	*C*	3.500	4.376	*D*	5.500	6.500	*E*
1.250	2.000	*A*	2.375	3.125	*C*	3.500	4.500	*D*	5.500	6.625	*E*
1.250	2.125	*A*	2.375	3.251	*C*	3.500	4.626	*D*	5.500	6.750	*E*
1.375	2.000	*B*	2.375	3.371	*C*	3.500	4.751	*D*	5.500	6.875	*E*
1.375	2.125	*B*	2.375	3.500	*C*	3.625	4.626	*D*	5.750	6.750	*E*
1.375	2.250	*B*	2.500	3.251	*C*	3.625	4.751	*D*	5.750	6.875	*E*
1.375	2.374	*B*	2.500	3.371	*C*	3.625	4.876	*D*	5.750	7.000	*E*
1.500	2.125	*B*	2.500	3.500	*C*	3.625	4.999	*D*	5.750	7.125	*E*
1.500	2.250	*B*	2.500	3.623	*C*	3.750	4.626	*D*	6.000	7.125	*E*
1.500	2.374	*B*	2.625	3.371	*C*	3.750	4.751	*D*	6.000	7.500	*E*
1.500	2.502	*B*	2.625	3.500	*C*	3.750	4.876	*D*			
1.625	2.250	*B*	2.625	3.623	*C*	3.750	4.999	*D*			
1.625	2.374	*B*	2.625	3.751	*C*						

D_s = shaft diameter
D_b = bore diameter
W = seal width
A = 5/16 in
B = 3/8 in w/o inner case; 1/2 in w/inner case
C = 7/16 in w/o inner case; 1/2 in w/inner case
D = 1/2 in
E = 9/16 in

from 0.25 to 1.00 mm (0.010 to 0.040 in). During the life of the seal, the shaft becomes smooth and burnished even though very little material is removed from the seal tip (Fig. 9.16).

Spiral machining grooves or inclined scratches, known as machine lead, can auger or pump the oil out from under the lip, resulting in early leakage. The spiral marks or nicks can also damage the lip during installation. The angle that the spiral grooves make with the shaft centerline is known as the lead angle. The lead angle can be determined by measuring the axial advance that a string will make when the shaft rotates [Eq. (9.1)].

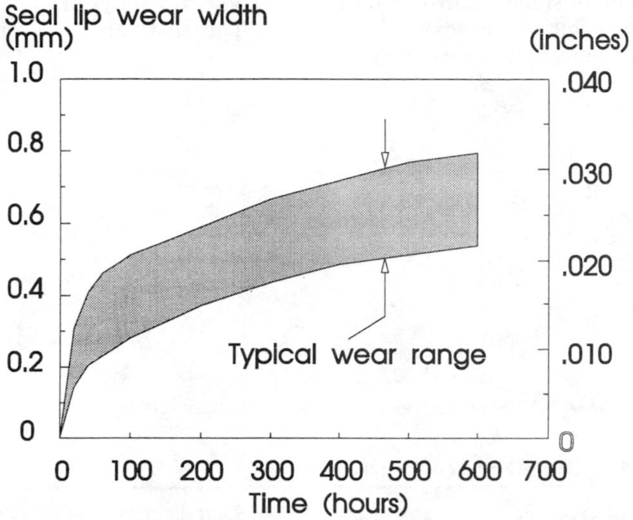

FIGURE 9.15 Seal wear track width versus running time.

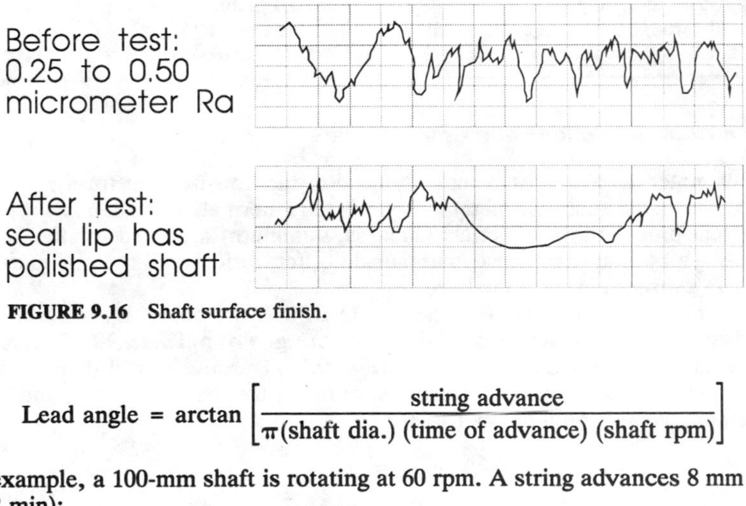

FIGURE 9.16 Shaft surface finish.

$$\text{Lead angle} = \arctan\left[\frac{\text{string advance}}{\pi(\text{shaft dia.})\,(\text{time of advance})\,(\text{shaft rpm})}\right] \tag{9.1}$$

For example, a 100-mm shaft is rotating at 60 rpm. A string advances 8 mm in 30 s (1/2 min):

$$\text{Lead angle} = \arctan\left[\frac{8\text{ mm}}{\pi(100\text{ mm})\,(1/2\text{ min})\,60\text{ rpm}}\right]$$

$$= 0.049° = 2\text{ min, }56\text{ s}$$

This shaft is just within the tolerance of the recommended lead angle of $0 \pm 0.05°$.

The best known method of obtaining the proper finish is plunge grinding. Shafts should be ground with mixed number rpm ratios between the work and the grinding wheel, with the wheel sparking out to avoid machine lead. Also, enough material should be removed during grinding to eliminate all machining grooves.

The shaft must have burr-free lead corners with a chamfer or radius to prevent seal damage during installation (Fig. 9.17). The shaft diameter must be held within the tolerances shown in Table 9.5.

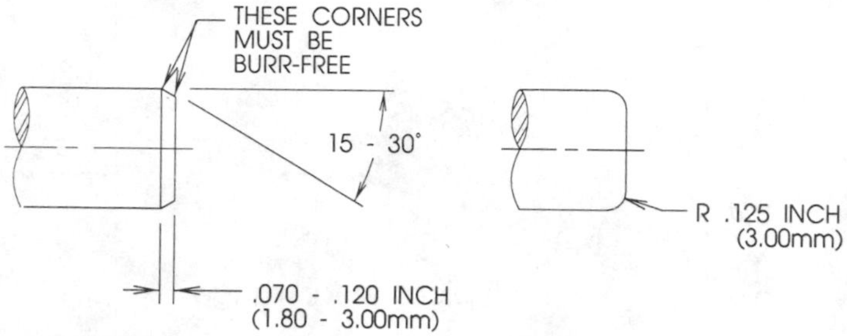

FIGURE 9.17 Specifications for shaft lead-in chamfer.

TABLE 9.5 Shaft Diameter Tolerances

Shaft diameter (ISO/R 286)	Tolerance, mm	Shaft diameter (SAE J946)	Tolerance, in
Up to and including 100	±0.08	Up to and including 4.000	±0.003
100.01 through 150	±0.10	4.001 through 6.000	±0.004
150.01 through 250	±0.13	6.001 through 10.000	±0.005

Bore Recommendations and Specifications

Ferrous materials are most commonly used in the housing that form seal bores. The conventional seal case material is steel; thus thermal expansion is not usually a problem unless other materials (such as aluminum) are used for the housing bore. When bore and seal case materials do differ, differential thermal expansion rates can cause leakage at the bore.

The proper press fit between the seal O.D. and the bore I.D. must be maintained to prevent leakage. Bore tolerances are given in Table 9.6. The leading edge of the bore should be chamfered (Fig. 9.18) to prevent seal damage during installation. The bore surface roughness, bore depth, chamfer length, and corner radius dimensions are given in Table 9.7.

TABLE 9.6 Bore Diameter Tolerances

ISO/R 286, bore diameter (D_b)	Tolerance, mm	SAE J946 bore diameter (D_b)	Tolerance, in
$D_b \leq 50$	+0.039, −0.0	$D_b \leq 3$	±0.001
$50 < D_b \leq 80$	+0.046, −0.0	$3 < D_b \leq 6$	±0.0015
$80 < D_b \leq 120$	+0.054, −0.0	$6 < D_b \leq 10$	±0.002
$120 < D_b \leq 180$	+0.063, −0.0	$10 < D_b \leq 20$	+0.002, −0.004
$180 < D_b \leq 300$	+0.075, −0.0	$20 < D_b \leq 40$	+0.002, −0.006
$300 < D_b \leq 440$	+0.084, −0.0	$D_b > 40$	+0.002, −0.010

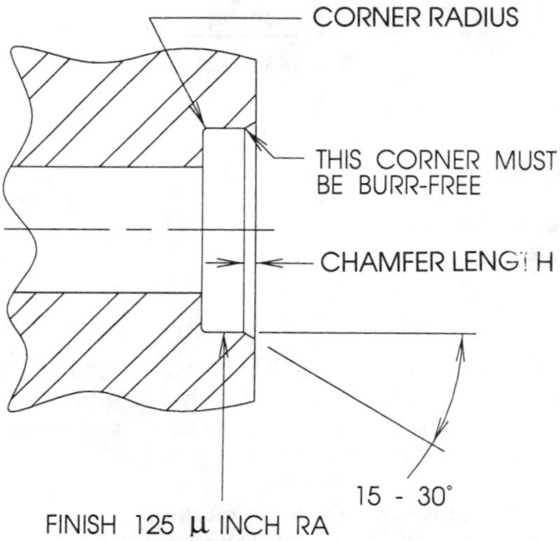

FIGURE 9.18 Bore dimensions.

TABLE 9.7

	Seal width (W)	Housing bore depth	Chamfer length	Maximum housing bore corner radius	Surface finish
ISO 6194/1, mm	$W \leq 10$	$W + 0.9$	0.70–1.00	0.50	Less than 3.2 μm RA
	$W > 10$	$W + 1.2$	1.20–1.50	0.75	
SAE J946, in	All	$W + 0.016$	0.06–0.09	0.047	Less than 125 μin RA

Seal Dimensional Specifications

The recommended tolerances for the primary lip diameter are given in Table 9.8. The tolerance for seal width is given in Table 9.9. The seal outer diameter is a critical dimension since this portion of the seal is press fit tightly into a housing bore. The seal O.D. is typically determined by averaging a minimum of three equally spaced measurements. The out of round (OOR) is defined to be the max-

TABLE 9.8 Primary Lip I.D. Tolerances

	Shaft diameter (D_s)	Tolerance
SAE J946, mm	$D_s \leq 75$	±0.50
	$75 < D_s \leq 150$	±0.65
	$150 < D_s \leq 250$	±0.75
SAE J946, in	$D_s \leq 3.000$	±0.020
	$3.000 < D_s \leq 6.000$	±0.025
	$6.000 < D_s \leq 10.000$	±0.025

TABLE 9.9 Seal Width Tolerances

	Seal width (W)	Tolerance
ISO 6194/1, mm	$W < 10$	±0.3
	$W > 10$	±0.4
SAE J946, in	$W < 0.400$	±0.015
	$W > 0.400$	±0.020

imum variation of these O.D. readings. The nominal press fit is the distance between the nominal seal O.D. and the nominal housing bore I.D. The tolerances for seal O.D., the maximum allowable out of round, and recommended press fit are given in Table 9.10 for steel O.D. seals and in Table 9.11 for rubber-covered seals.

The radial wall dimension (RWD) is the radial distance from the seal O.D. to the primary lip inner diameter. Radial wall variation (RWV) is the difference between the maximum and minimum values of RWD when measured around the circumference of the seal lip. The maximum allowable values for RWV are given in Table 9.12.

Radial load is the total force that the seal lip and spring exert on the shaft. The specifications for radial load appear on Table 9.13. Lip opening pressure (LOP) is a method used for quality control checks. It is defined as the pressure applied to the air side of the seal that will force the lip off of the shaft and allow 10,000 cm^3/min of air to blow by. The LOP tolerance is ±30 percent of the nominal pressure. The minimum range is 0.25 bar (4 psi).

TABLE 9.10 O.D. Specifications for Metal O.D. Seals

	Seal O.D. (D_o)	Nominal press fit	Seal O.D. tolerance	Seal O.D. OOR
ISO 6194/1, mm	$D_o \leq 50$	0.12	+0.20/+0.08	0.18
	$50 < D_o \leq 80$	0.14	+0.23/+0.09	0.25
	$80 < D_o \leq 120$	0.15	+0.25/+0.10	0.30
	$120 < D_o \leq 180$	0.17	+0.28/+0.12	0.40
	$180 < D_o \leq 300$	0.21	+0.35/+0.15	0.25% of O.D.
	$300 < D_o \leq 440$	0.28	+0.45/+0.20	
SAE J946, in	$D_o \leq 1$	0.004	±0.002	0.005
	$1 < D_o \leq 3$	0.004	±0.002	0.006
	$3 < D_o \leq 4$	0.005	±0.002	0.007
	$4 < D_o \leq 6$	0.005	+0.003/−0.002	0.009
	$6 < D_o \leq 8$	0.006	+0.003/−0.002	0.012
	$8 < D_o \leq 9$	0.007	+0.004/−0.002	0.015
	$9 < D_o \leq 10$	0.008	+0.006/−0.002	0.002 in per inch of seal O.D.
	$10 < D_o \leq 20$	0.008	+0.006/−0.002	
	$20 < D_o \leq 40$	0.008	+0.008/−0.002	
	$40 < D_o \leq 60$	0.008	+0.010/−0.002	

TABLE 9.11 O.D. Specifications for Rubber O.D. Seals

	Seal O.D. (D_o)	Nominal press fit	Seal O.D. tolerance	Seal O.D. OOR
ISO 6194/1, mm	$D_o \leq 50$	0.20	+0.30/+0.15	0.25
	$50 < D_o \leq 80$	0.25	+0.35/+0.20	0.35
	$80 < D_o \leq 120$	0.25	+0.35/+0.20	0.50
	$120 < D_o \leq 180$	0.32	+0.45/+0.25	0.65
	$180 < D_o \leq 300$	0.34	+0.50/+0.25	0.80
	$300 < D_o \leq 440$	0.38	+0.55/+0.30	1.00
SAE J946, in	$D_o \leq 1$	0.009	±0.003	0.010
	$1 < D_o \leq 3$	0.011	±0.003	0.014
	$3 < D_o \leq 4$	0.012	±0.003	0.020
	$4 < D_o \leq 9$	0.014	±0.004	0.025
	$9 < D_o \leq 10$	0.014	±0.004	0.031
	$10 < D_o \leq 20$	0.017	±0.005	0.039
	$20 < D_o \leq 40$	0.018	±0.006	0.045
	$40 < D_o \leq 60$	0.020	±0.007	0.050

TABLE 9.12 Specifications for Radial Wall Variation

	Shaft diameter (D_s)	Radial wall variation (RWV) maximum
SAE J946, mm	$D_s \leq 75$	0.6
	$75 < D_s \leq 150$	0.8
	$150 < D_s \leq 250$	1.0
SAE J946, in	$D_s \leq 3$	0.025
	$3 < D_s \leq 6$	0.030
	$6 < D_s \leq 10$	0.040

TABLE 9.13 Radial Load Specification

	Shaft diameter (D_s)	Radial load tolerance (range)
SAE J946, mm	$D_s \leq 75$	Nominal load ±45%
	$74 < D_s \leq 250$	Nominal load ±40%
SAE J946, in	$D_s \leq 3$	Nominal load ±45%
	$3 < D_s \leq 10$	Nominal load ±40%

Garter Spring Recommendations

Some radial lip seals use a garter spring (Fig. 9.19) to generate additional radial load when the seal is installed on the shaft. The spring also compensates for radial load changes that occur when the elastomer lip properties change due to exposure to heat and oil. It controls the seal lip finished I.D. by pulling the elastomeric lip in until the coils of the spring touch each other. The garter spring also adds additional radial stiffness which allows the seal to follow shaft runout.

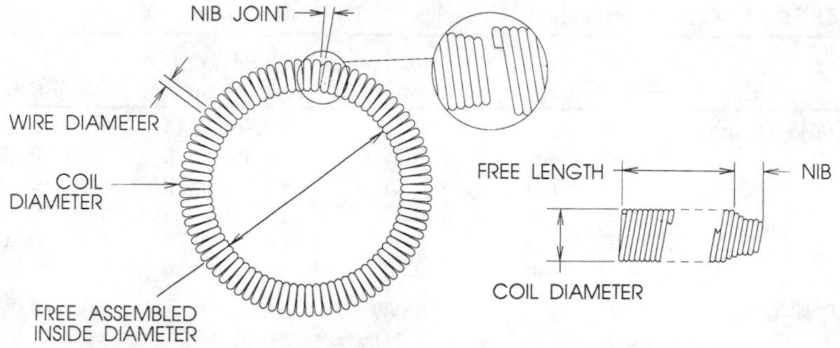

FIGURE 9.19 Garter spring features.

The spring should be designed to sit tightly in the spring groove and should not be dislodged during handling and installation.

Garter springs are usually made from carbon steel spring wire. The acceptable grades are SAE 1050 through 1095 (AISI,C1050 through C1095) and music wire. In high-temperature applications (up to 200°C, 392°F) or corrosive media, stainless steel springs are recommended (SAE 30302 through 30304; AISI type 302 through 304). Maximum corrosion resistance can be achieved by passivating the surface of the stainless steel wire spring.

The load versus the deflection curve for a garter spring is obtained by cutting the garter spring at a position approximately 180° from the nib, or connection point, attaching the ends to an extension device, and measuring the force as the spring is slowly extended. A typical load versus deflection curve for the extension spring is shown in Fig. 9.20. The initial spring tension must be exceeded before deflection occurs. After the initial tension has been exceeded, the increase in the spring load is directly proportional to the deflection. The change in spring tension per unit change in deflection is called the spring rate.

It is desirable to design the garter spring with the smallest possible wire diameter to ensure the highest possible initial tension within the allowable torsional stress limits for the material. This will give a spring with the lowest possible spring rate. The lower the spring rate, the less variance there will be in the total tension due to changes in the deflection of the spring. The high initial tension is

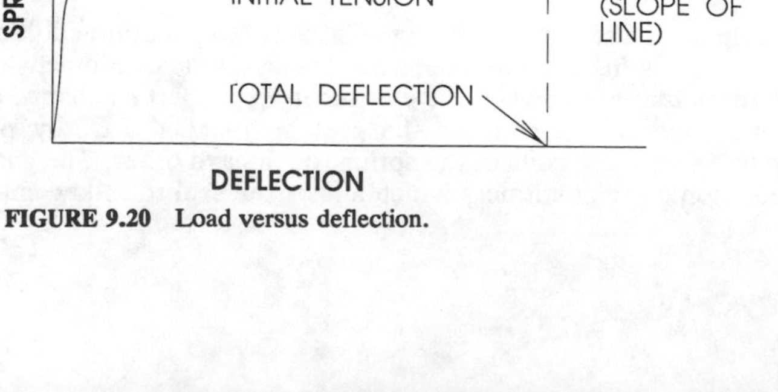

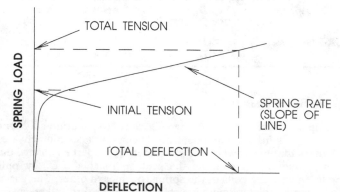

FIGURE 9.20 Load versus deflection.

desired to ensure that the coils of the spring bottom and touch when installed in the spring groove. This gives I.D. control of the finished seal. Normally, the initial tension is approximately 50 to 80 percent of the total load. The tolerance for spring load is ±0.14 N (±0.5 oz), or ±20 percent of the nominal load, whichever is greater. This tolerance is applied to the specified load at a given test length. The test length is usually the design stretch length.

Stress relieving is a heat treatment of the unassembled coiled spring to relieve stresses caused by the spring coiling process. It is intended to ensure that the spring force will not change in service due to exposure to heat. Stress relieving should only be specified when the spring is expected to function at temperatures exceeding 100°C (212°F). The temperature of stress relief must always be higher than the expected service temperature. The most common minimum stress relief temperatures are 200°C (392°F) for mild steel and 260°C (500°F) for stainless steel. It must be pointed out that stress relieving reduces the maximum obtainable initial tension. Because of this, care must be taken when specifying stress relieved springs to ensure that they can be economically manufactured. When stress relieving is specified, the recommended method of inspection is to subject an unassembled spring to the specified temperature for a minimum of 30 min and then, after cooling, measure the spring load. The spring must still meet the original load specification.

The recommended tolerance for coil diameter is ±0.13 mm (±0.005 in). The variation in coil diameter within a given spring should not exceed 0.08 mm (0.003 in). The wire diameter is usually specified as a reference dimension to give the spring manufacturer latitude to meet the initial tension requirements. If the wire diameter must be held, the recommended tolerance is ±0.03 mm (0.001 in). A coil to wire diameter ratio of 5 or more is recommended.

When the spring is assembled, the maximum allowable gap at the nib is three wire diameters. The spring may have a tendency to deform from a flat position into a figure 8. This results from wind-up during assembly and can be controlled or eliminated by back winding the spring ends before assembly. Springs that form a figure 8 are acceptable if they snap back to their original diameter when dropped approximately 300 mm (12 in) onto a flat, hard surface. The tolerances for the assembled inside diameter (AID) of the spring are given in Table 9.14. The gauge shown in Fig. 9.21 can be used to check both nib strength and AID.

A spring with an acceptable joint strength will pass over the largest dimension (1.35 times the nominal spring AID) without disassembly. This may be a destructive test since the spring may yield during stretching.

TABLE 9.14 Assembled Inner Diameter Tolerances

	Wire diameter	AID tolerance
SAE J946, mm	0.15–0.28	±0.20
	0.30–0.48	±0.30
	0.50–0.76	±0.40
	0.80–1.40	±0.50
SAE J946, in	0.006–0.011	±0.008
	0.012–0.019	±0.012
	0.020–0.030	±0.015
	0.031–0.055	±0.020

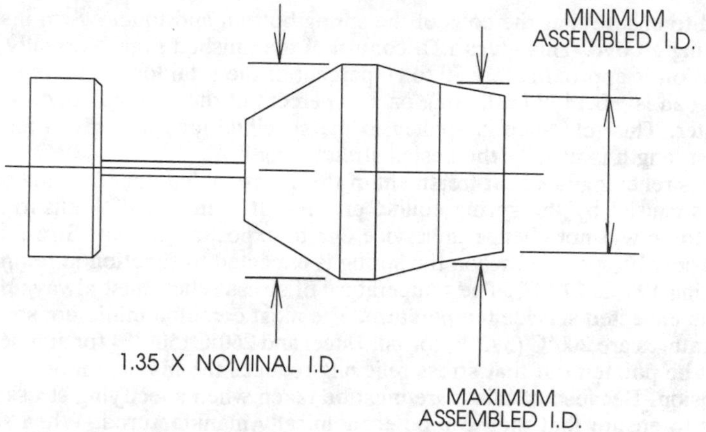

FIGURE 9.21 Spring AID and nib strength gauge.

Recommended Installation Procedures

Improper installation is one of the main causes of premature leakage. Because an elastomeric lip seal is a precise mechanical component, it must be assembled properly to obtain the reliability expected for the application. In high-volume production applications, the seal should be assembled with automatic or semiautomatic assembly tools. These tools will minimize assembly variables that may affect sealing efficiency. The following procedure should be followed before the seal is installed:

Check dimensions: If the shaft and bore dimensions do not match the dimensions specified for the selected seal, leakage will most likely occur. Replace the seal with one having the proper dimensions for the specified shaft and housing diameters.

Check seal: The seal should be examined for damage prior to installation. Leakage will also occur if the lip is turned back or nicked during installation. The seal outside diameter should be checked for dents, scores, or cuts. Replace a damaged seal. Never reuse a seal; always install a new one. Be sure the primary seal lip faces the same direction as the old one.

Housing bore: The housing edge must be deburred to prevent damage to the seal outside diameter. A chamfer or rounded corner should be provided whenever possible (see the section "Bore Recommendations and Specifications").

Check shaft: Remove surface nicks, burrs, and grooves that may damage the sealing lip. Examine the shaft for machine lead. Remove burrs or sharp edges from the shaft end. The shaft end should be chamfered (see the section "Shaft Recommendations and Specifications").

Prelubricate the sealing element: Carefully wipe the seal lip with a lubricant immediately before installation.

When the shaft edges are too severely nicked or burred to be properly prepared for seal installation, an installation cone is recommended. An assembly sleeve is required when the seal is installed over splines or keyways (Fig. 9.22). When the shaft is installed against the primary sealing lip, the use of double pilot assembly cones is recommended (Fig. 9.23).

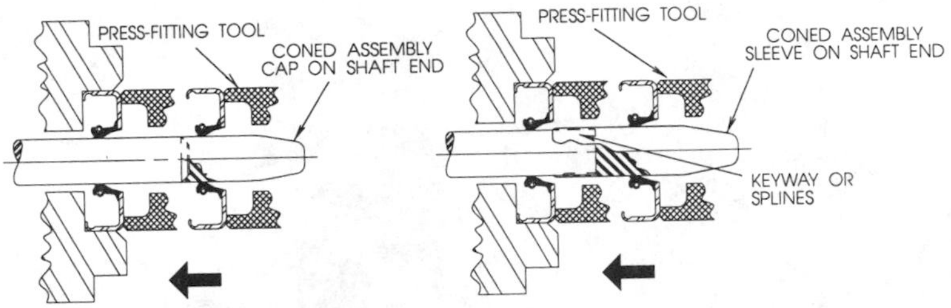

FIGURE 9.22 Assembly cones.

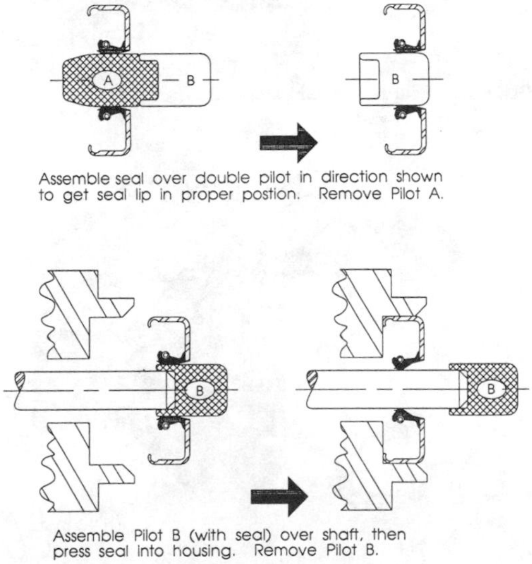

A
B
B
Assemble seal over double pilot in direction shown
to get seal lip in proper postion. Remove Pilot A.
B
B
Assemble Pilot B (with seal) over shaft, then
press seal into housing. Remove Pilot B.

FIGURE 9.23 Double pilot assembly tools.

Installation tools should always be used to install seals and should have an outside diameter that is smaller than the bore diamcter. To prevent seal distortion, apply pressure only at the seal outside diameter. To avoid cocking the seal in the bore, the installation tool must be designed to bottom the seal in the bore (Fig. 9.24) against the shaft (Fig. 9.25) or against the bore face (Fig. 9.26). If the shaft has been used before, position the seal to prevent the seal lip from running in the old wear track. After the installation is complete, check for other machine parts that may rub against the seal and cause friction and heat.

Proper seal installations are sometimes ruined by mishandling of the sealing area during normal maintenance. When machinery is painted, mask the seal area to keep paint off of the lip or the shaft where the lip rides. If the sump is vented to prevent pressure buildup, mask the vents to prevent clogging. If the paint must be baked or the mechanism exposed to other outside heat sources, take care to keep seal temperatures below material limitations. When cleaning or testing, do not allow cleaning fluids to contact seals. When testing or breaking in machinery, do not subject seals to conditions that exceed design recommendations. Seal damage may not be evident until much later.

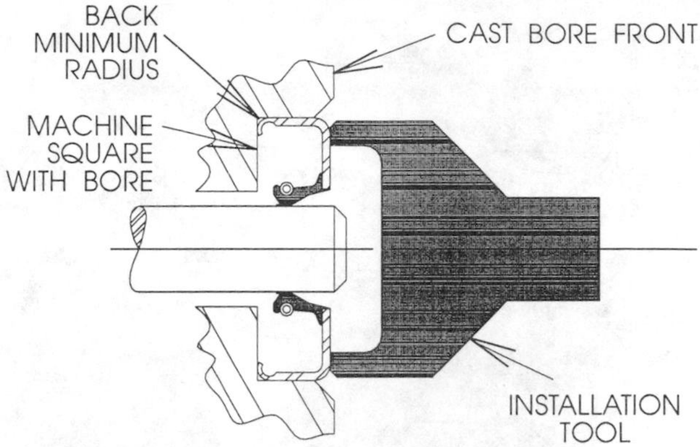

FIGURE 9.24 Seal bottoms in stepped bore.

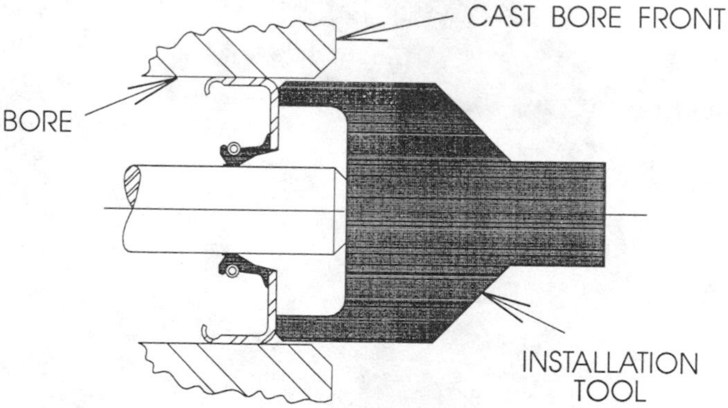

FIGURE 9.25 Through bore—installation tool bottoms on shaft face.

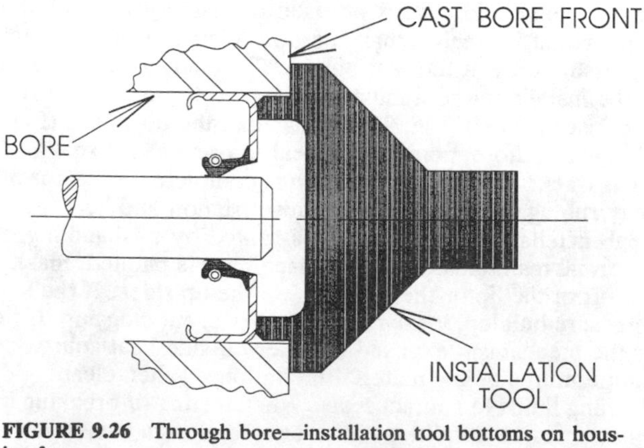

FIGURE 9.26 Through bore—installation tool bottoms on housing face.

PRINCIPLES OF SEAL OPERATION

Theory of Operation

The sealing concept appears to be simple. The flexing of the elastomeric lip and the spring as they are installed over the shaft generates a force between the lip and the shaft that acts as a dam to keep fluid in and contaminants out of the sump. In actual practice, the sealing mechanism is not understood even though elastomeric lip seals have been used since the 1940s. This lack of understanding leads to trial and error in both seal design and material compounding. The net result can be seals that leak instantly for no apparent reason. Some seal designs and material may have sporadic performance and have early leakers combined with long-lived seals. Trial and error can also result in a seal design and material that will function smoothly until the ultimate wearout mode is reached. This usually occurs when the material has lost its flexibility and can no longer follow the dynamics of the shaft. The many variables of lip design, material compounding, and application conditions combine to make a complete understanding of the sealing mechanism a difficult task. Fortunately, many researchers have been studying the phenomena since the mid-1950s,[22] and progress has been accelerating recently.

Early work[23] by Jagger demonstrated that the seal quickly established a wear track on the lip and the lip ran on a thin (0.001 to 0.002 mm) (40 to 80 $\times$ 10^{-6} in) film of oil. Jagger[24] then proposed that surface tension prevents the fluid from leaking through the sealing gap between the elastomeric lip and the shaft surface. He also observed that successful seals developed "microasperities" in the lip contact area, and the shaft surface became smooth. The influence of radial load and especially the correlation of bearing hydrodynamic film theory to oil seal film theory through the Petroff equation for fluid friction was provided by Brink.[5,25,26,27] Later work attempted to show how the oil film could support the lip load. Hirano and Ishiwata[28] estimated the pressure distribution under the lip and also confirmed foil bearing theory to predict seal frictional behavior. The surface tension concept was developed further by Iny and Cameron,[29] Rajakovics,[30] and Jagger and Wallace.[31] The surface tension concept proposed in Ref. 31 was partially agreed to by researchers Brink, Ishiwata, Field, Nau, Singleterry, Walker, and Upper. However, all felt that the sealing mechanism involved other contributing factors. Muller pointed out that successful seals could be made with reversed lip geometry when run under limited conditions of stress. Johnston suggested that supplementary assistance to the seal function comes from a pumping ability and observed that an oil droplet placed on the air side was soon transported across the lip into the sump. But it was the work of Robert L. Dega of General Motors Research Laboratories and his staff, under the overall direction of Dr. Greg Flynn, that showed how a seal could be made to pump oil by (1) cutting helical microgrooves in the shaft surface under the seal lip and/or (2) molding raised helical ribs on the seal surface.[32] Dega and Symons reported that oil could be transferred from the air side to the oil side of an oil seal lip, and the pump rate would increase with shaft speed.[33] Arai also reported that lip seals would pump small quantities of oil placed on the air side into the sump. This transfer occurred even when the sump was empty. He found that the pump rate increased with shaft speed and increased as viscosity was decreased. Pump rate increased slightly as dynamic shaft eccentricity increased. Johnston[34] developed a theoretical method and attempted to link measured seal torque to the dimensional characteristics of the rubber microasperities in the wear track of the seal lip. Kawahara and Hirabayashi conducted experiments which quantified the pumping effect.[35] Nakamura and Kawahara analyzed the contact pattern of the rubber lip on a glass shaft and confirmed the existence of microasperity contacts.[36]

Many researchers now agree that lip seal leakage can be minimized by a positive pumping action. Horve[37] established a relationship between microasperity formation, pumping ability, service reliability, and the formulation of the elastomeric material. Materials can be formulated that will not develop microasperities in the wear track of the sealing lip. They have poor pump rates and service reliability. When the wear tracks of such materials are deliberately roughened, the pump rates increase dramatically for a short time and then decay as the artificially induced microasperities are worn away. Materials can also be formulated to quickly develop an abundance of microasperities in the wear track. Seals made with these materials have high pump rates with excellent service reliability.

Other researchers have investigated different theories to explain the pumping phenomenon. Muller and Ott[38] described a pumping effect due to Gortler-Taylor vortices generated in the oil at the rotating shaft surface. These vortices created an inward pumping effect that equaled the hydrostatic pressure to prevent leakage. The phenomenon was demonstrated with a rigid bushing that had a wide gap between the rotating shaft and the bushing inner diameter.

Some researchers[35,39,40] have observed helical patterns of asperities in the wear track that act as miniature pumps. The shearing force of the rotating shaft displaces the rubber surface and forms a topography of ridges and valleys inclined to the axial direction. The rubber will take a set when exposed to heat and stress, and the ridges act as a microvisco pump. This would also explain the loss of pumping ability and in some cases leakage that occurs when the direction of shaft rotation is reversed.

Gawlinski[41,42] suggested that the radial lip sealing mechanism depends on an axial scrubbing action that results from shaft dynamic eccentricity. Tests by Horve[43] and others show that small amounts of reciprocating motion without rotary shaft motion will not induce the amount of pumping and capability observed when the shaft rotates. Prati[44] demonstrated that gross leakage will result when combinations of shaft dynamic eccentricity and shaft speed exceed the ability of the seal lip design and material combination to follow the shaft. Kalsi and Fazekas[45] reported that an O-ring that is slanted or cocked with respect to the shaft axis reduces leakage in high-pressure rotary applications. Martini reports that some O-rings have been known to seal shafts running at surface speeds as high as 1000 ft/min (305 m/sec),[46] although O-ring usage for rotating shafts normally should be avoided except for slow, reciprocating, or intermittent full rotation. Brink and Horve introduced a molded lip seal design with a wavy sealing edge that pumped and sealed more effectively than seal designs having a conventional straight sealing edge. Using infrared thermometry, Brink also found that the wavy lip presented a greater heat transfer area to the shaft, thereby reducing the lip operating temperature.[6] Increasing the amplitude of lip edge waviness and the number of waves increased the pump rate. This macroscopic pumping effect has been described mathematically by Horve.[43] Hermann and Seffler[47] reported that tilting the shaft from the horizontal resulted in a one-cycle wave seal contact band, which also resulted in increased pumping ability. One should be cautioned, however, that unless carefully designed, a cocked or tilted shaft or seal that is installed eccentrically can be a leading cause of seal failure. Certainly, small eccentric movements of the shaft during rotation can be of some benefit to the sealing mechanism in that as the shaft moves up, the upper part of the seal moves toward the oil sump while the lower section moves away from the sump, creating a dynamic wave action to the lip as the shaft makes a complete revolution. Shaft out-of-round conditions can produce a similar effect. Again, it must be stated that extremely small movements can produce favorable results, whereas something more than extremely small movements can produce failure, especially if some of the application conditions are combined. It is necessary, therefore, to work with the supplier of the oil seal to secure the most accurate information on application limits. Researchers continue to investigate and study the sealing mechanism. Answers will enable seal designers and material compounders to develop more reliable sealing products.

Measuring Seal Operating Characteristics. The oil seal is a friction device. Brink has shown its impact on fuel and energy consumption, and certainly more study remains to be done on the effect of seal leakage on the environment. No one needs to be reminded of the oil spots in our drives and parking lots, which is just a portion of the problem when an oil seal leaks. A radial force is exerted by the seal lip on the shaft. Brink shows that this force increases frictional heat, decreases seal life, and wastes world energy resources.[48,49] Brink shows the many application, seal design, material, and dynamic variables that can affect seal life[5] (Fig. 9.27). It is important to have methods of measuring key operating parameters and to be able to relate them to seal life in order to understand how life can be extended.

Seal life is very difficult to determine on equipment in the field because operating conditions are unknown, and they vary considerably. For this reason, seals are tested on special machines where variables can be closely controlled and monitored. Brink shows the interaction of some of the many variables on seal life. These are shown in Fig. 9.27.[5,49] The specifications for oil seal test machines appear in Table 9.15.[50–52] A typical test machine head is shown in Fig. 9.28.

It is recommended that seals be tested to life end and that Weibull statistics be used to analyze the results.[53] Seal failure criteria must be determined before testing begins and rigidly followed to eliminate subjective evaluations.

The radial load of an oil seal is defined as that load an oil seal exerts on the shaft, and it is the sum of the central acting forces around the seal lip.[5] The radial load is generally composed of two principal elements. They are the spring and the sealing element. The sealing element load is created by two distinct modes of stress within the material—the hoop tension created by stretching of the lip and the beam load created by radial deflection outward of the seal lip element (Fig. 9.29).

Various methods can be used to calculate the radial force that the seal lip exerts on the shaft.[1,5,54] Finite element analysis is one of several methods used to determine the total force and underlip pressure distribution (Figs. 9.30 and 9.31). Measurments of oil seal radial force are required to estimate seal performance and verify mathematical models. This force can be measured with an instrumented segmented or split shaft. In either case, a force transducer senses the seal load against the segment or split shaft.

The split shaft technique (Fig. 9.32) measures the total sum of the components of radial force acting perpendicular to the split line of the shaft. This resultant force P_D when multiplied by π is called the total radial load of a seal P_π. Equations for force calculations are shown in Eqs. (9.2), (9.3), and (9.4). Integration over a quarter shaft gives the following result:

$$\frac{P_D}{2} = \int_0^{\pi/2} pr_s \cos\theta d\theta = pr_s \sin\theta|_0^{\pi/2} = p_r \tag{9.2}$$

$$P_D = 2pr_s \tag{9.3}$$

$$P_\pi = 2\pi r_s p = \pi P_D \tag{9.4}$$

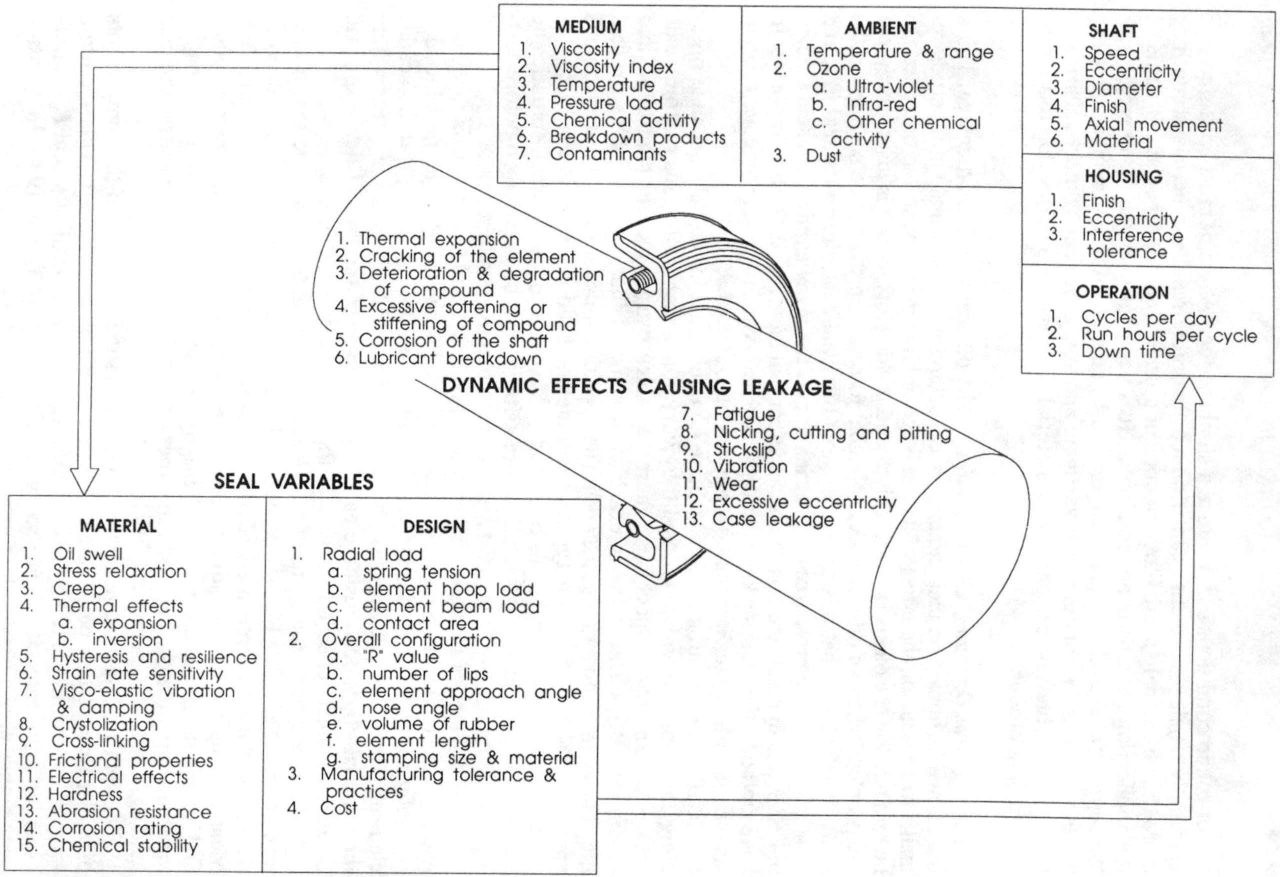

MEDIUM
1. Viscosity
2. Viscosity index
3. Temperature
4. Pressure load
5. Chemical activity
6. Breakdown products
7. Contaminants
AMBIENT
1. Temperature & range
2. Ozone
a. Ultra-violet
b. Infra-red
c. Other chemical activity
3. Dust
SHAFT
1. Speed
2. Eccentricity
3. Diameter
4. Finish
5. Axial movement
6. Material
HOUSING
1. Finish
2. Eccentricity
3. Interference tolerance
OPERATION
1. Cycles per day
2. Run hours per cycle
3. Down time
1. Thermal expansion
2. Cracking of the element
3. Deterioration & degradation of compound
4. Excessive softening or stiffening of compound
5. Corrosion of the shaft
6. Lubricant breakdown
DYNAMIC EFFECTS CAUSING LEAKAGE
7. Fatigue
8. Nicking, cutting and pitting
9. Stickslip
10. Vibration
11. Wear
12. Excessive eccentricity
13. Case leakage
SEAL VARIABLES
MATERIAL
1. Oil swell
2. Stress relaxation
3. Creep
4. Thermal effects
a. expansion
b. inversion
5. Hysteresis and resilience
6. Strain rate sensitivity
7. Visco-elastic vibration & damping
8. Crystolization
9. Cross-linking
10. Frictional properties
11. Electrical effects
12. Hardness
13. Abrasion resistance
14. Corrosion rating
15. Chemical stability
DESIGN
1. Radial load
a. spring tension
b. element hoop load
c. element beam load
d. contact area
2. Overall configuration
a. "R" value
b. number of lips
c. element approach angle
d. nose angle
e. volume of rubber
f. element length
g. stamping size & material
3. Manufacturing tolerance & practices
4. Cost

FIGURE 9.27 Interaction of seal and application variables.

TABLE 9.15 Oil Seal Test Machine Specifications

Test head	Level, rigid, vented to atmosphere
Spindle speed control	±3%
Spindle runout	±0.03 mm (±0.001 in)
	Up to 6000 rpm
Test shaft to bore alignment	±0.03 mm (±0.001 in)
	Through the operating temperature range
Test head temperature control	±3°C (±5°F)

where D_s = shaft diameter
r_s = shaft radius
θ = angle from center of shaft to any point on surface of the shaft
$d\theta$ = differential angle θ
p = incremental lip force per unit of circumference
P_D = force sensed by instrument
π = total circumferential radial load

Seal performance depends on temperature. The underlip temperature is higher than the sump when the shaft is rotating. Brink has accurately measured the underlip temperature with an infrared camera that is focused and scanned along the inside of a hollow shaft until the peak temperature is reached. The peak temperature is directly opposite the sealing lip.[5,26,49]

The pumping ability of a seal is determined by the time required to transfer a known quantity of oil for the air side of the contact zone to the oil side. Pump rates are a measure of reliability and can be used to compare different materials and designs. The method of measuring pump rate is illustrated in Fig. 9.33. The seal is installed in the normal sealing position facing the oil on a vertical pump machine. The secondary dirt lip is trimmed away to gain access to the primary lip, and the seal is run for 20 h while sealing a head of oil to allow the seal lip to establish a wear track on the rotating shaft. After the break-in period, a small drop (0.02 ml) of fluorescent-dyed oil is injected at the air side of the seal. The pump time is obtained with a stop watch by observing (under ultraviolet light) when the fluorescent oil meniscus on the air side of the seal disappears. Another method of obtaining pump time involves measuring seal torque (Fig. 9.34). The seal torque falls instantaneously when the oil drop is added to the air side and returns to the original value when the meniscus disappears. The pump time can be determined from the recorder strip. The pump time measurement is repeated for various shaft speeds, and a pump rate curve can be generated.

Seal power developed at the seal and shaft interface is calculated using the product of measured seal torque and shaft speed [Eq. (9.5)]. The coefficient of friction μ can be calculated using measured values of torque and radial load [Eq. (9.6)]. A Stribeck curve (Fig. 9.35) can be developed when μ is plotted versus the duty parameter G [Eq. (9.7)]. This dimensionless curve is useful when comparing various seal designs and materials operating under different conditions. Brink's development of dynamic radial load, underlip infrared thermometry, and dynamic torque measuring devices in 1966 showed that friction power developed at the seal interface could be calculated using Petroff's equation. Until accurate measurements could be made, it had been thought that the oil film interface was too thin (boundary lubrication) and variant to treat the seal application as a hydrodynamic bearing[26,27,49] [Eq. (9.9)]; (Fig. 9.36).

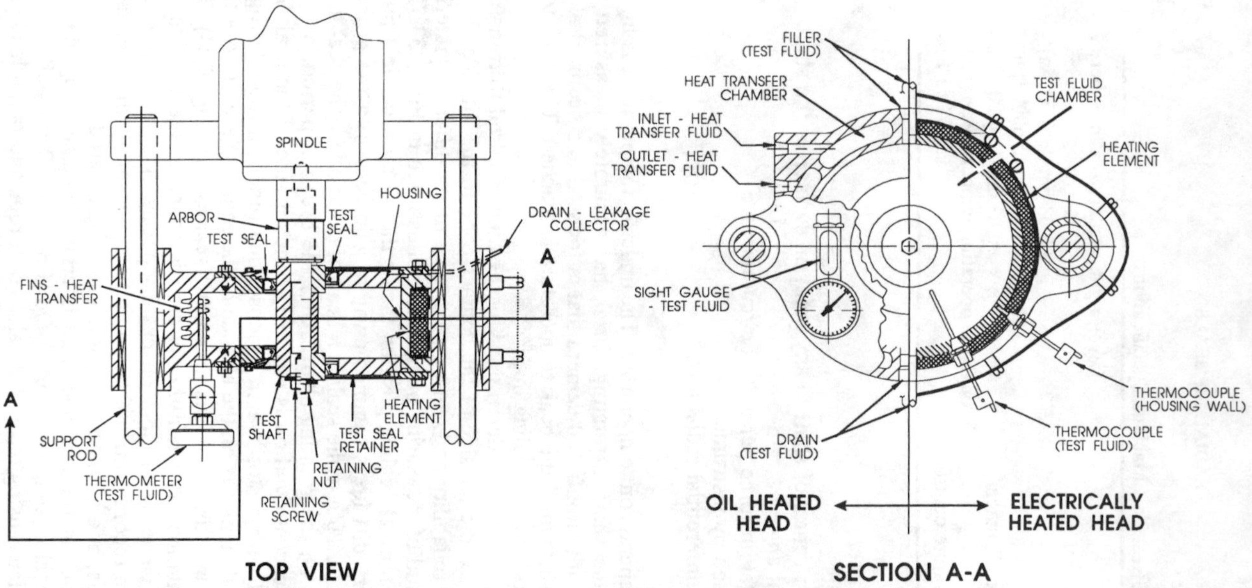

FIGURE 9.28 Schematic showing top view and cross section of standard oil or electrically heated seal test head.

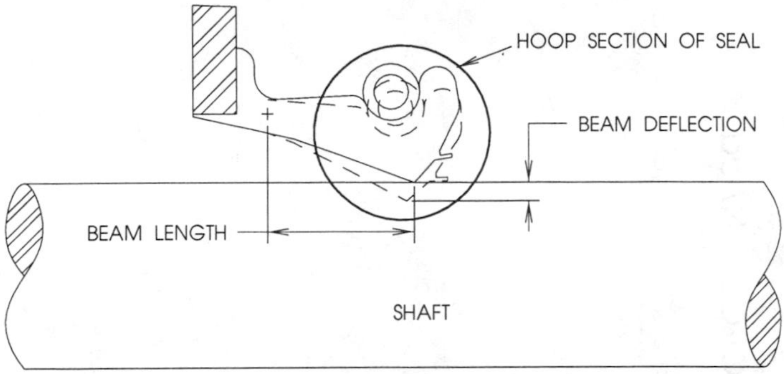

FIGURE 9.29 Seal lip cross section illustrating hoop and beam functions.

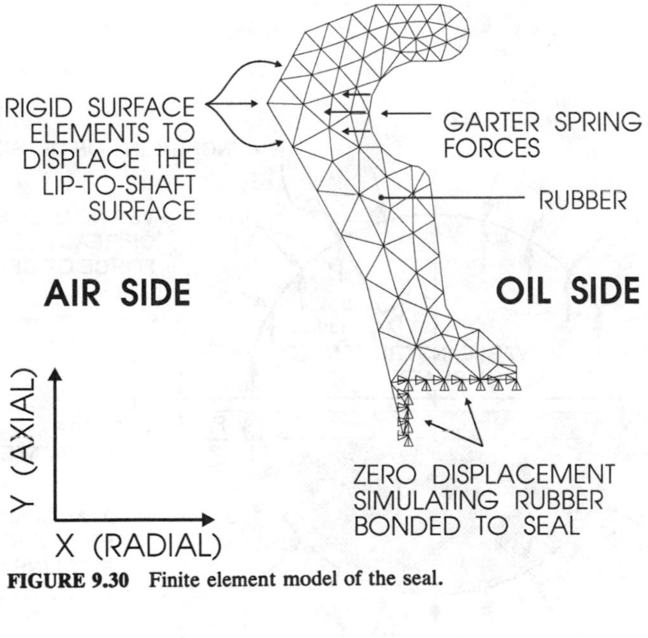

FIGURE 9.30 Finite element model of the seal.

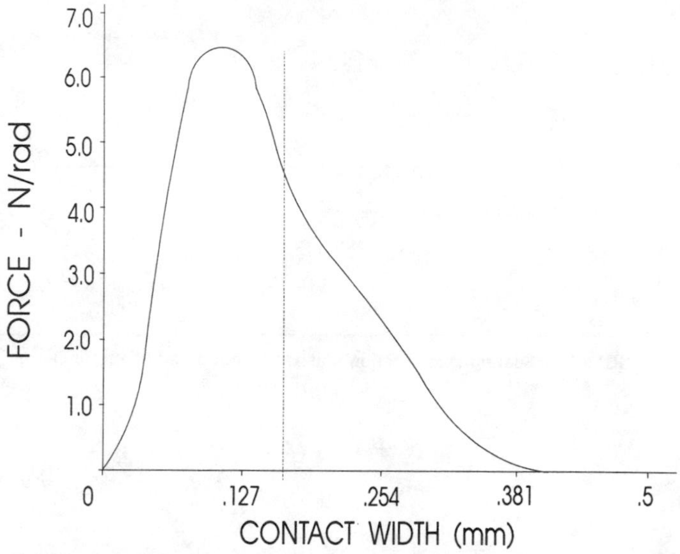

FIGURE 9.31 Pressure profile of the seal lip on the shaft.

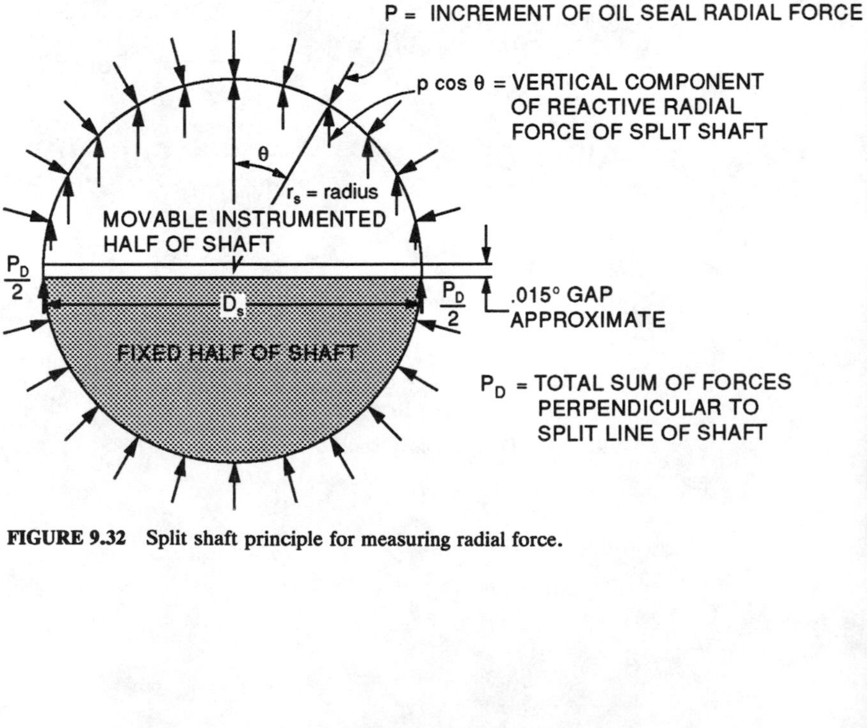

FIGURE 9.32 Split shaft principle for measuring radial force.

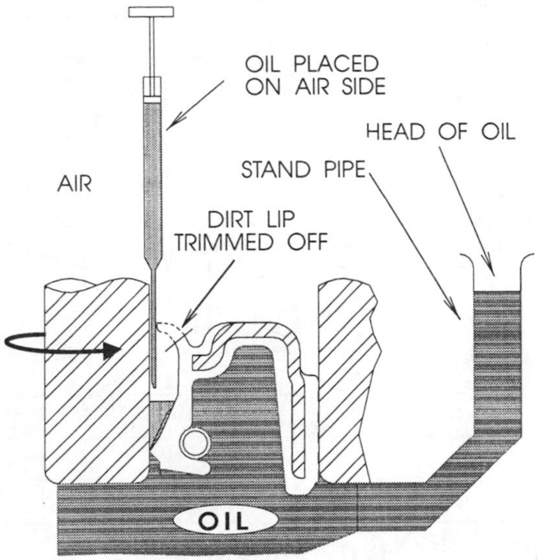

FIGURE 9.33 Oil seal pump test machine.

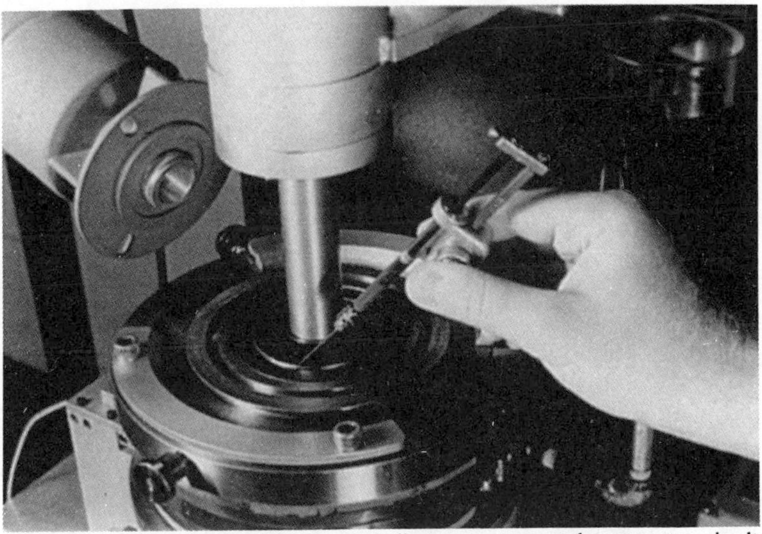

FIGURE 9.34 Measuring seal torque, underlip temperature, and pump rates simultaneously.

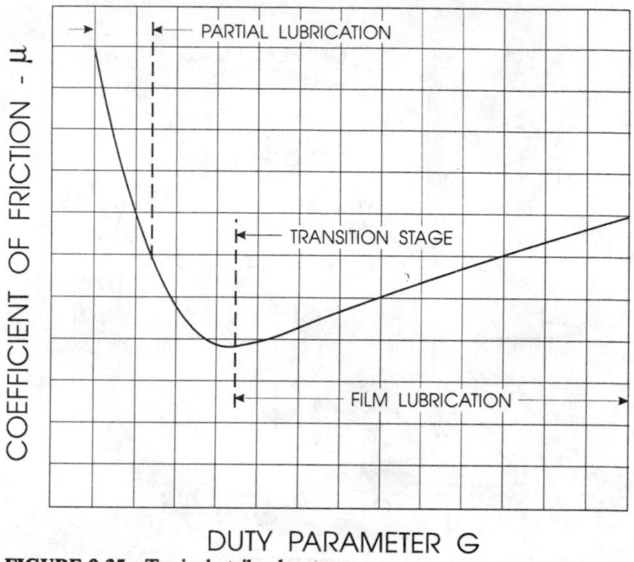

FIGURE 9.35 Typical stribeck curve.

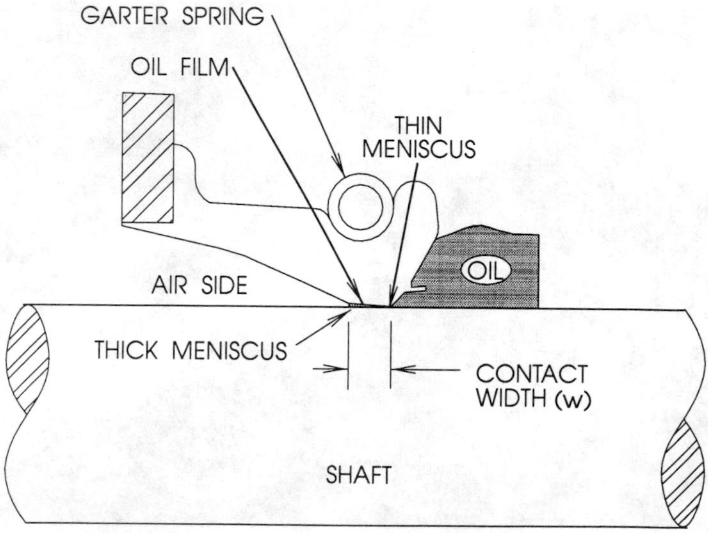

FIGURE 9.36 Lubricating film between seal lip and shaft.

$$P = 0.1046\, T_m N \quad \text{watts} \tag{9.5}$$

$$\mu = (1000)\frac{\text{torque}}{2\pi R_s^2 R_L} \tag{9.6}$$

$$G = 1.666 \times 10^{-11} \frac{Z_{cp} N_w}{R_L} \tag{9.7}$$

$$T_m = (1.666 \times 10^{-14}) \frac{(2\pi)^2 Z_{cp} R_s^3 N_w}{h} \quad \text{N} \cdot \text{m} \tag{9.8}$$

$$h = 1.666 \times 10^{-14} \frac{(2\pi)^2 Z_{cp} R_s^3 N_w}{\text{torque}} = \frac{2\pi R_s G}{\mu} \quad \text{mm} \tag{9.9}$$

where h_f = film thickness, mm
V_s = linear shaft velocity, mm/min
R_s = shaft radius, mm
Z_{cp} = fluid viscosity at underlip temperature, cP
N = shaft speed, rpm
w = contact width, mm
T_m = seal drag moment, N · m
R_L = oil soaked radial load at sump temperature, N/mm shaft circumference
μ = coefficient of friction
G = duty parameter

Sample calculations are as follows: A seal for an 80-mm- (3.150-in-) diameter shaft is operating in Shell SAE 30 at 5000 rpm with a sump temperature of 150°C (302°F). The torque under these conditions was measured at 0.5 N · m (70.8 oz · in) with an underlip temperature of 170°C (338°F). After test measurements show the seal contact width to be 0.300 mm (0.012 in) and the radial load at 150°C (302°F) is 0.0875 N/mm (0.5 lb/in). The viscosity of the oil under the lip at 170°C (338°F) is obtained from a viscosity chart (Fig. 9.37) and is about 4.5 cP.

$$P = 0.1046(0.5)(5{,}000) = 261.5 \text{ watts}$$

$$\mu = \frac{1{,}000\,(0.5)}{2\pi(40)^2(0.0875)} = 0.568$$

$$G = \frac{1.666 \times 10^{-11}(3.5)(5{,}000)(0.254)}{0.0875} = 0.8466 \times 10^{-6}$$

$$h_f = \frac{2\pi(40)(0.8466 \times 10^{-6})}{0.568} = 0.000375 \text{ mm} = 0.375\ \mu\text{m}$$

The Effect of Operating Parameters on Radial Lip Seal Performance

A series of tests were run by Horve to determine how variations in operating parameters affect seal underlip temperature, torque, power loss, pumping ability, and life.[57] All tests were run with nitrile (NBR) material using bench test equip-

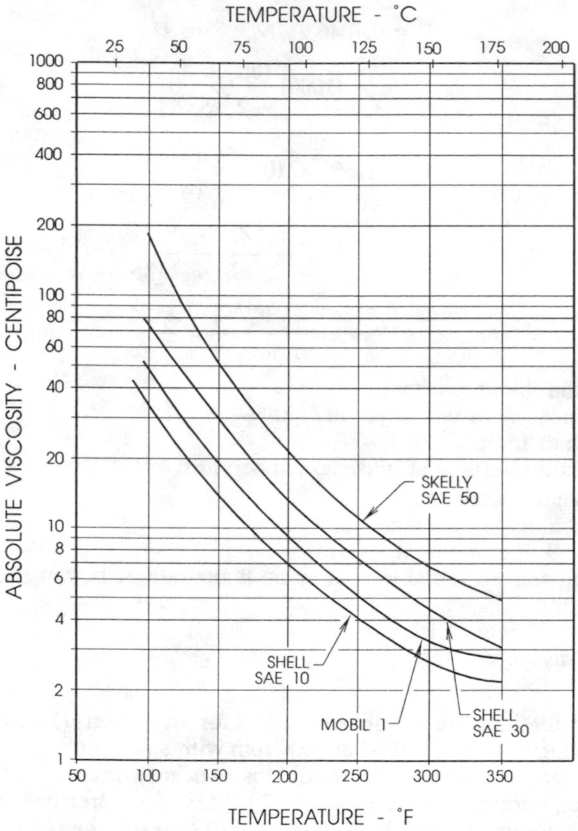

FIGURE 9.37 Absolute oil viscosity versus temperature.

ment. The base set conditions used in the test program are shown in Table 9.16. The effect of each key parameter on performance was determined by varying that parameter while keeping the others constant.

The underlip temperature increases as sump temperature and shaft speed increase (Fig. 9.38). Fluid viscosity drops (Fig. 9.37) as temperature increases,

TABLE 9.16 Base Line Test Conditions

Seal material	Nitrile
Shaft size	76.2 mm (3.000 in)
Dynamic runout	0.13 mm (0.005 in)
STBM	0.13 mm (0.005 in)
Seal cock	Zero
Shaft speed	2165 rpm, 8.63 m/s (1700 fpm)
Sump pressure	Zero
Fluid	SAE 30 engine oil
Sump temperature	93°C (200°F)
Fluid level	Centerline
Life test cycle	20 h, 4 h off

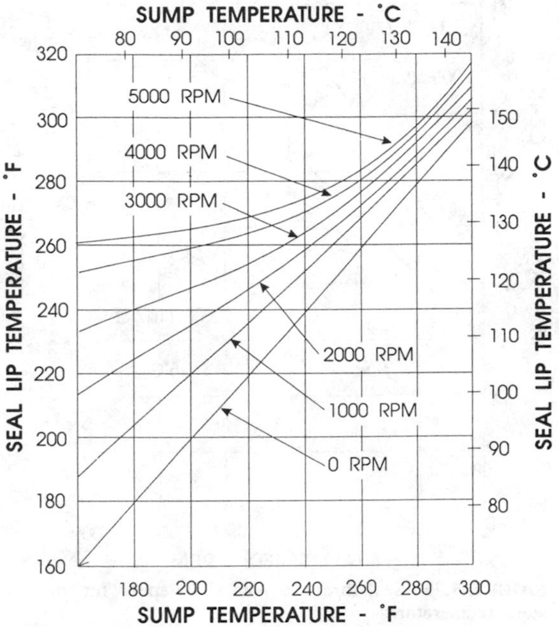

FIGURE 9.38 Seal lip temperature versus sump temperature for various shaft speeds.

which results in a reduction in seal torque (Fig. 9.39) and power consumption (Fig. 9.40). The pumping ability increases with shaft speed and sump temperature (Fig. 9.41).

Brink and Horve find excessive temperature is a prime culprit for seal failure.[5,48,49,58] As lip temperatures increase, the lubricating film becomes thinner and may break down, resulting in dry running and eventual failure. High underlip temperatures may break down the lubricants, creating sludge deposits, or produce carbonized abrasive particles. Increased chemical activity under the lip may cut a groove into the shaft. The typical failure mode for nitrile (NBR) and polyacrylate (PA) materials is the inability to follow shaft eccentricities as excessive temperatures cause the materials to harden and lose interference with the shaft. Silicone materials tend to soften, wear, and revert. Fluoroelastomer materials fail when oil sludge deposits, blisters, or cracks appear near the sealing surface. Oil sludge clogs the microasperities in the seal wear track and destroys the natural pumping ability, which allows leakage. Cracking appears at the lip-shaft interface since this is the hottest point of the sealing system. Increasing fluid temperatures may cause blistering of the elastomer. Blisters usually occur on the air side of the sealing lip where the temperature is hottest and the lubrication is poor. These blisters distort the sealing lip, promoting leakage. Thus, the net result of these temperature effects is a decrease in seal life. The life of nitrile seals decreases by a factor of 2 for every 14°C (25°F) increase in sump temperature. Methods of counteracting these temperature problems include reducing the sump temperature, improving the heat transfer in the seal area, specifying high-temperature lubricants with higher viscosities, and avoiding chemically active lubricants.

It is sometimes necessary to select premium materials in order to extend seal life at elevated sump temperatures. The life of polyacrylic and nitrile seals with identical

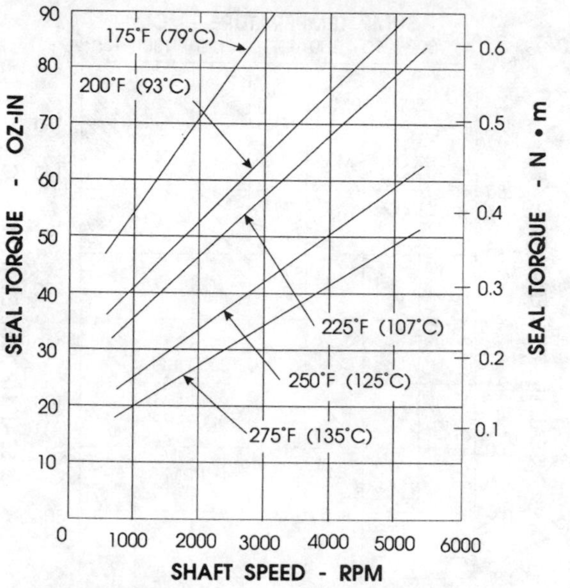

FIGURE 9.39 Seal torque versus shaft speed for various sump temperatures.

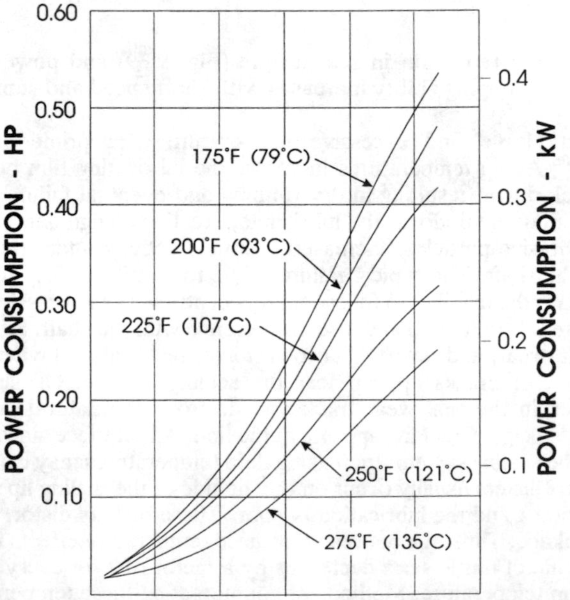

FIGURE 9.40 Power consumption versus shaft speed for various sump temperatures.

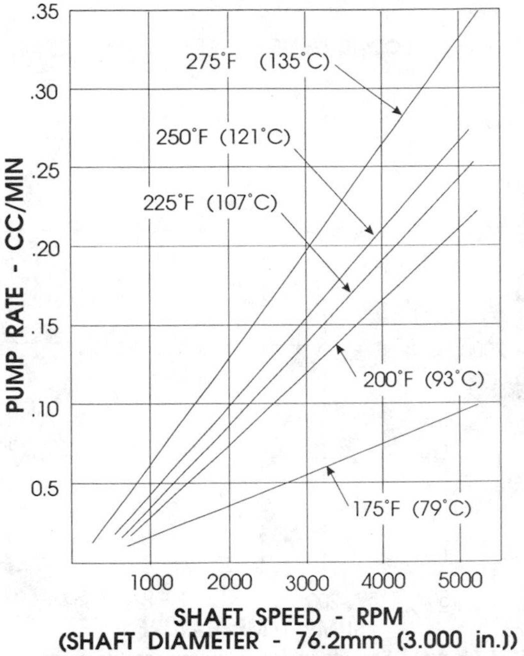

FIGURE 9.41 Pump rate versus shaft speed for various sump temperatures.

designs is compared in Fig. 9.42. At 135°C (275°F), the life of the nitrile seal is only 250 h. The polyacrylate material extends the life to approximately 3000 h under identical test conditions. Other more expensive materials such as silicone and fluoroelastomer will extend life even more. The underlip temperature increases as shaft speed increases. This results in a reduction of seal life (Fig. 9.43).

Two types of eccentricity affect seal performance. The shaft to bore misalignment (STBM) is the amount (in millimeters or inches) that the center of the shaft is offset with respect to the center of the bore. STBM usually exists in some degree due to normal machining and assembly inaccuracies. STBM is measured by attaching a dial indicator to the shaft and is indicated off of the seal bore as the shaft is slowly rotated. STBM should be held to 0.254-mm (0.010-in) offset or less. Dynamic runout (DRO) is the amount (in millimeters or inches) that the sealing surface of the shaft does not rotate about the true center. DRO is caused by misalignment, bending of the shaft, lack of shaft balance, shaft lobing, and other manufacturing inaccuracies. It is measured by the total movement of an indicator held against the side of the shaft while the shaft is slowly rotated. In general, the dynamic runout should not exceed 0.25 mm (0.010 in) total indicator reading (TIR). Sealing performance in eccentric conditions depends largely on the flexibility of the sealing element. As a general rule, typical spring-loaded seals will operate satisfactorily if the total eccentricity (combined indicator readings) does not exceed the maximums shown in Table 9.17. Special seal designs with convoluted sealing lips are recommended if eccentricities are excessive.

Changing the STBM and DRO from 0.0 to 0.5 mm (0.020 in) TIR had no effect upon underlip temperature rise, seal torque, power consumption, and pumping

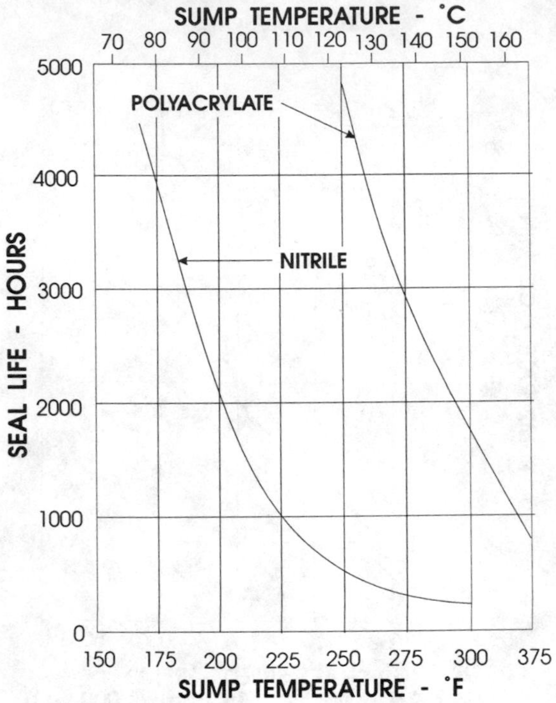

FIGURE 9.42 Seal life versus sump temperatures.

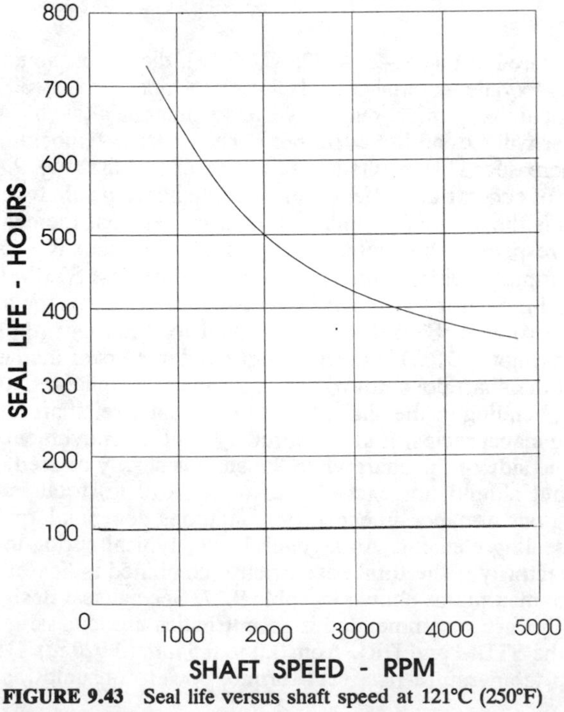

FIGURE 9.43 Seal life versus shaft speed at 121°C (250°F) sump temperature.

TABLE 9.17

Shaft speed, rpm	Maximum total eccentricity, STBM + DRO mm	in
100	0.635	0.025
200	0.508	0.020
500	0.457	0.018
1000	0.381	0.015
1500	0.330	0.013
2000	0.254	0.010
2500	0.229	0.009
3000	0.203	0.008
4000	0.178	0.007

ability. Increasing the STBM from 0.0 to 0.5 mm (0.020 in) had no effect on seal life. The life was dramatically effected by changes in the DRO (Fig. 9.44). Seals leaked instantly when the DRO exceeded 0.6 mm (0.024 in).

Seals are sometimes installed without the back face perpendicular to the centerline axis of the shaft. This condition is known as seal cock or angular misalignment. Cocked seals have regions of high and low pressure between the lip and the shaft. This results in increased underlip temperature, higher seal torque, and increased power consumption as seal cocking increases (Figs. 9.45, 9.46, and 9.47). Seal pump rate also increases with increasing seal cocking (Fig. 9.48). Since seals with high underlip temperatures have shorter lives than seals with low

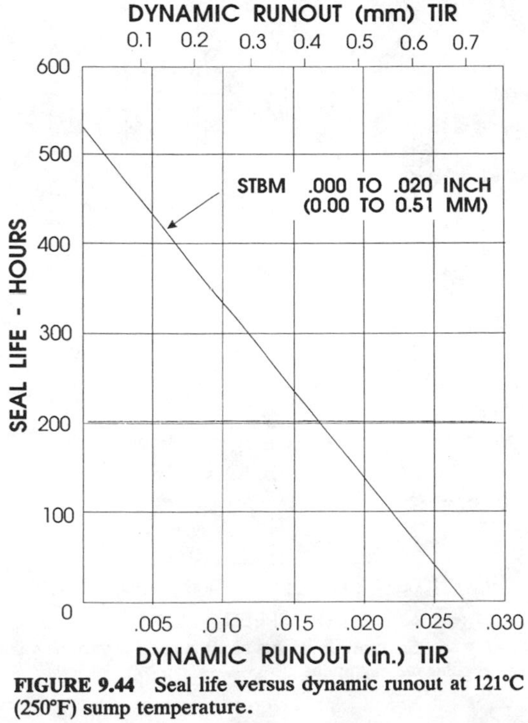

FIGURE 9.44 Seal life versus dynamic runout at 121°C (250°F) sump temperature.

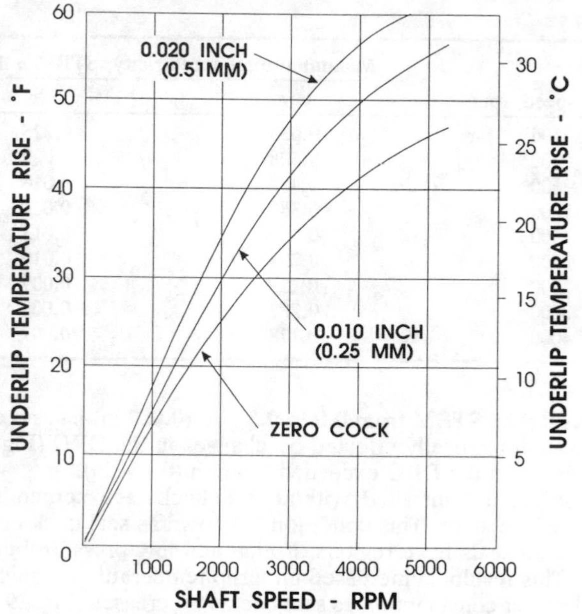

FIGURE 9.45 Underlip temperature rise versus shaft speed for various values of seal angular misalignment.

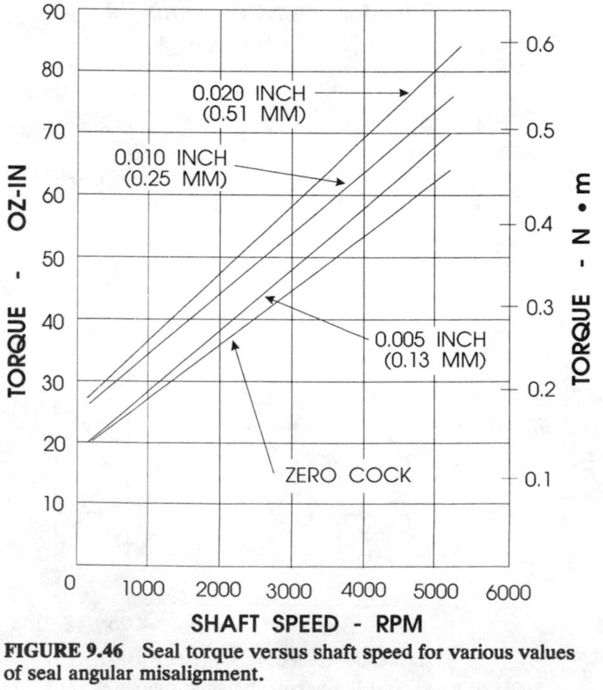

FIGURE 9.46 Seal torque versus shaft speed for various values of seal angular misalignment.

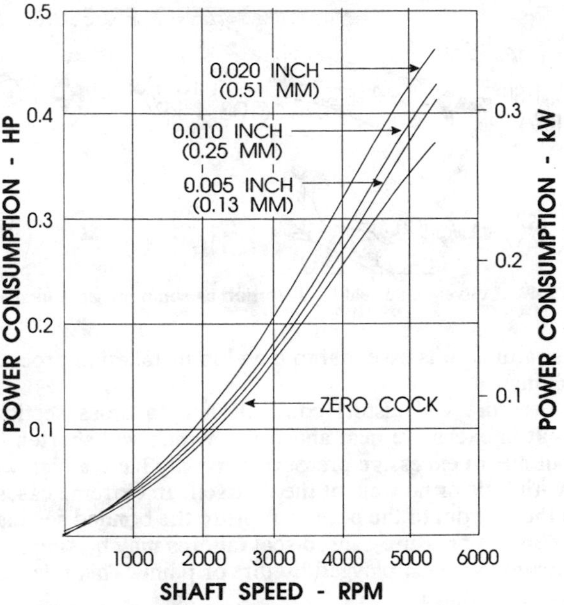

FIGURE 9.47 Power consumption versus shaft speed for various values of seal angular misalignment.

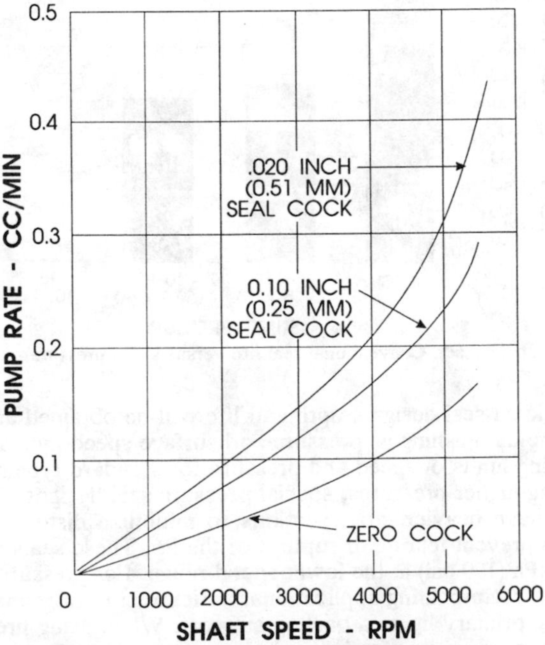

FIGURE 9.48 Pump rate versus angular misalignment.

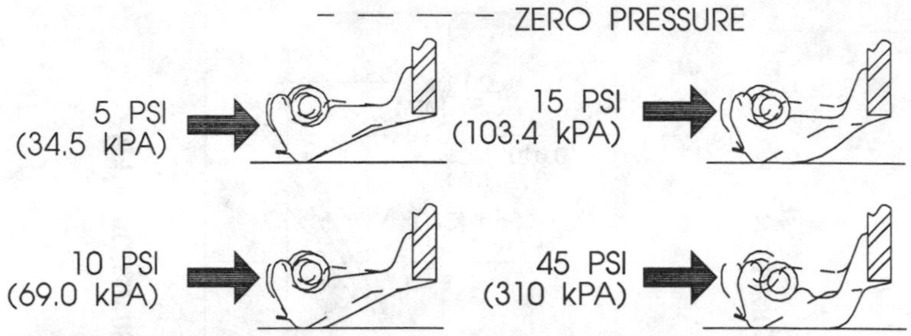

FIGURE 9.49 Conventional seal lip distortion as sump pressure increases.

underlip temperatures, it is essential to develop installation procedures that minimize seal cocking.

Excessive pressure will distort conventional lip cross sections (Fig. 9.49), which will result in excessive heat and wear, which will shorten life (Fig. 9.50). Seals which fail due to excessive pressure normally have a high wear area in the heel section, with little or no wear at the lip itself. In extreme cases, pressure can actually force the seal out of the bore or rupture the bonded sealing element from the case. The cure for pressure-caused seal failures may be simply to open the air vents, which may have been plugged by dirt or paint. This releases the pressure built up by operational heat.

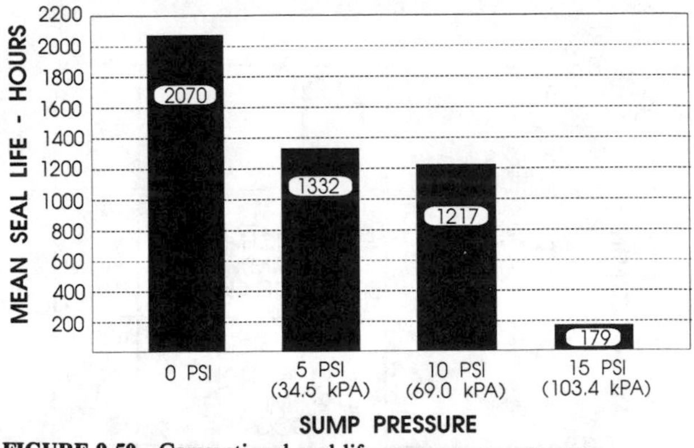

FIGURE 9.50 Conventional seal life versus sump pressure.

For conventional seal designs, optimum life will be obtained at "zero" pressure with the life diminishing as pressure and surface speed increase. Table 9.18 lists the operating limits of speed and pressure for standard type seals. In applications involving higher pressures, special pressure seal designs should be used. Pressure seals have heavier cross sections to minimize distortion and greater bonding area to prevent tearing or rupture of the lip. These seals can accommodate up to 700 kPa (100 psi) at the lower speed range. For pressures up to 10,350 kPa (1500 psi) in reciprocating applications, antiextrusion rings may be incorporated behind the primary lip as a backup support. When using pressure seals, a retainer washer, such as a snap-ring or bolt-on flange, must be inserted behind

TABLE 9.18 Operating Pressure Limits

Shaft speed, m/s	Maximum pressure, kPa	Shaft speed, ft/min	Maximum pressure, psi
0–5	48.3	0–1000	7
5–10	34.5	1000–2000	5
10–?	20.7	2000–?	3

*vs. = velocity of shaft.

the seal and mounted to the housing bore to prevent the seal from being forced out by pressure.

High-viscosity fluids generate higher torque (Fig. 9.51) with greater power losses (Fig. 9.52) than fluids with low viscosity. The pump rate is slightly higher with lower viscosity fluids (Fig. 9.53). Vassmer found that lubricant fill level also affects seal performance and life.[59] Lowering the sump fill level reduces torque and power consumption (Figs. 9.54 and 9.55). The heat transfer also decreases, which results in an increase in underlip temperature (Fig. 9.56). This results in lower seal life (Fig. 9.57).

Testing by Brink with high-speed photography and dynamic instrumentation to measure torque, radial load, and temperature demonstrated that lip seals should not run without lubrication for any prolonged period of time. As sump temperature or speed increases, the lubricating film becomes thin and may break down. In many applications, lubrication in the seal area is sparse, or nonexistent, for a brief period of time before the oil reaches the seal. Under these circumstances, direct contact between the lip and shaft increases, resulting in high wear

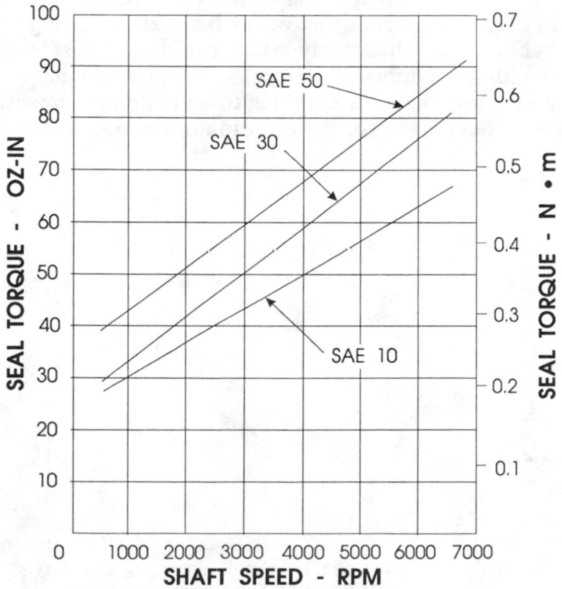

FIGURE 9.51 Seal torque versus shaft speed for various fluid viscosities.

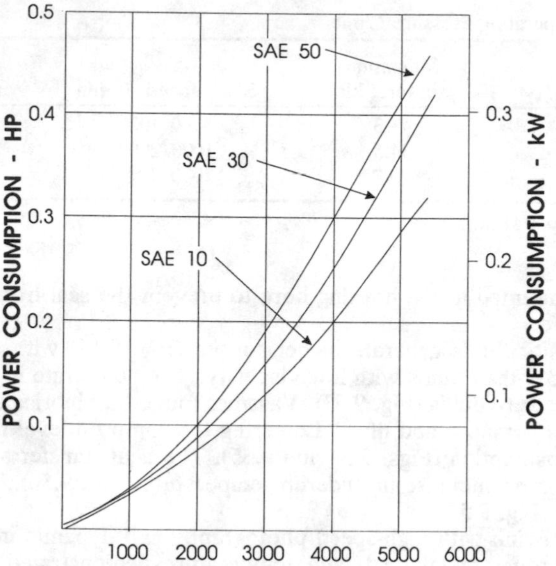

FIGURE 9.52 Power consumption versus shaft speed for various fluid viscosities.

or stick-slip, which may destroy the seal.[5] Stick-slip is a friction phenomenon that describes the jerky motion between two surfaces caused by alternate gripping and slipping of contact areas. This action produces a surface wave oscillation that can literally tear a seal apart or, at best, allow leakage. Different seal materials can behave quite differently when run dry. A test was run with the same design but with three different materials that illustrate the phenomena. Seal A ran smoothly for 5 min but after climbing to an underlip temperature of 475°F, went into stick-slip. Seal B ran for the duration of the test without stick-slip and

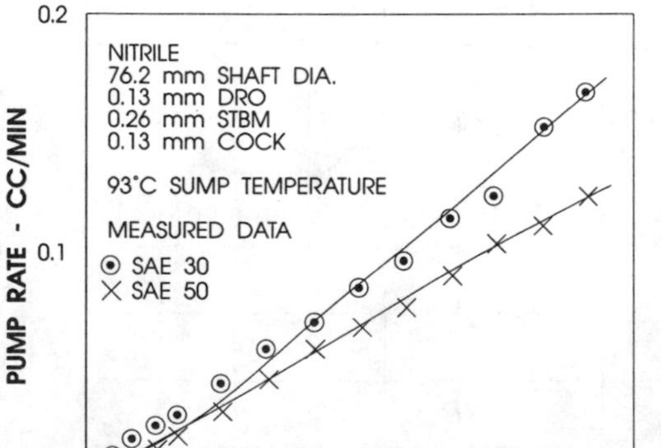

FIGURE 9.53 Pump rate versus fluid viscosity.

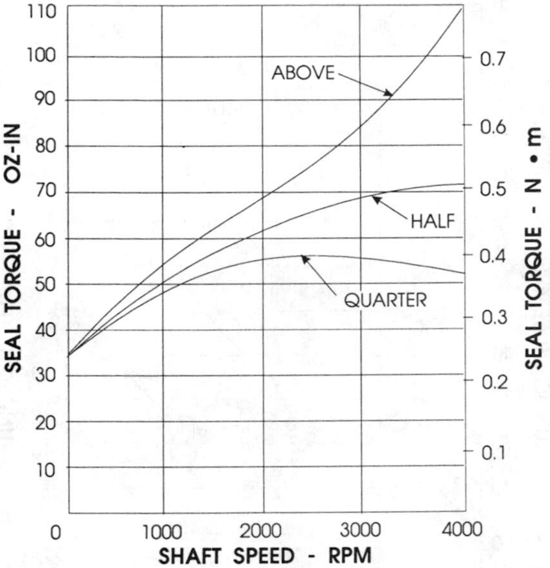

FIGURE 9.54 Seal torque versus speed for various sump fill levels.

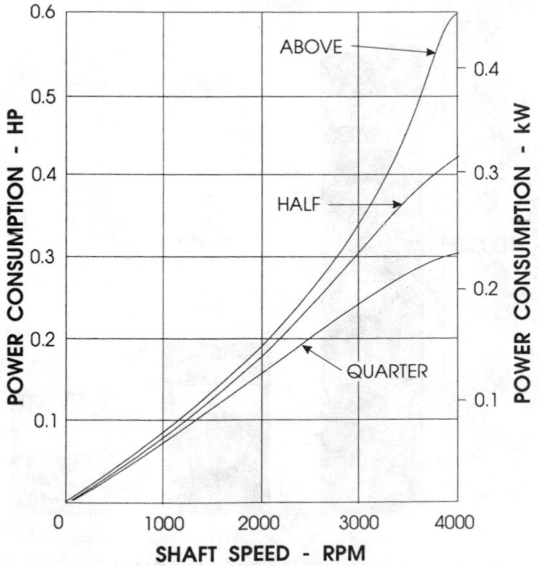

FIGURE 9.55 Power consumption versus speed for various sump fill levels.

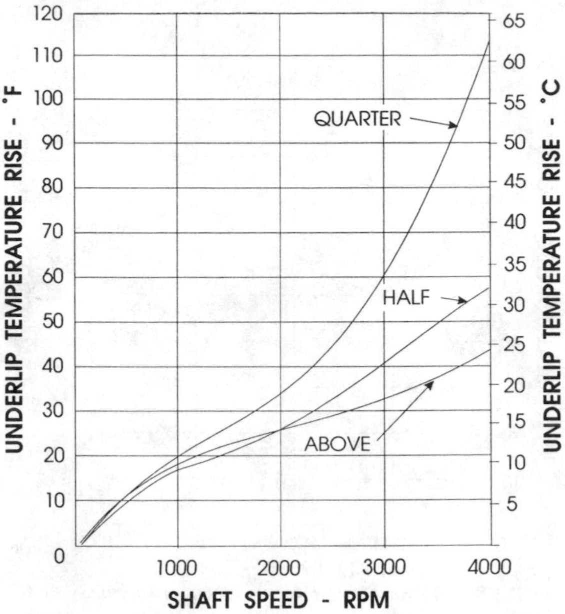

FIGURE 9.56 Underlip temperature rise versus shaft speed for various sump fill levels.

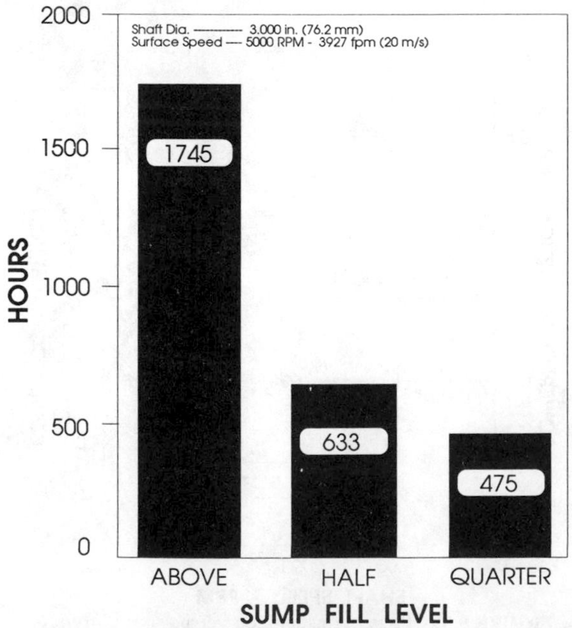

FIGURE 9.57 Seal life versus sump fill level.

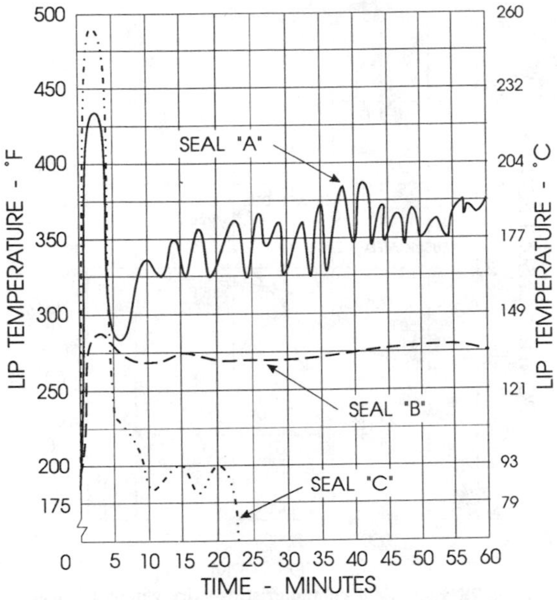

FIGURE 9.58 Dry run characteristics for three seal materials.

never reached an underlip temperature greater than 275°F. Seal C climbed to an excessively high temperature of 500°F and went into such violent stick-slip that it destroyed itself within 2 min (Fig. 9.58).

Seals compounded without dry lubricants and seals made of low modulus materials generally tend to stick-slip more than others. Some seals that run well dry may develop stick-slip when sparsely lubricated. Figure 9.59 demonstrates a seal that reaches a critical condition at 2500 rpm. The friction horsepower increases dramatically, which causes the lip temperature to escalate rapidly. Failure can result quickly. Presoaking silicone seals, which absorb oil, or the use of good-wearing nitriles or teflon materials will alleviate this type of problem in sparse lubrication.

The Effect of Design Parameters Upon Radial Lip Seal Performance

Seal torque, underlip temperature, power consumption, and pump rate increase as shaft size increases. At constant speed, the power loss and frictional torque are approximately directly proportional to the square of shaft diameter (Figs. 9.60, 9.61, and 9.62). Typical seal design practice increases the lip cross-section size as shaft diameter increases (Fig. 9.63). At a constant surface speed of 8.6 m/s (1700 ft/min) and a sump temperature of 93°C (200°F) seal life ranges from 1350 h for small shaft sizes (less than 38 mm) (1.50 in) to 2500 h for shaft sizes greater than 115 mm (4.50 in) (Fig. 9.64).

A garter spring is used to compensate for material changes and to provide a uniform load throughout the life of the seal. Seal life with no spring is about half that of seals using garter springs. When a spring is used, seal life and lip wear

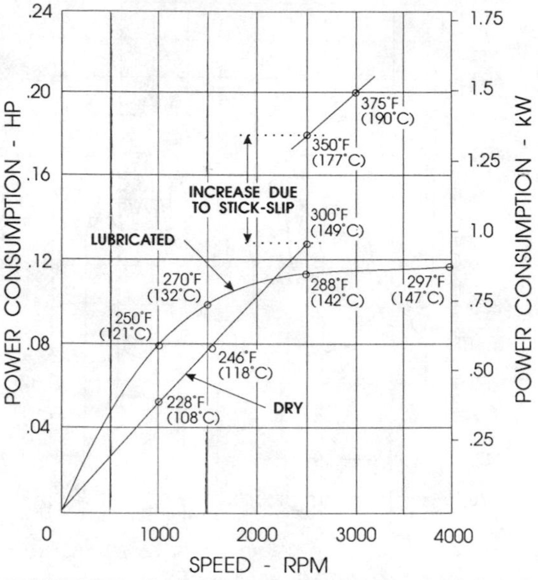

FIGURE 9.59 Seal characteristics—dry versus lubricated conditions.

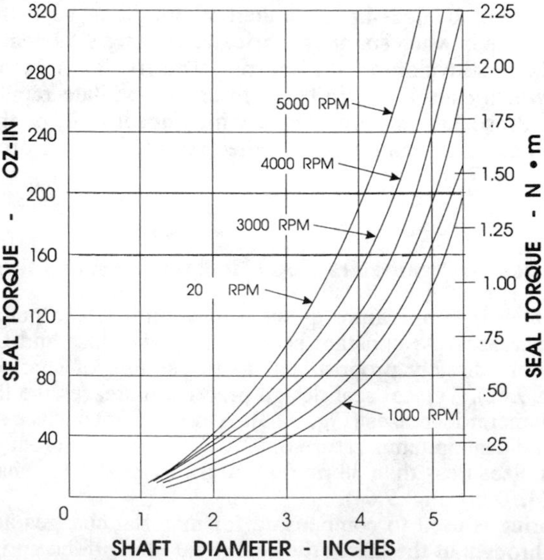

FIGURE 9.60 Seal torque versus shaft diameter.

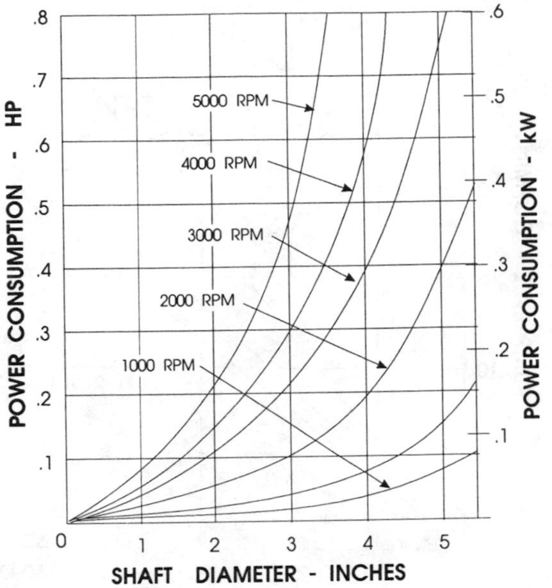

FIGURE 9.61 Power consumption versus shaft diameter.

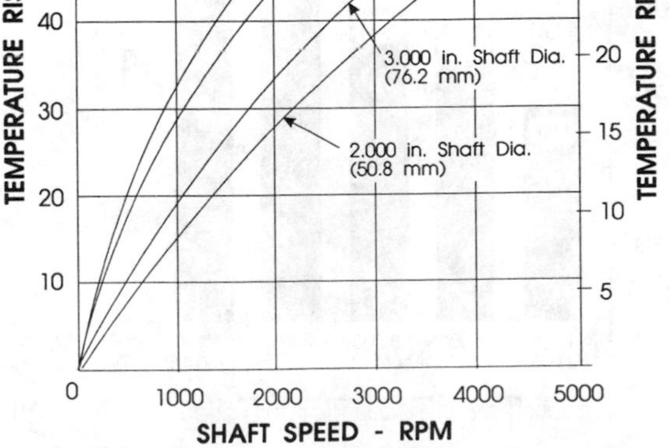

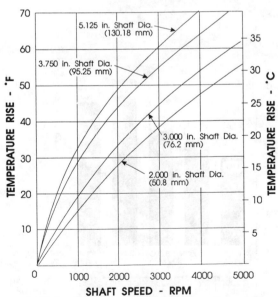

FIGURE 9.62 Underlip temperature rise versus shaft speed for various shaft sizes.

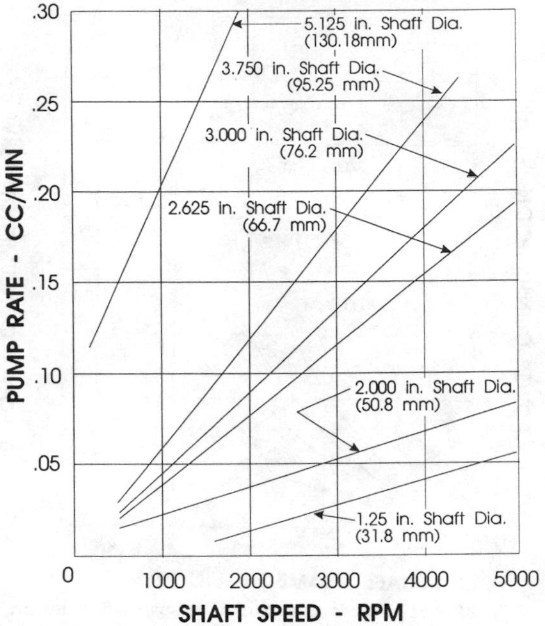

FIGURE 9.63 Pump rate versus shaft speed for various shaft sizes.

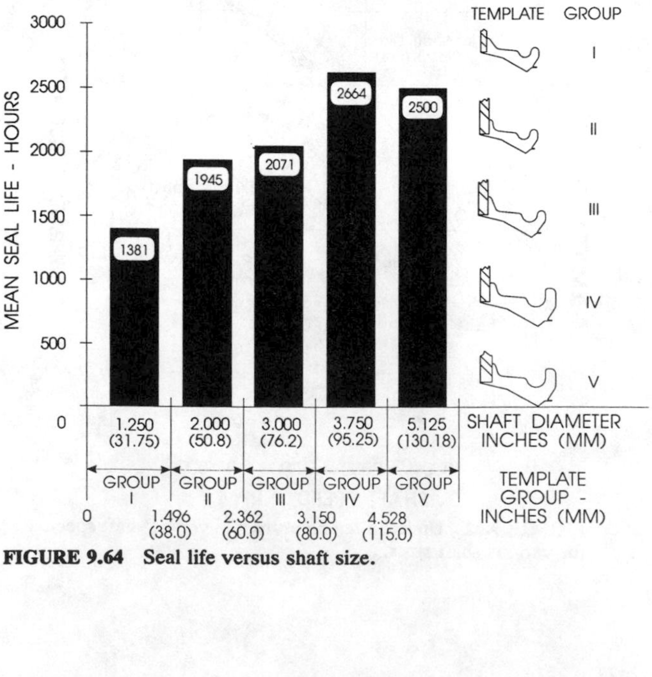

FIGURE 9.64 Seal life versus shaft size.

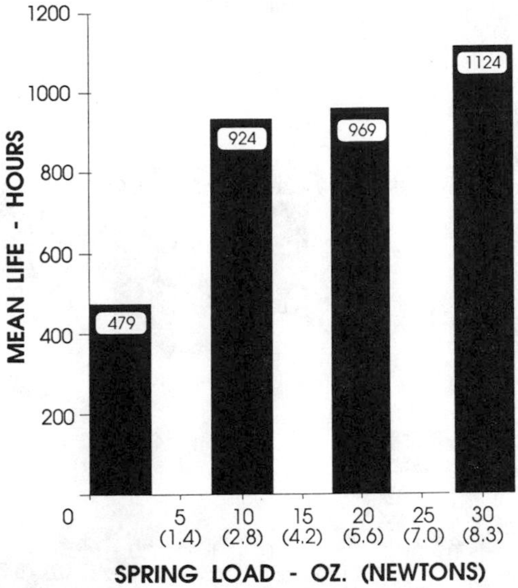

FIGURE 9.65 Seal life versus spring tension.

width are relatively independent of spring tension (Figs. 9.65 and 9.66). The underlip temperature, torque, and power consumption all increase with increasing spring tension (Figs. 9.67, 9.68, and 9.69). It is desirable to select a spring tension that provides good seal life without generating excessive power losses.

The R value is defined as the distance between the lip contact point and the centerline of the spring. Seals with negative on-shaft R values leak instantly. When the installed R value is positive, the seals function well. Seal life and lip wear are independent of the R value when the installed value is positive.

Nitrile seals will eventually fail when the material becomes too hard to follow shaft eccentricity. When this occurs, the sealing lip has lost much of the original lip to shaft interference. Increasing the sprung interference will provide greater seal life (Fig. 9.70). This increase in life at interference levels greater than 2.5 mm (0.100 in) becomes less pronounced because lip wear width (Fig. 9.71), radial load, underlip temperature, torque, and power consumption (Figs. 9.72, 9.73, 9.74, and 9.75) also increase with increasing interference. Since temperature increase and lip wear tend to reduce life, there is a leveling effect at the higher values of sprung interference. A sprung interference of approximately 2.5 mm (0.100 in) is optimum for general-purpose seals.

The Effect of Cold Temperature on Seal Performance

The exposure of machinery to subarctic temperatures has become more frequent within recent years. Special lubricants have been developed to cope with these hostile conditions since ordinary oils and greases can become as stiff as taffy candy at −40°C (−40°F). Cold temperatures harden most elastomeric materials and make them brittle enough to crack when subjected to impulse forces. The temperature at which cracking occurs depends on the dynamic runout of the

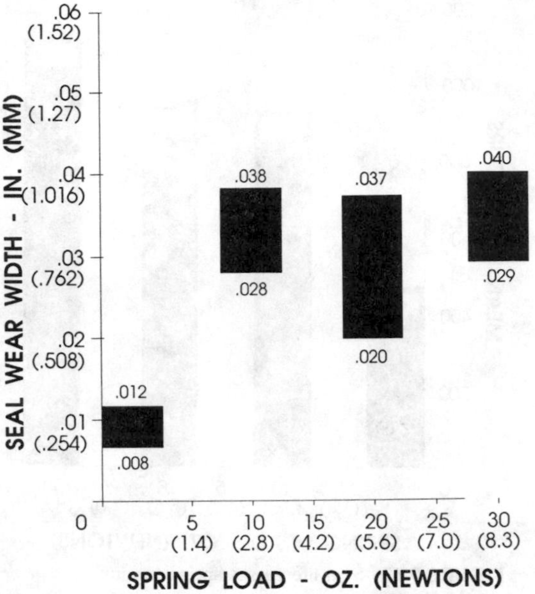

FIGURE 9.66 Seal wear width versus spring tension.

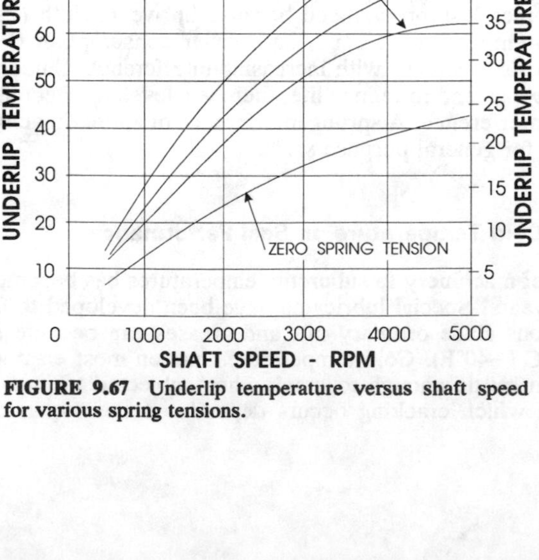

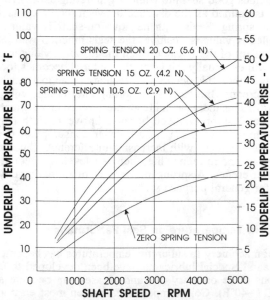

FIGURE 9.67 Underlip temperature versus shaft speed for various spring tensions.

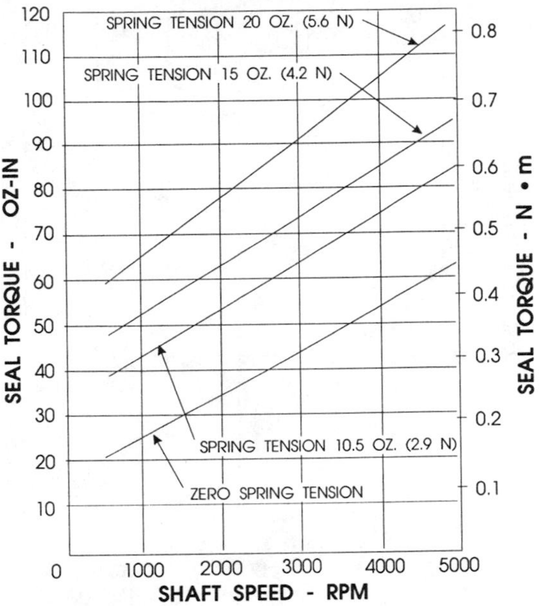

FIGURE 9.68 Seal torque versus shaft speed for various spring tensions.

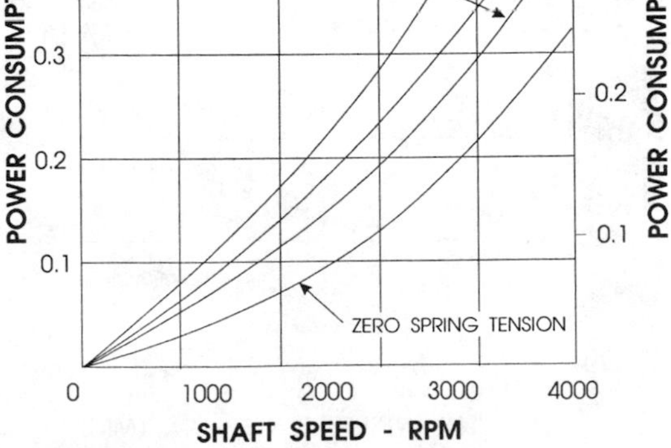

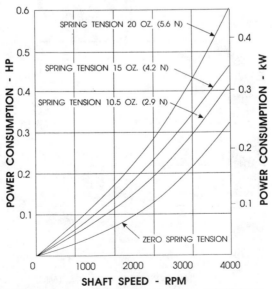

FIGURE 9.69 Power consumption versus shaft speed for various spring tensions.

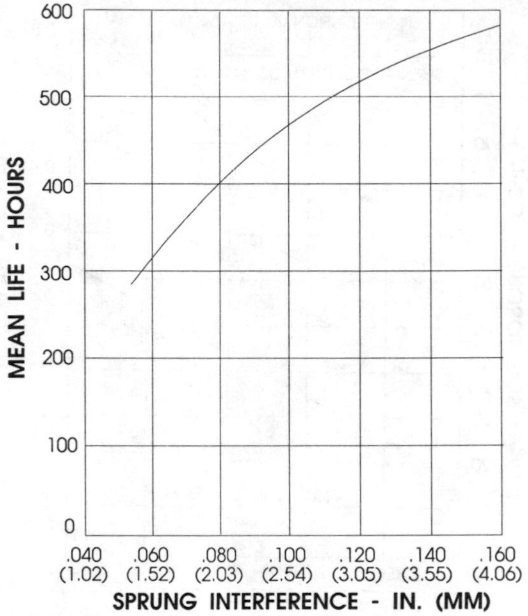

FIGURE 9.70 Seal life versus sprung interference.

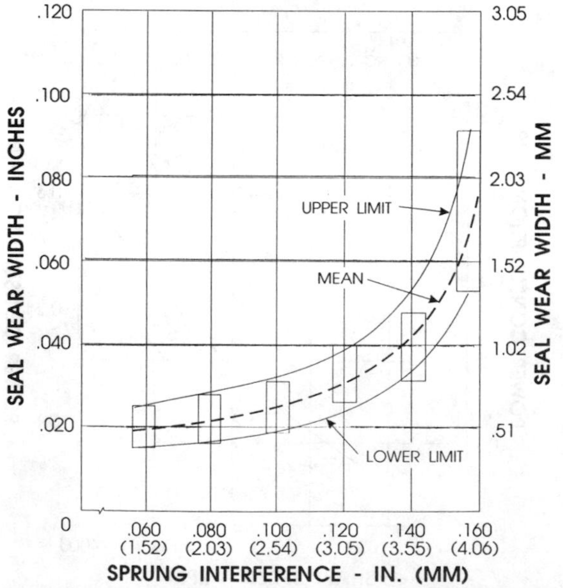

FIGURE 9.71 Seal lip wear width versus sprung interference.

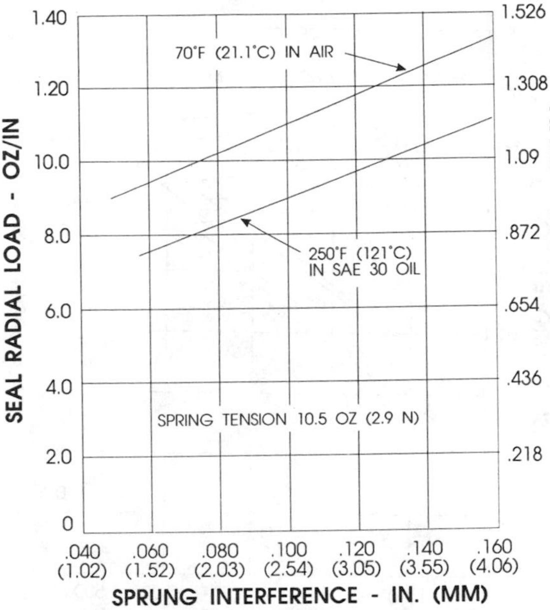

FIGURE 9.72 Radial load versus sprung interference.

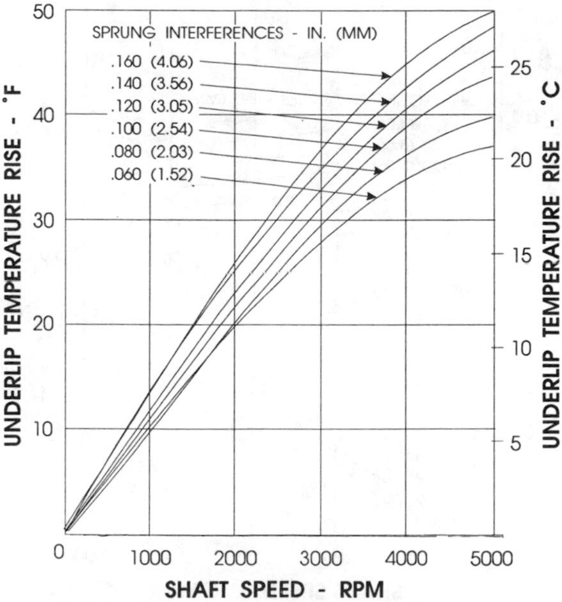

FIGURE 9.73 Underlip temperature rise versus shaft speed for various sprung interference levels.

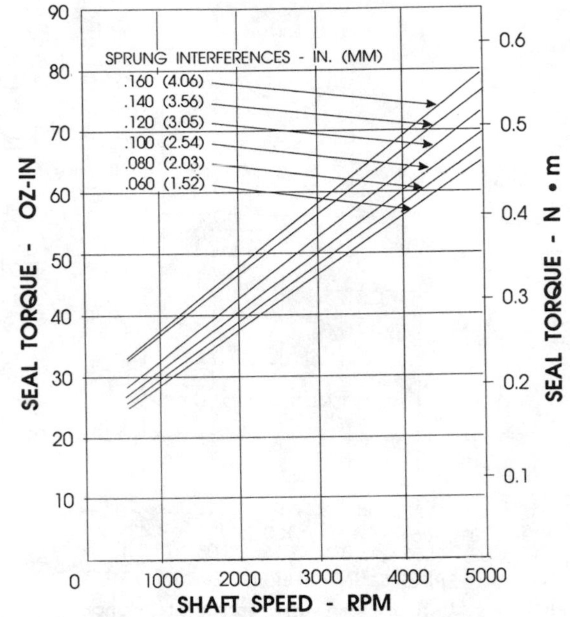

FIGURE 9.74 Seal torque versus shaft speed for various sprung interference levels.

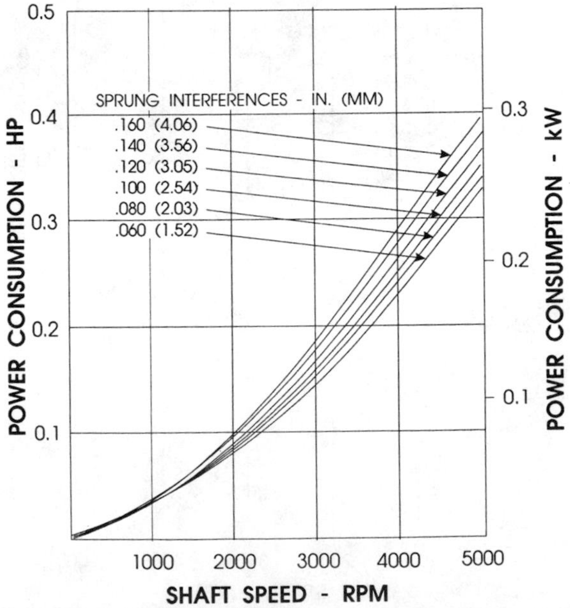

FIGURE 9.75 Power consumption versus shaft speed for various sprung interference levels.

shaft. As the runout increases, the force required to move the sealing lip at start-up increases dramatically, and fracture will occur at higher temperatures. The fracture temperature is relatively independent of seal design parameters. Polyacrylate seals made with a standard lip design, a design with a long flex section, and a design with a thick flex section all fracture at about the same temperature for a given dynamic runout.

The cold fracture limits of elastomeric seals are highly dependent on material and compounding techniques. Three polyacrylate compounds were tested under cold conditions and compared. Compound C was the poorest and A was the best. At a runout of 0.76 mm (0.030 in), the fracture of C occurred at −18°C (0°F). Compound A did not fracture until the test equipment was cooled to −40°C (Fig. 9.76).

Seal leakage can occur at low temperatures even if the lip does not fracture. Material modulus increases as temperature decreases. The lip becomes stiff and may be unable to follow shaft eccentricities. This may not be a problem with conventional fluids since they become very viscous at low temperatures and will not leak through the small gap between the seal lip and the shaft. Synthetic fluids do not thicken at low temperatures and leakage may occur. Friction will warm the elastomeric lip, followability of shaft eccentricity will return, and leakage will stop. The higher material modulus at low temperatures can cause excessive wear since the radial load between the seal and the shaft increases and lubrication is poor. If the seal case and housing bore are made from different materials, differential thermal contraction can result in bore leakage. If the seal is subjected to temperatures of −20°C (−4°F) or lower, the low-temperature properties of the elastomer must be considered. The seal case material should be matched to the housing material. If this is not possible, rubber O.D. seals should be used to minimize thermal contraction effects. The equipment designer should also minimize shaft dynamic runout.

External Contaminants

Contaminants such as water, dust, and mud can enter the bearing area and shorten the seal's life by causing corrosion or abrasion. Heavy-duty exclusion seals can be used to prevent entry of these contaminants by facing the sealing lip toward the atmosphere. These seals can be kept lubricated and clear of dirt by purging them with grease; however, grease can be forced through them only if the lips point outward.

If a minimal amount of contaminants are present at a seal which is retaining a fluid, a conventional secondary radial dirt lip design can be used for exclusion. However, if the secondary lip contacts the shaft and runs dry at speeds over 1000 fpm, the frictional heat generated may cause more harm than good.

Measured underlip temperature profiles by Brink[49] reveal that a nonlubricated, contacting secondary lip can raise the underlip temperature of the lubricated primary lip by almost 10°C (18°F) over a seal without the secondary lip (see Fig. 9.77). After a short period, the secondary lip has hardened and is slightly larger than shaft size. It is recommended that noncontacting secondary radial dirt lips be used if high shaft speeds are expected.[60,61]

A series of tests was performed to determine the optimum dust exclusion design.[62] Dust testing was conducted using a conventional seal test machine with a dust box attached. The dust in the box was constantly agitated with a fan. It was found that test results are inconsistent if the fan is not driven with an exter-

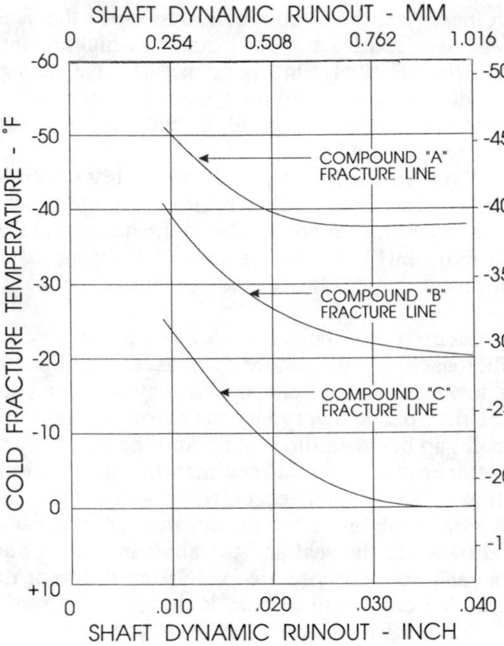

FIGURE 9.76 Cold fracture limits for various seal designs and polyacrylate materials.

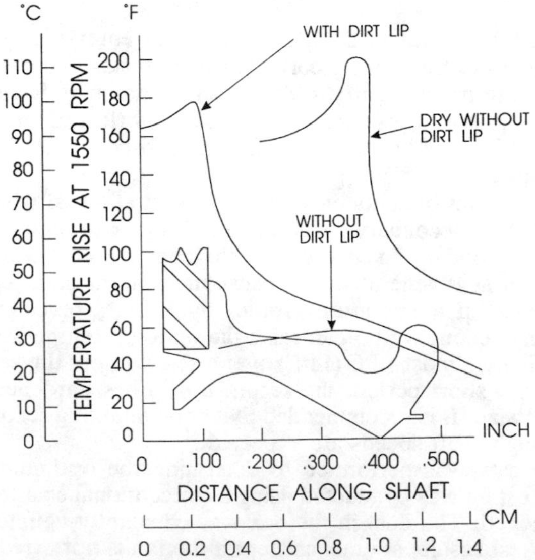

FIGURE 9.77 Seal temperature profiles.

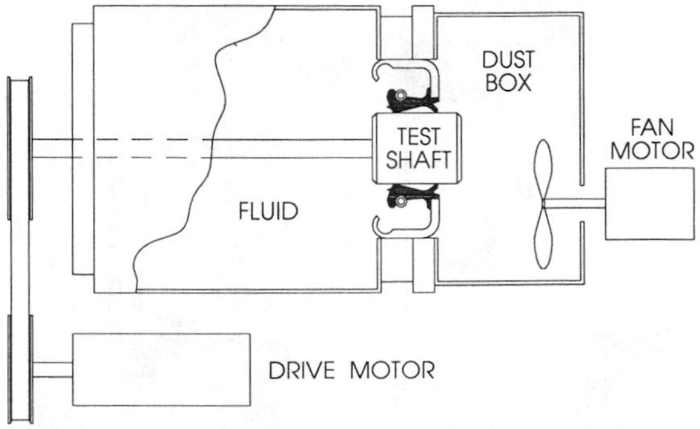

FIGURE 9.78 Dust test machine.

TABLE 9.19 Test Procedure

1. Run the test without dust for 20-h break-in
2. Add 800 ml of Arizona coarse dust.
3. Adjust the paddle speed to 700 rpm.
4. Run the test 20 h on, 4 h off.
5. Check seals for leakage at the end of each 20-h period.
6. Continue the test until leakage occurs.

nal motor (Fig. 9.78). Recommended test procedures appear in Table 9.19 and the dust test parameters are given in Table 9.20.

A common material (fluoroelastomer) and a common helix seal design were evaluated. Only the dirt exclusion lips (Figs. 9.79 and 9.80) were changed and compared. A single lip helix seal (no dust lip) was also tested to serve as a baseline. The best results were obtained with the axial dust lip design shown in Fig. 9.80. The second best results occurred for the control lip seal without any dust exclusion lip. All of the dirt exclusion designs shown in Fig. 9.79 had performance inferior to that of the control seal without a dust lip. Only the axial dirt lip seal design is recommended if high shaft speeds are encountered in the application.

TABLE 9.20 Dust Test Parameters

Dust box	203-mm (8-in) diameter, 203-mm (8-in)-wide cylinderical cavity
Dust paddle	104-mm (5.525-in) diameter
Paddle speed	700 rpm
Dust	Arizona coarse
Amount of dust	800 ml
Seal	Various
Shaft diameter	76 mm (3.000 in)
Shaft speed	2165 rpm (1700 fpm) (10 m/s)
DRO	0.1 mm (0.005 in) TIR
STBM	0.25 mm (0.010 in) TIR
Fluid temperature	107°C (225°F)
Test fluid type	SAE 30

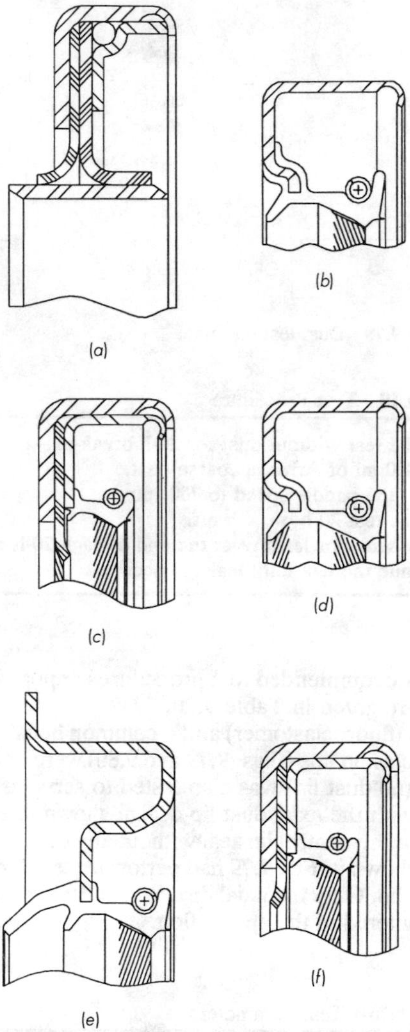

FIGURE 9.79 Seal radial dust lip designs. (*a*) No dirt lip; contacting dirt lip; noncontacting dirt lip; (*b*) dirt lip with helix; (*c*) nylon/felt dirt lip; (*d*) PTFE dirt lip; (*e*) inverted PTFE dirt lip with helix; (*f*) vented and nonvented extended dirt lip.

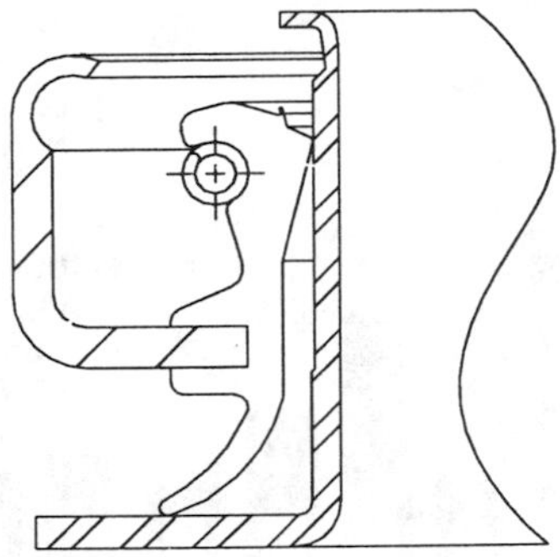

FIGURE 9.80 Axial dirt lip.

ESTIMATING SEAL LIFE

As discussed earlier, there are many variables that affect seal life. Seal laboratory tests under controlled conditions provide the best data to compare designs and materials. Estimating the life of a sealing product requires a solid baseline of information and control of the many variables that can influence the outcome of a test. Brink[5,27,49] has developed a general life equation that can be used under controlled test conditions and is based on the thermodynamic principle of work and waste energy. The concept assumes that the work done at the seal and shaft interface produces wear and heat. It is the seal's function to minimize the generation of heat and wear, and, of course, the seal's capacity to absorb the effects of wear and heat will ultimately determine how long the seal will survive. This logic is valid for any seal device in which wearout and thermal degradation due to friction are the failure modes.

The working life of a seal, therefore, is governed by the capacity of the seal to absorb the effects of the work done during its operational period. It will be shown that:

> *The total number of cycles or shaft revolutions a seal can absorb in its lifetime* (N_T) *is proportional to the cyclic rate (shaft speed* n*), the total work done* (W_T) *and inversely proportional to the rate at which work is done* (W_r) (Fig. 9.81).

For a given set of operating conditions, the seal life (t) is proportional to the total work done (W_T) and inversely proportional to the rate at which work was done (W_r) [Eq. (9.10)]. A common life factor for a family of seals can be determined making it possible to transfer the knowledge of one test application to another without first having to test. That life factor (B_W) is defined as *the total revolutions or cycles* (N_T) *at work rate* (W_r) *per inch or millimeter of shaft diameter* (D). Since,

$$t = \frac{W_T}{W_r} \tag{9.10}$$

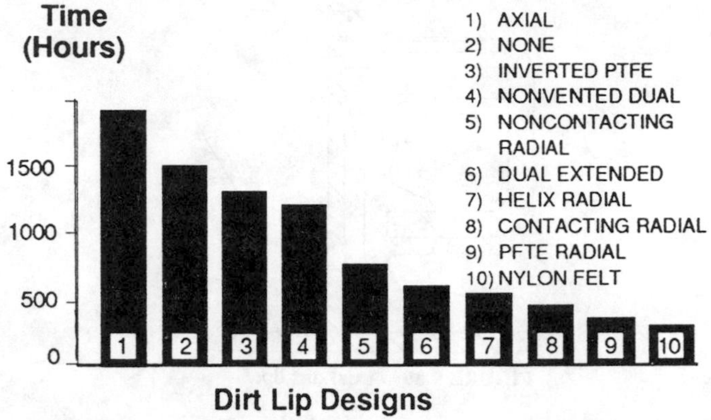

FIGURE 9.81 Life test results.

and

$N_T = nt$ and n = shaft speed, revolutions/hour

then

$$N_T = n\frac{W_T}{W_r} \tag{9.11}$$

To determine the number of life cycles for any shaft size, divide both sides of the equation by D, and

$$\frac{N_T}{D} = \frac{nW_T}{DW_r} \tag{9.12}$$

Therefore,

$$\frac{(W_T)n}{D} = \frac{N_T(W_r)}{D} \quad \text{which is the definition for } B_W$$

$$B_W = \frac{(W_T)n}{D} = \frac{N_T(W_r)}{D} = nt\frac{W_r}{D} \tag{9.13}$$

and therefore t for any other seal of the same family operating in the same oil but at different speeds, torque, and shaft size can be determined since

$$t = \frac{B_w D}{nW_r} \tag{9.14}$$

For example, a seal in NBR is operating at 5000 rpm and 100°C (212°F) in SAE 30 on a 50-mm shaft. The measured torque at these conditions is 0.30 N · m and the measured life is 2000 h. It is desired to predict life of the same seal design and material operating on a 75-mm shaft at 100°C in SAE 30 oil at 3500 rpm. The measured torque of this seal at these conditions is 0.525 N · m:

$$B_W = n\frac{W_r}{D}$$

Since the relationships will be relative, there is no need to convert to absolute units. Then,

$$B_W = 5000\,(2000)(0.30)\,\frac{5000}{50}$$
$$B_W = 300(10^6)$$

and

$$t = \frac{B_w D}{nW_r}$$

Therefore,

$$t = 300\,(10^6)\,\frac{75}{3500}\,(0.525)(3500)$$
$$t = 3498 \text{ h}$$

Here is another example: An oil seal design made of a nitrile material was found to provide a mean life of 1300 h when tested under the following conditions:

Shaft size	76.2 mm (3.00 in) having a surface finish of 15 centerline average (CLA)
Fluid	SAE 10W-30 motor oil
Fluid temperature	107°C (225°F)
Eccentricity	0.25 mm (0.010 in) TIR dynamic runout [total indicator reading; 0.13 mm (0.005 in) offset]
Speed	1840 rpm
Torque	0.42 N · m (3.7 lb · in) (measured after 1 h of stabilization)

Keep in mind as you work through the examples that work is force through a distance, therefore:

$$2\pi\,\frac{\text{radians}}{\text{revolution}}\,(3.7 \text{ lb} \cdot \text{in}) = \text{work per revolution}$$

Two new applications for a seal of the same template (design) and material are being considered. The question is: How long will the new seals operate? The conditions are for Application I:

Shaft size	12.7 mm (0.500 in) diameter (15 CLA)
Fluid	SAE 10W30 motor oil
Fluid temperature	107°C (225°F)

Eccentricity	0.076 mm (0.003 in) TIR dynamic runout; 0.051 mm (0.002 in) offset
Speed	4200 rpm
Torque	0.025 N · m (0.22 lb · in) (measured after 1 h of test stabilization)

The conditions for Application II are:

Shaft size	44.5 mm (1.75 in) diameter (15 CLA)
Fluid	SAE 10W30 motor oil
Fluid temperature	107°C (225°F)
Eccentricity	0.13 mm (0.005 in) TIR dynamic runout; 0.052 mm (0.002 in) offset
Speed	3450 rpm
Torque	0.26 N · m (2.3 lb · in) (measured after 1 h of test stabilization)

The oil seal life capacity (B_W) can be calculated for the known seal life of the 76.2-mm (3.00-in) shaft size. Thus,

$$B_W = \frac{(1840 \times 60)(1300)(1840 \times 60 \times 2\pi \times 3.7)}{3.00}$$

$$B_W = 12{,}278 \times 10^{10} \frac{\text{revolutions inch-pounds}}{\text{inch-hour}}$$

B_W is a constant since the operating conditions are the same for all three applications. Therefore, for the 13-mm (½-in) seal

$$t = \frac{(12{,}278 \times 10^{10})(0.500)}{(4200 \times 60)(2\pi \times 4200 \times 60)}(0.22)$$

$$t = 699 \text{ h}$$

For the 44.5-mm (1¾-in) seal,

$$t = \frac{(12{,}278 \times 10^{10})(1.75)}{(3450 \times 60)(2\pi \times 3450 \times 60)}(2.3)$$

$$t = 362 \text{ h}$$

Here is a third example: For the 76.2-mm (3.00-in) seal of the previous example, it is necessary to set up short-term quality audit tests. It has been determined that 100-h tests can be economically justified. What shaft speed should be used (assuming constant seal torque to achieve a mean life of 100 h?

$$n^2 = \frac{(12{,}278 \times 10^{10})(3.00)}{(100)(2\pi)(3.7)}$$

$$n^2 = 15.8517 \times 10^{10}$$

$$n = 398143 \text{ revolutions per hour}$$

or

$$n = 6636 \text{ rpm}$$

It is important to understand from Eq. (9.13) that seal factor B_W, when multiplied by the shaft diameter, is the total work W_T the seal can absorb in a lifetime at shaft speed n. Figure 9.82 illustrates a typical life relationship for three different shaft sizes. The measured torque after 1 test hour has been a reasonable time value to use when making relative life comparisons.

The life equation will predict the effect of shaft diameter, shaft speed, and frictional torque upon seal life. However, this expression alone will not predict the effect of changing fluids, sump temperature, seal design, seal materials, or seal pressure. The work function B_W can be modified by these parameters and, therefore, they either must be controlled or accounted for to make the testing meaningful. For example, $B_W = B_L(C_F C_S C_M C_T C_p C_E)$ shows some of the major variables that could affect the test results if not controlled,

where B_W = oil seal life factor in desired application
B = seal life factor in a given application
C_F = fluid correction factor
C_S = seal design correction factor
C_M = material correction factor
C_T = fluid temperature correction factor
C_p = pressure correction factor
C_E = eccentricity correction factor

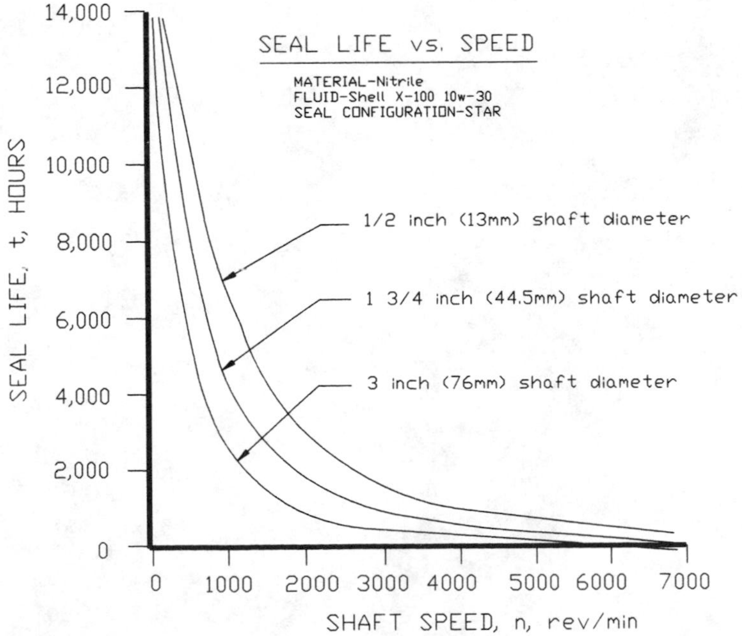

FIGURE 9.82 Life decreases rapidly with speed.

The above information is not presented to discourage one from using the life equation but rather to make one cognizant of other factors that can affect the seal's performance if they are not controlled.

SEALING SYSTEM FAILURE ANALYSIS

The radial lip seal is only one part of the sealing system. The shaft (or running surface) and the bore are other components of the sealing system that are equally as important as the seal itself. It is not possible to provide an accurate sealing system failure analysis unless all three components of the system are studied. Symptoms, probable causes for failure, and recommended corrective action, appear in Table 9.21. The basic steps in conducting an analysis appear in Table 9.22. A comprehensive study has been prepared by the RMA which has descriptions and pictures of clues to look for when conducting a detailed analysis.[62] The SAE manual of the Fluid Sealing Committee has also prepared a *Radial Lip Seal Handbook* that concerns itself with some of the practical problems encountered by the seal user.[63]

TABLE 9.21 Sealing System Failure Analysis

Symptom	Possible cause	Corrective action
1. Early lip leakage	a. Nicks, tears or cuts in seal lip	Examine shaft. Eliminate burrs and sharp edges. Use correct mounting tools to protect seal lip from splines, keyways, or sharp shoulders. Handle seals with care. Keep seals packaged in storage and in transit.
	b. Rough shaft	Finish shaft to 10–20 μin RA or smoother.
	c. Scratches or nicks on surface	Protect shaft after finishing.
	d. Lead on shaft	Plunge-grind shaft surface.
	e. Excessive shaft whip or runout	Locate seal close to bearings. Ensure good, accurate machining practices.
	f. Cocked seal	Use correct mounting tools and procedures.
	g. Paint on shaft or seal element	Mask seal and adjacent shaft before painting.
	h. Turned-under lip	Check shaft chamfer for roughness. Machine chamfer to 32 μin RA or smoother, blend into shaft surface. Check shaft chamfer for steepness. Use recommended lead chamfer. Use correct mounting tools and procedures.
	i. Damaged or "popped out" spring	Use correct mounting tools and procedures to apply press-fit force uniformly. Protect seals in storage and transit.
	j. Damaged or distorted case	Use correct mounting tools and procedures to apply press-fit force uniformly. Protect seals in storage and transit.
	k. O.D. sealant on shaft or lip element	Use care in applying O.D. sealant. Purchase precoated seals.
2. Lip leakage, intermediate life	a. Excessive lip wear	Check seal cavity for excessive pressure. Provide vents to reduce pressure. Provide proper lubrication for seal. Check shaft finish. Make sure finish is 10–20 μin RA.
	b. Element hardening and cracking	Reduce sump temperature if possible. Upgrade seal material. Provide proper lubrication for seal. Change oil frequently. Change seal design.
	c. Element corrosion and reversion	Check material-lubricant compatibility. Change material.
	d. Excessive shaft wear	Check shaft hardness. Harden to Rockwell C30 minimum. Change oil frequently to remove contaminants. Use dust lip in dirty atmosphere.
3. O.D. leakage	a. Scored seal O.D.	Check housing machining. Use 125 μin RA. Check edges on housing bore. Use recommended chamfer. Remove burrs.
	b. Damaged seal case	Use correct mounting tools and procedures to apply press-fit uniformly. Protect seals in storage and transit.

TABLE 9.22 Basic Steps in Analyzing Sealing System Failures

Step 1 Inspect the seal application before removal			
Amount of leakage	☐ Slight	☐ Seal area damp	☐ Heavy leakage
Condition of area	☐ Clean	☐ Dusty	☐ Mud packed
Leakage source	☐ Between lip & shaft ☐ At retainer gasket ☐ At retainer bolt holes	☐ Between O.D. & bore ☐ Between elements of seal ☐ Between wear sleeve & shaft	
Step 2 Wipe area clean & inspect			
Check conditions found	☐ Nicks on bore chamfer ☐ Seal cocked in bore ☐ Seal installed improperly ☐ Shaft to bore misalignment	☐ Seal loose in bore ☐ Seal case deformed ☐ Paint spray on seal ☐ Other ________	
Step 3 Rotate shaft if possible			
Check conditions	☐ Excessive end play	☐ Excessive runout	
Step 4 If the location of leakage cannot be confirmed at this point, either introduce ultraviolet dye into the sump or spray area with white powder, operate for 15 min, and check for leakage with ultraviolet or regular light.			
Step 5 Mark the seal at the 12 o'clock position & remove it carefully			
	☐ Retain an oil sample		
Step 6 Inspect the application with seal removed			
Check conditions found	☐ Rough bore surface ☐ Shaft clean ☐ Coked lube on shaft ☐ Shaft damaged	☐ Flaws or voids in bore ☐ Shaft corroded ☐ Shaft discolored	
Step 7 Inspect the seal			
Primary lip wear	☐ Normal ☐ None	☐ Excessive	☐ Eccentric
Primary lip condition	☐ Normal ☐ Soft (flexible)	☐ Damaged	☐ Hardened (stiff)
Seal O.D.	☐ Normal	☐ Axial scratches	☐ Damaged rubber
Spring	☐ In place ☐ Corroded	☐ Missing	☐ Separated

REFERENCES

1. Brink, R. V., J. Brady, R. Daly, W. W. Rasmussen, "Seal Technology Assures Reliability," ASME Annual Meeting, November 1967 (WA/LUB-13).
2. Horve, L. A., "Achieving Dimensional Control with Molded Lip Seals," ASLE, May 1975 (75AM-7B-1).
3. OS-2, RMA Technical Bulletin, "Alpha Beta 'R' Method for Defining Oil Seal Cross-Sections," Rubber Manufacturers Association, Washington, D.C.
4. SAE J111C, "Terminology of Radial Lip Seals," Society of Automotive Engineers, Warrendale, PA.
5. Brink, R. V., "Oil Seal Life—Good Loading or Good Luck," SAE 650656 1965 Annual Meeting.

6. Brink, R. V., and Leslie A. Horve, "Wave Seals—A Solution to the Hydrodynamic Compromise," *ASLE Lubrication Engineering,* 29, 6, 265-270 (1973).

7. Jackowski, R. A., R. Keller, and D. Strubel, "Advanced Elastomeric Seal Design and Material for Automotive Crankshaft Applications," SAE 87057, February 1987.

8. OS-14, RMA Technical Bulletin, "Recommended Practices for Measuring Thickness and Bond Strength for O.D. Coatings for Radial Lip Type Oil Seals," Rubber Manufacturers Association, Washington, D.C.

9. Peisker, Glenn W., "A Unitized Heavy Duty Axle Oil Seal," SAE 852348, December 1985.

10. Jackowski, R. A., and J. B. Wagner, "Unitized Bonded PTFE Wafer Radial Lip Seal," SAE 880302, February 1988.

11. Mims, S., "Development of an Inner Rack Seal with External Dirt Lip and Integral Oil Contaminant Filter," SAE 900332, February 1990.

12. Mims, S., "Development of an Engineering Thermoplastic Retainer and Rear Crankshaft Seal Assembly for Automotive Applications," SAE 890663, February 1989.

13. Horve, L. A., "The Effect of Humidity upon Polyacrylate Radial Lip Seal Parameters," SAE 840184, February 1984.

14. SAE J946, "Application Guide to Radial Lip Seals," August 1989.

15. RMA Technical Bulletin, *Oil Seal Radial Load,* September 1970.

16. RMA OS-1, "Shaft Finishing Techniques for Rotating Shaft Seals," 1985.

17. RMA OS-4, "Application Guide for Radial Lip Type Shaft Seals," 1984.

18. RMA OS-5, "Garter Springs for Radial Lip Seals," 1984.

19. RMA OS-7, "Storage and Handling Guide for Radial Lip Type Shaft Seals," 1973.

20. ISO 6194/1, "Rotary Shaft Lip Type Seals—Part 1: Nominal Dimensions and Tolerances," 1982.

21. ISO 6194/3, "Rotary Shaft Lip Type Seals—Part 3: Storage, Handling and Installation," 1988.

22. Lein, J., "Mechanical Investigation of Oil Seals for Rotating Shafts," *Konstruction Magazine,* Karlsruhe, Germany, October 1954.

23. Jagger, E. T., "Study of the Lubrication of Synthetic Rubber Rotary Shaft Seals," *Proceedings Conference Lubrication & Wear,* 1957, #409.

24. Jagger, E., "Rotary Shaft Seals—The Sealing Mechanism of Synthetic Rubber Seals Running at Atmospheric Pressure," *Proceedings of the Institute of Mechanical Engrs.,* 1957, #171.

25. Brink, R. V., and W. W. Rasmussen, "The Influence of Oil Seal Design Parameters on Lip Opening Pressure," SAE 895A, September 1964.

26. Brink, R. V., and D. E. Czernik, "Research in Engine Sealing Products," *Diesel & Gas Turbine Progress,* April 1966.

27. Brink, R. V., "The Working Life of a Seal (An Elementary Theory)," *Lubrication Engineering* (ASLE) paper presented at 25th ASLE Annual Meeting in Chicago, May 4–8, 1970, pp. 375–380.

28. Hirano, F., and H. Ishiwata, "The Lubricating Condition of a Lip Seal," *Proceedings of the Institute of Mechanical Engineers,"* 1965–66 (Pt. 3B), #187.

29. Iny, E. H., and A. Cameron, "The Load Carrying Capacity of Synthetic Rubber Rotary Shaft Seals," *Proceedings of the 1st International Conference on Fluid Sealing,* BHRA, 1961, #C1.

30. Rajakovics, G. E., "On the Sealing Mechanism of Fluid Seals," *Proceedings 5th International Conference on Fluid Sealing,* BHRAA, 1971, Paper A6.

31. Jagger, E. T., and D. Wallace, "Further Experiments on the Sealing Mechanism of a

Synthetic Rubber Lip Type Seal Operating on a Rotary Shaft," *Proceedings Institute of Mechanical Engineers,* 1973, 127, 29.

32. Dega, R. L., General Motors Seal Symposium, 1964.
33. Dega, R. L., "Zero Leakage, Results of an Advanced Lip Seal Technology," *Journal of Lubrication Engineering Paper* WA/LUB-11, 1967.
34. Johnston, P. E., "Using the Frictional Torque of Rotary Shaft Seals to Estimate the Film Parameters and the Elastomer Surface Characteristics," *8th International Conference on Fluid Sealing* (BHRA), September 1978.
35. Kawahara, Y., and H. Hirabayashi, "An Analysis of Sealing Phenomena on Oil Seals," *Transactions ASLE,* vol. 23 (1978) 1.
36. Nakamura, K., and Y. Kawahara, "An Investigation of Sealing Properties of Lip Seals through Observations of Sealing Surfaces under Dynamic Conditions," *Procedures of the 10th International Conference on Fluid Sealing,* BHRA, 1984, Paper C1.
37. Horve, L. A., "The Correlation of Rotary Shaft Radial Lip Seal Service Reliability and Pumping Ability to Wear Track Roughness and Microasperity Formation," SAE 910530, February 1991.
38. Muller, H. K., and G. W. Ott, "Dynamic Sealing Mechanism of Rubber Rotary Shaft Seals," *Procedures, 10th International Conference on Fluid Sealing,* BHRA, 1984, Paper K3.
39. Kammuller, M., "Zur Abdichtwirkung von Radial-Wellendichtringen," Thesis 1986, Univ. of Stuttgart.
40. Muller, H. K., "Concepts of Sealing Mechanism of Rubber Lip Type Rotary Shaft Seals," *11th International Conference on Fluid Sealing,* BHRA, April 1987, Paper K1.
41. Gowlinski, M. J., "Lip Motion and Its Consequences in Oil Lip Seal Operation," *Proceedings, 9th International Conference on Fluid Sealing,* BHRA, 1981, Paper D2.
42. Gawlinski, K., "Dynamic Analysis of Oil Lip Seals," *10th International Conference on Fluid Sealing,* BHRA, April 1984.
43. Horve, L. A., "A Macroscopic View of the Sealing Phenomenon for Radial Lip Oil Seals," *11th International Conference on Fluid Sealing,* BHRA, April 1987, Paper K2.
44. Prati, E., "A Theoretical-Experimental Method for Analyzing the Dynamic Behavior of Elastomeric Lip Seals," ACHS Meeting, New York, April 1986.
45. Fazekas, G. A., and M. S. Kalsi, "Feasibility Study of a Slanted O-Ring as a High Pressure Rotary Seal," ASME 72WA/DE/4, November 1972.
46. Martini, L. J., *Practical Seal Design,* Marcel Dekker, New York and Basel, 1984.
47. Hermann, W., and H. Seffler, "New Knowledge of the Sealing Mechanism of Radial Shaft Sealing Rings," *Auto. Tech. Zeitschrift,* 87, no. 9, 1985.
48. Brink. R. V., "Shaft Seal Friction," *National Conference on Fluid Power,* 1974.
49. Brink, R. V., "The Heat Load of an Oil Seal," *6th Int. Conf. on Fluid Sealing,* Munich, Germany, BHRA, February 1973.
50. SAE J110, Testing of Radial Lip Seals," June 1985.
51. ISO 6194-4, "Rotary Shaft Lip Type Seals—Performance Test Procedures."
52. OS-18, Rubber Manufacturing Association Technical Bulletin, "Environmental Exclusion Test Procedures for Radial Shaft Seals," 1990.
53. Horve, L. A., "Statistical Interpretation of Shaft Seal Performance," SAE 741044, October 1974.
54. Horve, L. A., "The Calculation of Shaft Seal Steady State Radial Loads with Non-Uniform Cross-Section and Local Stretch Forces Included," *Transactions of the ASCF,* 13, 288–294, October 1970.
55. Sahiouni, J., "Finite Element Analysis as a Design Tool in the Radial Lip Seal Industry," SAE 900341, March 1990.

56. OS-6, Rubber Manufacturers Association Technical Bulletin, "Radial Force Measurement," 1984.

57. Horve, L. A., "The Effect of Operating Parameters upon Radial Lip Seal Performance," SAE 841145, October 1984.

58. Horve, L. A., "The Operation of Elastomeric Radial Lip Seals at High Temperatures," SAE 750810, September 1985.

59. Vassmer, G., "The Effect of Lubricant Level on Radial Lip Seals," SAE 890611, March 1989.

60. Horve, L. A. "Evaluating Dust Exclusion for Rotating Shaft Seal Applications," SAE 780402, March 1978.

61. Peisker, G., and J. Sahiouni, "Performance of Various Shaft Seal Dust Lips in a Dust Environment and Their Ability to Form a Vacuum Between the Primary and Secondary Lips," SAE 891853, September 1989.

62. OS-17, Rubber Manufacturers Association Technical Bulletin, "Sealing System Leakage Analysis Guide," 1990.

63. *SAE Fluid Sealing Handbook—Radial Lip Seals,* HS J1417, 1982. Prepared by members of the Fluid Sealing Manual Committee, R. V. Brink, chairman.

TABLE OF NOMENCLATURE*

α	Flow resistance coefficient
α_h	Helix angle, degrees
β	Velocity ratio (V_s/V_c)
β_{hs}	Hydrostatic pressure factor
$(\beta_{hs})_{CRIT}$	Critical hydrostatic pressure factor
ΔP	Pressure drop, g/cm^2
ΔP_c	Pressure gradient generated by centrifugal force, g/cm^2
$(\Delta P_c)_{CRIT}$	Critical pressure gradient generated by centrifugal force, g/cm^2
γ	Pressure ratio (P_o/P_i)
ρ	Fluid density, g/cm^3
ρ_i	Upstream (system) gas density, g/cm^3
ρ_o	Downstream (outlet) gas density, g/cm^3
θ	Angle, degrees
μ	Coefficient of friction
ν	Kinematic viscosity, cm^2/s
π	Pi = 3.14159
a	Axial land width, cm
A	Ratio defined to be $b/(a + b)$
A_f	Interface area, mm^2
b	Axial groove width, cm
B	Ratio defined to be $(h + hg)/h$

*Because the alphabet is limited, the same letter might be used to represent more than one concept. Because each symbol is defined when first used, confusion should not occur.

B_a	Balance ratio
B_e	Balance ratio (externally pressurized)
B_i	Balance ratio (internally pressurized)
B_L	Seal life factor in a specific application
B_w	Seal life factor, watt · revolutions/mm (hp · revolutions/in)
D_{ave}	Average face diameter, mm (in)
D_b	Bore diameter, cm (in)
D_{fi}	Inside interface diameter, mm (in)
D_{fo}	Outside interface diameter, mm (in)
D_o	Seal outer diameter, cm (in)
D_s	Shaft diameter (effective sealing diameter), cm (in)
D_{sb}	Effective sealing diameter for bellows, mm (in)
e	Shaft eccentricity, mm (in)
f	Fluid resistance coefficient
F_{cl}	Closing force, N (lb)
F_f	Friction force, N (lb)
F_h	Hydraulic force, N (lb)
F_{hs}	Hydrostatic force, N (lb)
F_{net}	Net force, N (lb)
F_s	Spring force, N (lb)
g	Gravitational constant, 980 cm/s^2 (32.17 ft/s^3)
G	Duty parameter ($Z_{cp}N/P$)
G_{LC}	Leakage rate, laminar compressible fluid (concentric), g/s (oz/s)
G_{LEC}	Leakage rate, laminar compressible fluid (eccentric), g/s (oz/s)
G_{TC}	Leakage rate, turbulent compressible fluid (concentric), g/s (oz/s)
G_{TEC}	Leakage rate, turbulent compressible fluid (eccentric), g/s (oz/s)
G_s	Flow rate of stream, g/s (oz/s)
h	Radial clearance, cm (in)
h_f	Film thickness, cm (in)
h_g	Groove depth, cm (in)
K	Exponent of adiabatic expansion for gas
K_T	Tooth constant for labyrinth
L	Bushing length, cm (in)
m	Multi ring factor
n	Operating speed, rpm or rph
n_e	Ratio (eccentricity/clearance) e/h
N	Force, N (lb)
N	Operating speed, rpm or rph
N_s	Number of stages in labyrinth
N_T	Total number of lifetime revolutions or cycles
p	Incremental radial force, N (lb)

p	Pressure at radial position *r*, g/cm^2 (oz/in^2 or psi)
P	Average underlip pressure, g/cm^2 (oz/in^2 or psi)
P_D	Split shaft load measurement, N (lb)
P_i	Upstream (system) pressure, g/cm^2 (oz/in^2 or psi)
P_o	Downstream (outlet) pressure, g/cm^2 (oz/in^2 or psi)
P_π	Total radial load, N (lb)
Pow	Power, watts
P_s	Standard pressure, g/cm^2 (psi)
P_V	Pressure velocity factor, MPa m/s
Q	Flow rate, cm^3/s (in^3/s)
Q_L	Laminar flow rate (concentric), cm^3/s (in^3/s)
Q_{LE}	Laminar flow rate (eccentric), cm^3/s (in^3/s)
Q_T	Turbulent flow rate (concentric), cm^3/s (in^3/s)
$R_1 Q_{TE}$	Turbulent flow rate (eccentric), cm^3/s (in^3/s)
$R_2 R_e$	Reynolds number (vh/v, $Q/2\pi Rv$)
R_L	Radial force per unit of shaft circumference, N/cm (lb/in)
R_m	Mean radius of annulus, cm (in)
R_r	Ring resistance factor (ML/h)
R_s	Shaft radius, cm (in)
t	Seal life, h
T	Total radial force of seal, N (lb)
T_i	System temperature, kelvin
T_s	Standard temperature, kelvin
T_L	Seal life, h
torque	Seal torque, N · m (lb)
V	Linear velocity, cm/s (in/s)
V_c	Critical fluid velocity, cm/s (in/s)
V_i	Specific volume of steam, cm^3/g (in^3/lb)
V_s	Linear shaft velocity, cm/s (in^3/s)
W	Seal width, cm (in)
W_r	Work rate
W_T	Total work, N · m (lb · ft)
Z_{cp}	Absolute fluid viscosity, gm · s/cm^2 (lb · s/in^2)

P • A • R • T • 3

FLUIDS AND LUBRICANTS

CHAPTER 10
LUBRICANTS

LUBRICATION THEORY

One of the functions of a seal is to retain lubricants. These lubricants have different formulations to perform different functions for many applications of modern machinery. These fluids also lubricate the seals used in the applications and may react adversely with the elastomeric materials used in the seal. It is important for the seal designer and the end user to have a basic understanding of lubricants and lubricant theory to assist in the selection of the proper seal design and material. The seal and lubricant are exposed to a variety of operating conditions that result in several levels of seal lubrication. These lubrication regimes depend on the application load (P), operating speed (N), temperature, and resistance of the lubricant to flow (viscosity, Z_{cp}). They can be defined by studying a Streibeck curve, which is obtained by plotting the coefficient of friction (μ) versus the dimensionless duty parameter G ($Z_{cp}N/P$). The lubricating film thickness also changes with G. The coefficient of friction is determined by measuring torque. The fluid viscosity must be measured at the application temperature since the viscosity of most lubricants decreases with increasing temperature (see Fig. 10.1).

In the full-film lubrication regime (Fig. 10.2), the oil film is thick enough to separate the moving surfaces completely. Friction in this region is linearly dependent on $Z_{cp}N/P$. Low pressure is generated in the contact zone. To maintain full-film lubrication, it is necessary to maximize speed and lubricant viscosity while minimizing load. This is usually not practical with most machine operations. As a result, full-film lubrication is rarely obtained.

In contrast to full-film lubrication, lubricant viscosity in the boundary lubrication regime plays little or no part in defining the friction coefficient. Most of the applied load is carried by microscopic peaks on the surfaces of gears and bearings, called asperities, rather than as a continuous oil film. Under conditions of shock loading, low speed, high torque, and critically low viscosity, the full oil film fails to support the load and direct contact between moving surfaces occurs. Improperly formulated lubricants allow rapid wear under these conditions. Chemically reactive boundary or extreme pressure (EP) additives are believed to prevent wear by forming metal-salt films that have low shear strength but a high melting point.

The mixed film lubrication regime falls between the two extremes of hydrodynamic and boundary lubrication. The load is partially supported by contacting

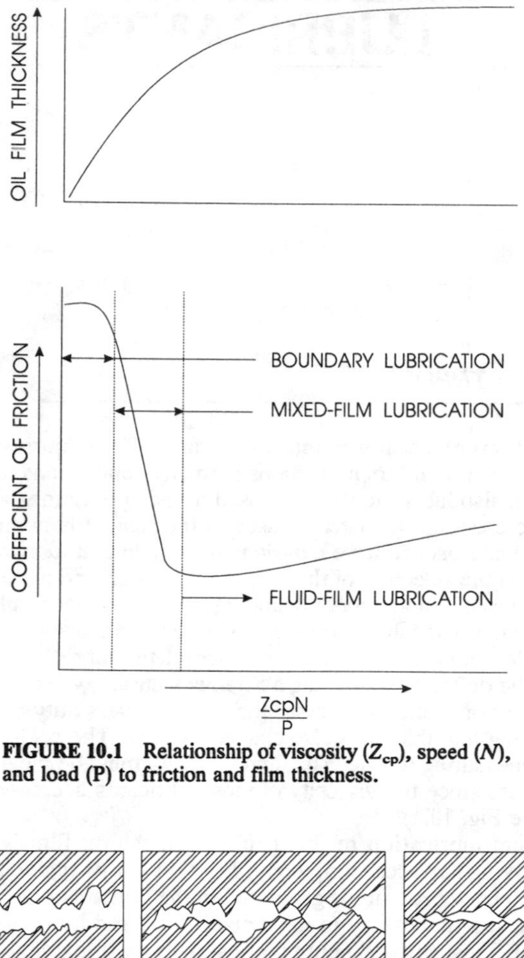

FIGURE 10.1 Relationship of viscosity (Z_{cp}), speed (N), and load (P) to friction and film thickness.

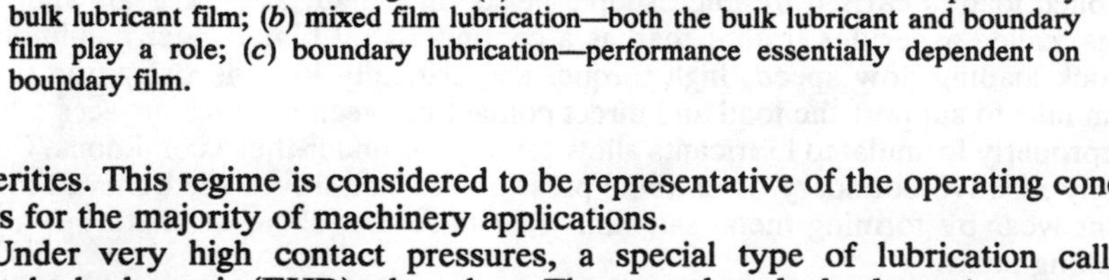

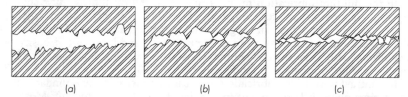

FIGURE 10.2 Lubrication regimes. (*a*) Fluid film lubrication—surfaces separated by bulk lubricant film; (*b*) mixed film lubrication—both the bulk lubricant and boundary film play a role; (*c*) boundary lubrication—performance essentially dependent on boundary film.

asperities. This regime is considered to be representative of the operating conditions for the majority of machinery applications.

Under very high contact pressures, a special type of lubrication called elastohydrodynamic (EHD) takes place. The term *elastohydrodynamic* means deformation without permanent distortion (elasto-) in a full film (hydrodynamic) regime of lubrication. EHD lubrication is believed to occur commonly during meshing gear operations. As the lubricant is exposed to high contact pressures, it undergoes a rapid viscosity increase. This thin, high-viscosity film maintains a separation of the mating surfaces and prevents metal to metal contact. The EHD film actually becomes more rigid than the metal and causes elastic deformation of the metal surfaces.

Hypoid gears have high relative sliding velocities. As a result, metal to metal contact is more prevalent with hypoid gears than other gear types. The high-viscosity EHD film is sheared and heat is generated, which causes a decrease in the mean film viscosity, which reduces film thickness. The reduced film thickness allows increased metal to metal contact, often resulting in scoring of the gears. Failure can be prevented with oil additives of sufficient quality.

BASE OILS

The performance characteristics of a lubricant depend on the physical and chemical characteristics of that fluid. In turn, the physical and chemical characteristics of a lubricant depend on the properties of the base oil used and the additives employed. A number of tests are used to determine the properties of the base oil and thus help to characterize the lubricant. These tests are summarized in Table 10.1.

The four constituents of a base oil determined by the clay-silica gel column test are asphaltenes, polars, aromatics, and saturates. Asphaltenes are very reactive, resinous-type materials having very poor stability. Modern base oils generally do not contain significant quantities of asphaltenes. Polars are hetrocyclic

TABLE 10.1 Measurement of Base Oil Properties

Oil property	Description and units	Recommended test
Viscosity (resistance to flow)		
Kinematic	Centistokes (mm^2/s)	ASTM D445
Saybolt universal second (SUS)	Empirical unit: time in seconds to flow through standard orifice	
Viscosity index	Empirical dimensionless unit: Measures the effect of temperature change by comparison to standard oils	ASTM D2270
Gravity or density	Mass per unit volume of material	ASTM D287 or D1298
API gravity	(141.5/specific gravity) − 131.5	
Viscosity gravity constant (VGC)	Formula calculation—removes viscosity variable	ASTM 2501
Flash point	Temperature where open flame will ignite oil vapors	ASTM D92 and D93
Pour point	Temperature at which oil can no longer be poured	ASTM D99
Color	Color compared to standards	ASTM D1500
Refractive index	Ratio of the velocity of specified wave length of light in air to the velocity in the oil	ASTM D1218
Analine point	Temperature where a 50/50 mixture of oil and aniline form a single phase	ASTM D611
Molecular analysis	Characterizes oil in terms of four constituents using clay-silica gel columns	ASTM 2007

molecules with either a sulfur, nitrogen, or oxygen molecule replacing one of the carbon atoms of the ring system. The ring systems containing a polar molecule tend to be aromatic and are largely removed from the oil in the solvent extraction test.

High-viscosity oils usually contain a high number of polar molecules. Aromatic molecules contain multiple unsaturated rings and are volatile. Saturates have saturated rings (napthlenes) and varying degrees of side chains called paraffins. These materials are not absorbed by the test columns. Since these molecules are saturated and thus relatively inert, they are stable to heat and light degradation.

Base oil selection for a lubricant depends on the requirements the lubricant must fulfill. It must maintain a lubricating film between moving surfaces to prevent excessive wear, and it must minimize friction and absorb and transmit the heat generated away from the moving surfaces. The lubricant must be stable over a broad range of operating conditions, and it should be cost effective.

The most important oil parameter is the viscosity. High-load applications require high-viscosity oils to maintain an oil film between the moving surfaces. High-viscosity oils require more force to move or shear them, thus more heat is generated. A balance is required to select the lowest oil viscosity that will still provide adequate lubrication. The International Organization for Standardization (ISO) has developed a viscosity classification system for industrial fluid lubricants (see Table 10.2).

The temperature the oil encounters in operation will determine its viscosity. An oil that thins out too much at high temperature will not maintain a safe film. Higher temperatures of operation require a higher-viscosity oil. Prolonged high-temperature operation may cause the oil to thicken and a buildup of carbon and other products may cause bearing and seal failure. Additives are often used to improve stability at high temperatures. Highly saturated oils provide better heat

TABLE 10.2 Viscosity Classification System for Industrial Fluid Lubricants

ISO viscosity grade numbers	Viscosity grade ranges, cSt at 40°C	
	Minimum	Maximum
2	1.98	2.42
3	2.88	3.52
5	4.14	5.06
7	6.12	7.48
10	9.00	11.0
15	13.5	16.5
22	19.8	24.2
32	28.8	35.2
46	41.4	50.6
68	61.2	74.8
100	90.0	110
150	135	165
220	198	242
320	288	252
460	414	506
680	612	748
1000	900	1100
1500	1350	1650

TABLE 10.3 Comparison of Properties for Three Classes of Base Oils

Base oil property	ASTM method	Paraffinic 64742-54-7	Naphthenic 64741-96-4	Aromatic 64742-03
Viscosity, cSt@40°C	D 445	40	42	36
Viscosity, SUS@100°F	D 2161	205	222	192
Viscosity, cSt@100°C	D 445	6.2	6.2	4.0
Viscosity, SUS@210°F	D 2161	46.9	46.9	40.0
Viscosity index	D 2270	100	−15	−185
Specific gravity	D 287	0.8628	0.9100	0.9826
API gravity	D 287	32.5	24.0	12.5
Viscosity-gravity constant		0.807	0.862	0.957
Flash point, COC°C	D 92	229	180	160
Aniline point, °C	D 611	107	78	17
Pour point, °C	D 97	−15	−36	−24
Color	D 1500	L 0.5	2.5	D 8.0
Molecular weight	D 2503	440	320	246
Refractive index	D 1747	1.4755	1.4997	1.5503
Clay-gel analysis	D 2007			
% polars		0.2	2	8
% aromatic		8.5	36	80
% saturates		91.3	62	12

stability, and in extreme cases, synthetic oils may be required. Low-temperature operations require a low viscosity oil and a low pour point to prevent thickening.

Naphthenic oils are often used in metalworking applications because the additives required for the application can readily be dissolved. Many of these additives will not stay in solution with paraffinic oils. If the oil is required to be water soluble, naphthenic oils are preferred since paraffinic oils are not water soluble. Naphthenic oils are made from wax-free crude oils and have better low-temperature properties than paraffinic oils. The highly saturated straight chain characteristics of paraffinic oils are more heat stable than naphthenics, although some highly refined naphthenics are also highly saturated. Paraffinic oils are less volatile than naphthenics; thus losses due to vaporization are low. For those applications that require as little change in viscosity with temperature as possible, the high VIs of paraffinic oils are a necessity.

In general, paraffinic oils cause some seal materials to shrink, which can cause leakage, particularly in O-ring applications. Naphthenic oils will cause the elastomeric materials to swell, which can reduce seal load in radial lip seal applications, which can result in seal leakage. Table 10.3 compares the three classes of base oils.

ADDITIVES

Additives are used in virtually all finished lubricants for every application from the smallest chain saw engine—under 1 hp—to the largest marine engines—up to 50,000 hp. In addition, additives play an important role in industrial oils, transmission fluids, and turbine lubricants.

Lubricant additives are laboratory chemical compounds used at dosages from

TABLE 10.4 Typical Dosages for Additive Packages

0.5–2 wt %	Industrial lubricants
4–7 wt %	Tractor hydraulic fluids and gear oils
3–16 wt %	Automotive engine oil
7–20 wt %	Zinc-free railroad oils
10–30 wt %	Marine cylinder oils

under 1 to over 33 percent, depending on the application (Table 10.4). They are added to refined base oils—generally petroleum oil, but sometimes synthetic fluids—to enhance their properties. In this way, additive-treated lubricants contribute to longer equipment life, lower maintenance, and improved field performance.

Lubricant additives were first used in the mid 1800s to thicken grease. In the early 1900s, the primary use of additives was to prevent oxidation. World War II accelerated the development of additive and lubricant technology. Today throughout the world, over 700 million gallons of additives are used annually in nearly 10 billion gallons of finished lubricant.

The largest end use of additives is in automotive crankcase lubricants, followed by railroad and marine engine oils. Areas of smaller volume include transmission fluids, two-stroke cycle, and industrial oils.

Dispersants

Fuel combustion products create sludge and varnish that contaminate gasoline engine oils and settle out in the engine oil passages and on operating parts until engine function becomes impaired. In low-temperature operations, water condensation increases sludge formation. High-speed, warm weather highway driving creates less sludge in the engine oil than low-speed, cold weather local driving.

Many other oils exposed to high temperatures require dispersing agents to minimize sludging. Ashless dispersants, based on succinimides, function like soap in water to disperse the contaminants and are particularly effective for low-temperature operation where water can condense in machinery cavities. The ashless dispersants have polar chemical heads attached to rather large hydrocarbon groups. The sludge and water become attached to the polar heads and the hydrocarbon groups enable the entire molecule, complete with sludge, to dissolve in the oil. The ashless dispersants are usually made from polybutenes of about 1000 to 10,000 molecular weight, which are connected to a polymine derivative or occasionally to an alcohol group. When a well-dispersed oil is discarded at drain time, virtually all the sludge and other contaminants go with it, leaving the machined surfaces relatively clean.

Detergents

The chemically and thermally deteriorated oil that accumulates in the ring-belt area of an engine piston eventually leaves hard carbonlike deposits behind the piston rings. These deposits are usually found near the top ring, which is hottest, and they interfere with ring function. Detergents are soaplike compounds which, when used with alkaline agents, keep the ring-belt area clean. They function by lifting the carbon deposits from the metal surfaces and holding them in suspension. The most common detergents are alkaline-earth soaps (most often calcium), which are made by treating petroleum lube stocks with sulfuric acid, alkylphenol sulfides, alkysalicylic acid, and alkylphosphonic acids. Detergents are generally available as liquid blends that contain about 50 percent of the active ingredient.

Corrosion Inhibitors

Oxidized oil contains acids which can chemically corrode metal surfaces. Some EP agents can also attack metal if not properly formulated. Corrosion inhibitors can protect either ferrous (iron-containing) or soft metal (copper alloys). Ferrous metal corrosion inhibitors or rust inhibitors coat the surfaces, especially during downtime, to prevent attack by moisture or oxygen. Typical rust inhibitors are basic sulfonates and fatty amines.

Soft metal corrosion inhibitors are highly surface active materials used at very low levels in the performance package. Their limited solubility allows them to coat and protect copper alloys (i.e., bronze) from the effect of acids and active sulfur. Effective compounds are heterocucles containing nitrogen and/or sulfur. Bench tests used to evaluate corrosion protection are ASTM D130 (copper corrosion), ASTM D665 (rust), and ASTM D1748 (humidity chamber).

Oxidation Inhibitors

Oxidation inhibitors prevent breakdown of the base oil, which results in oxidative thickening and sludge formation. Oxygen, especially at elevated temperatures, reacts at selected sites on oil molecules, forming hydroperoxides. These unstable molecules can lead to base oil polymerization, greatly increasing viscosity and preventing flow to critical surfaces. Thick oil can also reduce fuel economy and inhibit engine starting during cold weather. Effective inhibitors include hindered phenols, aromatic amines, some sulfur containing compounds, and zinc dialkyldithiophosphate.

Friction Modifiers

Friction modifiers are added to oils to reduce the static coefficient of friction between contacting metal surfaces. These modifiers have long, straight hydrocarbon chains with a polar end such as an acid, alcohol, amide, or ester. Oleic acid is often used as a friction modifier. These compounds generally do not react but are absorbed or plated onto the metal. This reduces the frictional heating at the oil-metal interface and allows for improved efficiency under boundary lubricating conditions.

The plating process is often enhanced by pressure. In some cases, the modifier will plate out on the shaft surface directly under the loaded portion of the sealing lip. These particles may accelerate lip wear, causing early leakage. Functional testing of seals in fluids with friction modifiers is recommended if no historical data is available.

Friction Reducers

Power loss results from the viscous drag of the lubricant that is contained between moving surfaces. This hydrodynamic drag can be reduced by decreasing the lubricant viscosity without destroying the oil film between the surfaces. Polymers that incur a temporary viscosity loss when the molecules are aligned by shearing are often added to oils to reduce power loss during machinery start-up. The viscosity of the oil and additive package combination will approach that of the base oil which provides the lowest practical viscosity.

Viscosity Improvers (VIs)

Straight mineral oils are not recommended for a broad temperature range since its viscosity changes dramatically with temperature. Low viscosity at high temperatures results in a thin unstable oil film between the moving surfaces, and high viscosity at low temperatures results in high drag and power consumption. Oil-soluble polymers are added to a low-viscosity base oil to change the shape of the viscosity-temperature curve. At low temperatures the polymer molecules are in a coiled configuration contributing very little to oil viscosity. At high temperatures, the configuration is more open and linear and makes a significant contribution to the solution viscosity.

Some polymer types that are used include: polymethacrylate esters, hydrogenated styrene-isoprene copolymers, hydrogenated styrene-butadiene copolymers, and ethylene-propylene copolymers (OCP). These polymers are dissolved in a low-viscosity oil to form a solution with 10 to 30 percent polymer. Generally, about 1 to 5 percent of undiluted polymer or about 10 percent of diluted polymer is required to make a multigrade oil such as 10W30 (Fig. 10.3).

Pour Point Depressants (PPDs)

Although most of the wax is removed from base oils during refining, enough remains to impede the flow since large lacy crystals are formed at low tempera-

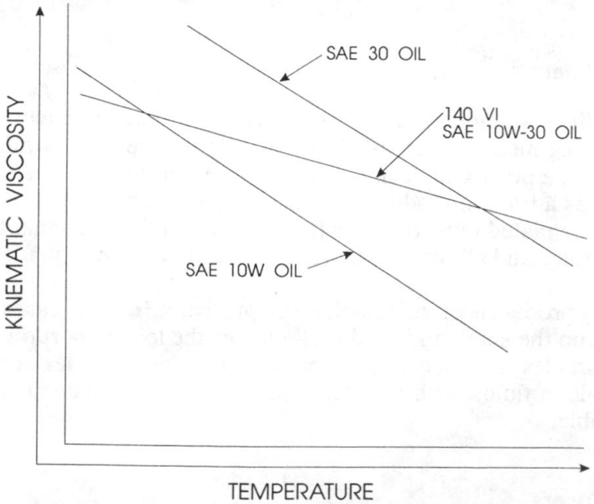

FIGURE 10.3 Typical viscosity temperature relationship for SAE 10W30 oil.

tures. This results in starting difficulties for machinery in cold weather. Fortunately, the crystallization of wax can be modified to alleviate this problem to an appreciable extent. This is accomplished by adding small amounts of materials which are not waxes but are polymers with waxlike segments in their structures. These pseudo-waxes co-crystallize with the wax and form structures that improve flow. Wax modifiers must come out of solution at the same temperature the wax crystals are forming.

Polymethyacrylate esters of waxy alcohols and also alkylated naphthalene are typically used as wax modifiers. They are dissolved in light oil, and the resulting solutions contain about 50 percent active product. In the finished oil, 0.5 to 1.0 percent of the diluted product may lower the pour point (flow temperature) of most finished oils by 10 to 20°C.

Antiwear Agents

Zinc dialkydithiosphates (ZDTPs) are used to inhibit wear of moving surfaces in boundary lubrication. The ZDTPs break down at high temperatures and deposit films on the metallic surfaces that provide lubrication. Modern engines require relatively unstable ZDTPs for emission systems and valve train wear protection. More stable ZDTPs are needed to prevent ring sticking in fastburn engines that have higher piston temperatures. These conflicting needs often result in formulations containing more than one kind of ZDTP.

Extreme Pressure Agents

These compounds generally contain phosphorus, sulfur, chlorine, or boron, and they react with the metal at elevated pressure and temperatures. They are believed to form metal-salt films which shear more easily than the metal alone to prevent damaging stress and wear. The effectiveness of an EP agent depends on its temperature of activation. Since activation temperatures vary, combinations of EP agents are used to provide a wider range of protection. EP agents are found in most gear oils.

Antifoaming Agents

The churning action that occurs in machinery can cause air entrainment, or foaming, in the oil. Foaming can result in poor lubrication of moving surfaces and loss of lubricant through vent or filler tubes. Foam inhibitors, or antifoaming agents, reduce the surface tension of the oil and allow the air bubbles to collapse. Common foam inhibitors are polyacrylates and polydimethysiloxanes (silicones).

Seal Compatibility and Seal Swell Additive

Additives that improve lubricant properties may often be incompatible with the seal material. For example, disulfide additives give lubricants antiwear properties, but they also cure or harden the sealing element. Many of the additives in EP lubricants become chemically more active as they heat up, affecting the rubber far more at 110°C (230°F) than at 82°C (180°F). Some seal materials revert and fall apart. When the sealing member softens excessively with use, or whenever there is not enough frictional heat present in the application to explain the excessive hardening of the seal element, the lubricant and seal material are probably chemically incompatible. The remedy lies in proper seal and fluid selection.

Additives can also affect other parts of the sealing system. Radial lip seals made with NBR had equivalent life curves (Fig. 10.4) when run under similar conditions in engine oil and gear oil with EP additives. The shaft wear rate was dramatically different. At 93°C (200°F) the shaft wear rate in the EP gear lube was about double the wear rate observed when engine oil was used. Increasing temperature to 107°C (225°F) did not change the shaft wear rate in engine oil, but the shaft wear rate in gear oil increased by a factor of 4 (Fig. 10.5). The shaft wear in the gear oil is accelerated by the EP additives that cause a chemical etching process. These additives are not present in the engine oil.

Nitrile rubbers are sensitive to fluid composition and tend to shrink in the presence of an all-paraffinic base oil. This has led to the use of special seal swell additives in ATF which are designed to give a small amount of swell to nitrile rubber and keep it soft and pliable. These additives may be phosphate esters, phenols, aromatic esters, sulfones, or lactones. Polyacrylic and silicone elastomers are also used in transmission application, and they will swell in paraffinic oils. Bench tests are usually run with these materials to make sure that additive components in the ATF have no detrimental effect on the oil seals. Full-scale transmission tests permit another examination of fluid and seal compatibility. Any detrimental result would eliminate that fluid and/or the seal material from commercial acceptance.

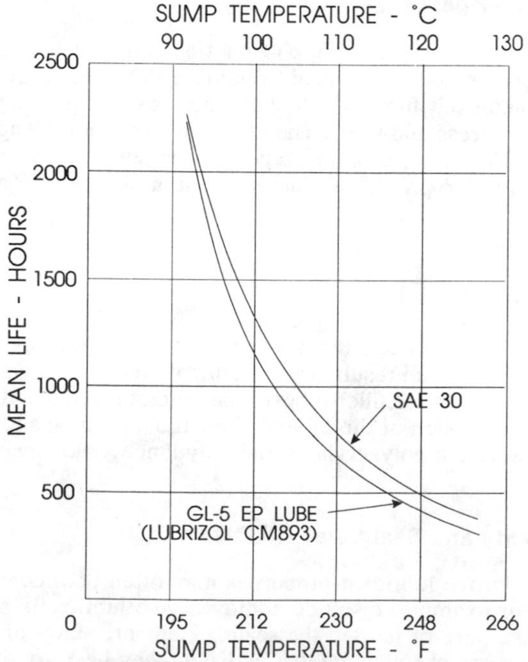

FIGURE 10.4 Seal life versus fluid type.

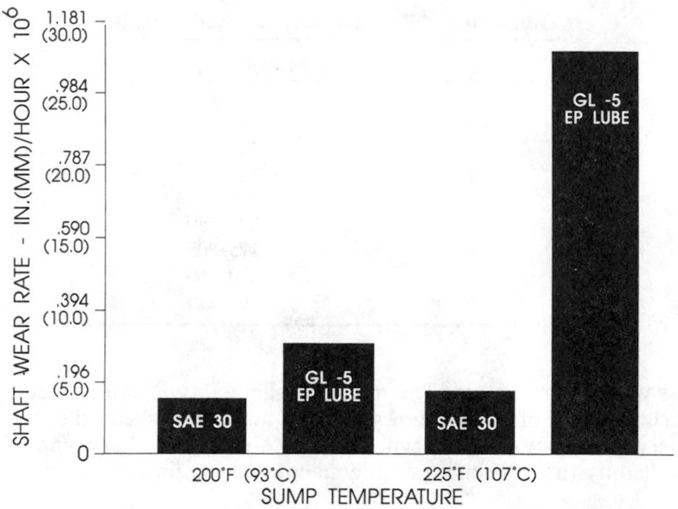

FIGURE 10.5 Shaft wear rate versus fluid type.

LUBRICATING GREASES

Lubricating grease is defined by the American Society for Testing and Materials (ASTM) and The National Lubricating Grease Institute (NLGI) as a solid to semisolid dispersion of a thickening agent in a liquid lubricant. Other ingredients in the form of an additive package may be added to improve performance. The thickener is usually a soap which is the reaction of a long chain fatty acid and a metallic hydroxide. The lubricant is usually mineral oil, and the additive system may contain oxidation inhibitors, wear inhibitors, and other chemicals.

The high-viscosity inherent with grease reduces runoff that results if oils are used. Grease also provides a barrier to prevent external contaminants from entering the gear box cavity. They prevent corrosion and are also used to fill gear box cavities to dampen noise in certain applications.

Consistency

The most frequently measured property of lubricating greases is the consistency. This is a measure of the hardness as displayed by the depth in tenths of a millimeter that a standard cone will penetrate a sample under given conditions. The NLGI classifies grease in grades dependent on the penetration, from 000 through 6, with the 6 being the hardest (Table 10.5). The most commonly used grades are NLGI 1, 2, and 3. For a given soap type, the consistency is influenced most by the thickener content. More thickener provides a heavier grease with lower penetration.

Mechanical Stability

The mechanical stability of a grease may be significant in an application. This is a measure of the ability of the grease to resist changes in consistency as the

TABLE 10.5 NLGI Consistency Grade Numbers

NLGI no.	ASTM worked penetration
000	445–475
00	400–430
0	355–385
1	310–340
2	265–295
3	220–250
4	175–205
5	130–160
6	85–115

grease is worked or sheared. The mechanical stability is determined by comparing the consistency of a sample of grease that has been sheared extensively with the initial consistency. Greases will generally soften with shear, and this property is controlled by the thickener and the manufacturing process used.

Drop Point

The dropping point of a grease is defined as the temperature at which the thickener transforms from a solid to a liquid. The dropping point for calcium grease is about 93°C (200°F). Lithium 12-hydroxystearate, one of the most common greases, will exhibit drop points near 190°C (375°F).

Additives

It may be desirable to incorporate oxidation resistance into a lubricating grease if it is to be in service for an extended period of time or if it is to be exposed to high temperatures. Oxidation is a chemical deterioration of the lubricant and is inhibited by adding chemical antioxidants.

Corrosion protection is a desirable property in many applications. Some greases, such as those using sodium-soap thickeners, are soluble in water and prevent corrosion of ferrous materials. Chemicals must be added to water-insoluble greases to achieve the same corrosion resistance. If other metals, such as copper, are present, different chemicals are required to prevent corrosion.

Additives are also used in grease to reduce friction and reduce wear. Metallic or sulfur compounds are often used.

Apparent Viscosity

The final grease property that is sometimes specified is the apparent viscosity. The term *apparent* is used because, since greases are a combination of a solid and a liquid, their behavior is "nonnewtonian." The viscosity of a newtonian fluid (such as water or light oil) is independent of the rate of shear. The viscosity of a nonnewtonian fluid such as grease will vary with shear rate. With no shear applied, the grease is a solid. Under high shear, such as in a high-speed bearing, the viscosity of a grease approaches that of the base fluid. The apparent viscosity of

TABLE 10.6 Relative Importance of Lubricating Grease Properties for Automotive Uses Shown

	Wheel bearings	Univ. joints	Chassis	ELI* chassis	Multipurpose applications
Structural stability (incl. mech. stability)	H	M	L	H	H
High dropping pt. (High-temp. serv.)	H	M	L	M	H
Oxidation resist.	H	M	L	M	H
Protection against friction and wear	M	H	M	H	H
Protection against corrosion	M	M	L	H	M
Protection against washout	M	M	M	H	M

*Extended lubrication interval.
H = highest; M = moderate; L = least.

a grease is affected by the thickener type, thickener content, and base fluid used and can be related to pumpability or the ease with which a grease can flow through dispensing lines.

Application Guide

The Society of Automotive Engineers (SAE) has provided guidelines that show the relative importance of grease properties for specific applications (Table 10.6). A grease application guide prepared by NLGI compares the properties of common greases (Table 10.7).

ENGINE OILS

Engine oils are required to lubricate moving parts; remove soot, varnish, water, and other contaminants from the engine; and prevent combustion gases from escaping through gaps between rings, pistons, and cylinder walls. Consumers expect engine oils to function well in engines manufactured by different companies, in different climates, under different driving conditions, for long periods of time at a low price. These requirements result in engine oils that are especially formulated with specialized ingredients.

Engine Oil Viscosity

The SAE has established a system (SAE J300) of defining and classifying engine oil viscosity to aid engine manufacturers in selecting the right oils for their applications. This system has 10 classifications and is designed to indicate both the high- and low-temperature viscosities for the oil (Table 10.8).

The grades with the letter *W* are often referred to as "winter" grades since

TABLE 10.7 Grease Application Guide

Properties	Aluminum	Sodium	Calcium—conventional	Calcium—anhydrous	Lithium	Aluminum complex	Calcium complex	Lithium complex	Polyurea	Organo-clay
Dropping point, °F	230	325–350	205–220	275–290	350–400	500+	500+	500+	470	500+
Dropping point, °C	110	163–177	96–104	135–143	177–204	260+	260+	260+	243	260+
Max. usable temp, °F	175	250	200	230	275	350	350	350	350	350
Max. usable temp, °C	79	121	93	110	135	177	177	177	177	177
Water resistance	Good to excellent	Poor to fair	Good to excellent	Excellent	Good	Good to excellent	Fair to excellent	Good to excellent	Good to excellent	Fair to excellent
Work stability	Poor	Fair	Fair to good	Good to excellent	Good to excellent	Good to excellent	Fair to good	Good to excellent	Poor to good	Fair to good
Oxidation stability	Excellent	Poor to good	Poor to excellent	Fair to excellent	Fair to excellent	Fair to excellent	Poor to good	Fair to excellent	Good to excellent	Good
Protection against rust	Good to excellent	Good to excellent	Poor to excellent	Poor to excellent	Poor to excellent	Good to excellent	Fair to excellent	Fair to excellent	Fair to excellent	Poor to excellent
Pumpability (in centralized systems)	Poor	Poor to fair	Good to excellent	Fair to excellent	Fair to excellent	Fair to good	Poor to fair	Good to excellent	Good to excellent	Good

Oil separation	Good	Fair to good	Poor to good	Good	Good to excellent	Good to excellent	Good to excellent	Good to excellent	Good to excellent	Good to excellent
Appearance	Smooth & clear	Smooth & fibrous	Smooth & buttery	Smooth & buttery	Smooth & buttery	Smooth & buttery	Smooth & buttery	Smooth & buttery	Smooth & buttery	Smooth & buttery
Other properties		Adhesive & cohesive	EP grades available	EP grades available	EP grades available, reversible	EP grades available, reversible	EP and antiwear inherent	EP grades available	EP grades available	
Production volume and trend*	No change	Declining	Declining	No change	The leader	Increasing	Declining	Increasing	No change	Declining
Principal uses†	Thread lubricants	Rolling contact bearings	General uses for economy	Military multiservice	Multiserv. automotive & industrial	Multiserv. industrial	Multiserv. automotive & industrial	Multiserv. automotive & industrial	Multiserv. automotive & industrial	High temp. (frequent relube)

*Lithium grease over 50 percent of production and all others below 10 percent.
†Multiservice includes rolling contact bearings, plain bearings, and others.

TABLE 10.8 SAE Engine Oil Viscosity Classification

SAE viscosity grade	Maximum viscosity (cP) at temp., °C	Maximum borderline pumping temp., °C	Stable point max., °C	Viscosity (cSt), °C	
				Min.	Max.
0W	3250 at −30	−35		3.8	
5W	3500 at −25	−30	−35	3.8	
10W	3500 at −20	−25	−30	4.1	
15W	3500 at −15	−20		5.6	
20W	4500 at −10	−15		5.6	
25W	6000 at −5	−10		9.3	
20				5.6	<9.3
30				9.3	<12.5
40				12.5	<16.3
50				16.3	<21.9

they are most often recommended for cold temperature operation. These grades are based on a maximum low-temperature viscosity and maximum borderline pumping temperature, as well as a minimum viscosity at 100°C (212°F).

Three test procedures are necessary to define the SAE viscosity grades containing the letter *W*. The low-temperature viscosity test procedure (modified ASTM D2606) uses an apparatus called the cold cranking simulator. Viscosities measured by this method have been found to correlate with the engine speeds developed during low-temperature cranking. It defines the maximum fluid viscosity at a given temperature which allows the engine to crank at a sufficient speed to start. As an example, SAE 10W must have a maximum viscosity of 3500 cP or less at −20°C (−4°F).

The second test to define the SAE viscosity grades which contain the letter *W* is the borderline pumping temperature, or BPT (ASTM D3829), test. This procedure uses a minirotary viscometer apparatus to simulate conditions which correlate with actual cold temperature engine performance. The BPT for an oil and engine combination is defined as the temperature at which the minimum system pressure at any time after 1 min of test is 138 kPa (20 psig). The BPT is a measure of an oil's ability to flow to the engine oil pump inlet and provide adequate engine oil pressure during the initial stages of operation.

The third test used to define the winter SAE viscosity grades is a high-temperature viscosity test (ASTM D445). Viscosities measured by this procedure are useful as a guide in selecting the proper viscosity oil for use under normal engine operating temperatures. Winter grade oils define only a minimum viscosity at 100°C (212°F).

SAE viscosity grades which do not contain the letter *W* are based only on the minimum and maximum viscosity at 100°C (212°F). As an example, SAE 30 must have a viscosity between 9.3 and 12.5 cSt at 100°C (212°F). These viscosity grades are not evaluated for cold temperature performance and are, therefore, not typically recommended for cold winter climates. These grades have been referred to as "summer" grades.

An engine oil which qualifies for only one of the primary 10 SAE viscosity grades is referred to as a single-grade engine oil. Table 10.8 is actually a list of single-grade engine oils. A car owner living in the northern part of the United

States would use a viscosity grade containing the letter *W* in the winter and then change to a viscosity grade without the letter *W* in the summer.

The SAE viscosity classification system is also designed to define multigrade engine oils. A multigrade, or multiviscosity, oil is one whose low-temperature viscosity and borderline pumping temperature satisfy the requirements for one of the *W* grades and whose 100°C (212°F) viscosity is within the prescribed range of one of the summer grade classifications. This is possible since the grades containing the letter *W* define a minimum viscosity at 100°C.

It is possible to add viscosity index improvers to engine oils and obtain a formulation with both excellent cold temperature flow properties and sufficient high-temperature viscosities to protect the engine at high operating temperatures. Multigrade engine oils are labeled as such by including both the SAE grade containing the letter *W* (defined at cold temperatures) and the SAE grade not containing the letter *W* (defined at high temperature). The most popular multigrade oils are SAE 10W40 and SAE 10W30.

API Service Classification

The API classification system attempts to rate engine oils by performance levels or application categories (SAE J183). Oils primarily designed for gasoline engines have a designation beginning with *S* (Table 10.9). The *S* stands for service and is used by service stations, garages, and new car dealers. There are seven classes and each must meet specific requirements of the automotive industry.

SF oils provide the quality necessary to meet the 1980–1989 warranty requirements of the automotive manufacturers. SF oils provide protection against sludge, varnish, rust, wear, and high-temperature oil thickening. Oils in this cat-

TABLE 10.9 API Engine Oil Service Classification System

Letter designation	API engine service description	ASTM engine oil description
SA*	Formerly for utility gasoline and diesel engine service	Oil without additive
SB	Minimum-duty gasoline engine service	Provides some antioxidant and antiscuff capabilities
SC	1964 gasoline engine warranty maintenance service	Oil meeting 1964–1967 requirements of the automotive manufacturers
SD	1968 gasoline engine warranty maintenance service	Oil meeting 1968–1971 requirements of the automotive manufacturers
SE	1972 gasoline engine warranty maintenance service	Oil meeting 1972–1980 requirements of the automotive manufacturers
SF	1980 gasoline engine warranty maintenance service	Oil meeting 1980 requirements of the automobile manufacturers
SG	1989 gasoline engine warranty maintenance service	Oil meeting 1989 requirements of the automobile manufacturers

*S = service (service stations, garages, etc.).

egory provide increased oxidation stability and improved antiwear performance relative to engine oils which meet only the minimum requirements for API service SE. SF oils may be used where API service SE, SD, or SC are recommended. In 1989, SG oils, which provide additional protection against sludge and viscosity buildup, were introduced (Fig. 10.6).

The *C* classification stands for commercial and is intended for fleets, contractors, farmers, etc. This designation describes oils intended primarily for diesel engines (Table 10.10).

It is worthy to note that the performance levels do not usually consider the effect of the oils on shaft seals. It is quite possible that a seal material that functions well in an SF oil may be adversely affected by the additives that are placed in an SG oil. Seal manufacturers and users are advised to test seal materials in any oils new to their applications to ensure against adverse effects on the seal material.

Fuel Economy

After the federal government imposed Corporate Average Fuel Economy (CAFE) requirements on the automobile industry, it became obvious that every means of reducing fuel consumption was going to be needed to comply. One area where some savings can be achieved is by reducing engine friction with friction modifiers added to the engine oil. An oil is classified as "energy conserving" if testing shows a 1.5 percent reduction in fuel consumption.

The Environmental Protection Agency (EPA) has developed a test procedure to determine if an engine oil reduces fuel consumption. Five automobiles representing a cross section of available engines are run on a chassis dynamometer through a driving cycle which is a modified version of a trip through Los Angeles, CA. The car's fuel consumption is measured precisely, first using a reference SAE 20W30 reference oil. The fuel economy performance of the candidate oil is

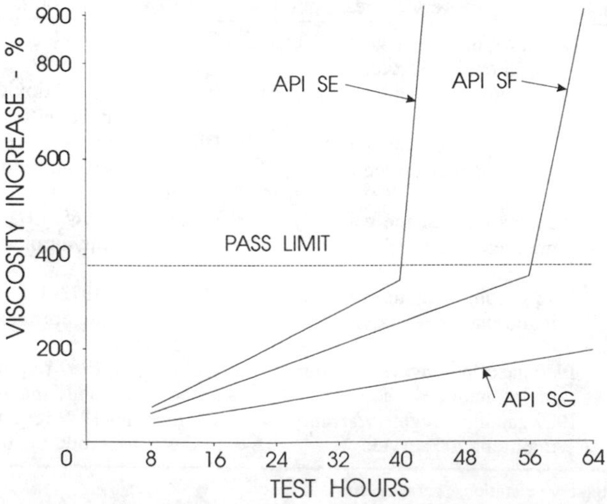

FIGURE 10.6 Sequence IIIE viscosity control comparison.

TABLE 10.10 API Engine Oil Service Classification System

Letter designation	API engine service classification description	ASTM engine oil description
CA*	Light-duty engine service; high-quality fuels.	Oil meeting the requirements of MIL-L-2104A (1954 vintage).
CB	Light to moderate engine service; lower-quality fuels.	Same as above but diesel engine run using high-sulfur fuel.
CC	Moderate- to severe-duty diesel and gasoline engine service.	Oil meeting requirements of MIL-L-2104B (1964 vintage).
CD	Severe-duty diesel engine service.	Oil meeting Caterpillar's certification requirements (1955 vintage).
CD-II	Severe-duty two-stroke cycle diesel engine service.	Oil meeting the performance requirement described in category CD by the Cat. 1G2 and CRC L-38 tests, plus Detroit Diesel 6V-53 T test to address deposits and wear in turbocharged and supercharged two-stroke cycle diesel engines.
CE	Service typical of turbocharged or supercharged heavy-duty diesel engine manufactured since 1983 and operated under both low-speed, high-load and high-speed, high-load conditions.	Oil meeting all requirements of category CD and having additional performance as described by the Cummins NTC 400 engine test for oil consumption control and the Mack EO-K/2 engine oil specification.
CF-4	This category was adopted in 1991 and describes oil for use in high-speed, four-stroke cycle, diesel engines.	API CF-4 oils exceed the requirements of the CE category, providing improved control of oil consumption and piston deposits. These oils should be used in place of CE oils. They are particularly suited for on-highway, heavy-duty truck applications. Based on CRC L-38, Cat 1K, Mack T-6 & T-7, Cummins NTC 400 (rev.).

*C = commercial (fleets, contractors, etc.).

compared to the reference oil and is reported as a percentage improvement versus the reference oil. If the candidate gives a statistically significant fuel consumption of at least 5 percent, the oil is said to be energy conserving.

API Service Classification Symbol

There are currently 14 API service designations and 10 SAE viscosity grades. These items show up in various locations on the engine oil cans (or bottles) and in hundreds of different combinations. The API's Fuels and Lubricants Committee has established the Engine Service Classification Symbol (commonly called the API doughnut). The symbol has been divided into three areas that display one specific characteristic of the oil (Fig. 10.7). The upper half displays the appropriate API service category. The categories are restricted to current API service categories such as SF, SG, CD, etc., and/or any combinations thereof. The center of the symbol is reserved for the SAE viscosity grade or grades for multiviscosity oils. The lower section displays the energy-conserving features of an oil, if applicable. The symbol is the manufacturer's warranty that the oil within the container complies with the designated service classification category. A formal API licensing system for use of the symbol has been established.

Engine Oil Additives

The typical engine oil consists of 73 to 85 percent high-quality lubricating oil, 8 to 15 percent viscosity index improvers, and 7 to 12 percent additives for enhancing performance. Most commercial additives perform more than one function, and all the functions shown above can be represented by five or six separate additives. More than one chemical version of a given additive may be used to provide protection over a wide range of operating conditions. Table 10.11 gives two examples of additive formulations for top grade oils, one for use in an SF/CD SAE30 grade heavy-duty diesel oil and the other for use in an SF/CC 10W30 passenger car oil.

To a minor extent, engine oil additives are sold as commodities to oil blenders who then combine them with other additives to make their own finished oils. However, most additives are sold as packages or blends of several additives which, in the customer's base oil, meet the test specifications of the performance level selected by the customer.

Additives are reactive chemical agents. Many physical and chemical interactions occur among these additives and the base oil that affect the finished oil performance, clarity, viscosity, and other properties. These additives may have an adverse effect on elastomers. Testing of seals in new oil formulations is recommended.

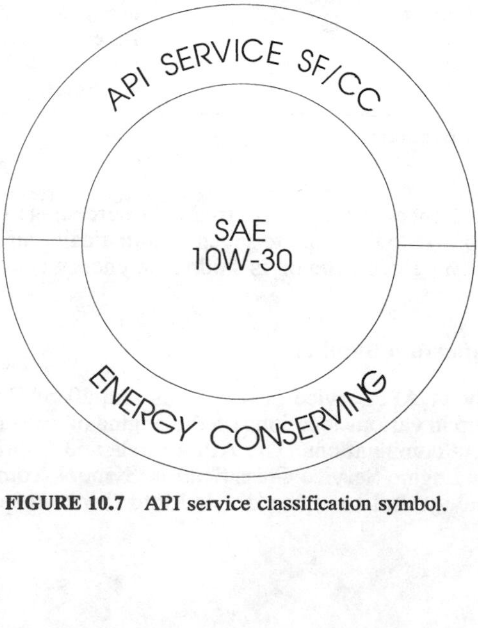

FIGURE 10.7 API service classification symbol.

TABLE 10.11 Engine Oil Additives

	SF/CD additive (SAE 30) treatment for diesel service		SF/CC additive (10W30) treatment for passenger car service	
	Ingredient	% weight in oil	Ingredient	% weight in oil
Dispersant	Polymer-amine	6.0	Polymer-amine	4.0
Primary detergent	Low-base sulfonate	1.5	Low-base sulfonate	0.5
Detergent alkaline agent	High-base phenate	0.5		
Antioxidant	Low-base phenate	1.5	Sulfurized hydrocarbon	0.5
Rust inhibitor alkaline agent	High-base sulfonate	0.5	High-base sulfonate	1.0
Antioxidant antiwear	Zinc dialkyldithiophosphate	1.5	Zinc dialkyldithiophosphate-A	0.8
Friction reducer			Zinc dialkyldithiophosphate-B	0.5
Viscosity modifier			Sulfurized fat	0.7
			Ethylene-propylene copolymer	10.0
Total		11.5		18.0

GEAR OILS

Gear Box Failures

Power is usually transmitted from the source to the end use through a gear box that can change speeds, direction, and torque. Lubricants for gear boxes are formulated to prevent premature component failure (gears, bearings, cross-shafts), assure reliable operation, and increase equipment service life. The objectives are accomplished through a number of vital lubricant functions that minimize friction and wear, inhibit corrosion, reduce operation noise, improve heat transfer, and wash wear particles from contact zone.

The most critical function is the minimization of friction and wear to extend equipment service life. Gear and bearing failures can be traced to mechanical problems or lubricant failure. The reasons for lubricant-related failures are usually contamination, collapse of the oil film, inadequate additive level, and use of the wrong oil type for the application.

The most common lubricant-related failures are caused by the presence of foreign matter in the gear oil. Off-road equipment is especially prone to contamination by dust and water. Dust particles are highly abrasive and act as cutting tools under thin film conditions, resulting in "plowing" wear, or ridging. Water contamination can cause rust on working surfaces, destroying metal integrity. Even the accidental addition of a small quantity of an engine oil may lead to failure. Additives of the motor oil can react with those of the gear lubricant, gelling and negating their protective actions. Good housekeeping and, in certain applications, the use of oil filters can reduce the likelihood of contamination failures.

An insufficient oil film thickness can be caused by an incomplete oil charge or by the use of an oil with a viscosity which is too thin for the application. Another major problem occurs when the oil film is interrupted by a wear particle passing through a gear mesh. This results in damage to the tooth surface and leads to fatigue failures. Many truck manufacturers recommend an oil change after a short break-in period to prevent this from occurring. Use of an inadequate additive level or improper additive type can cause gear failure. A motor oil will not provide sufficient EP protection under the boundary conditions experienced by many gear applications.

SAE Viscosity Grades

Gear lubricants must flow freely when the axle is cold, yet have sufficient thickness or viscosity to lubricate at elevated operating temperatures. Axle and transmission lubricant viscosity is defined through the SAE J306 recommended practice. As indicated in Table 10.12, each viscosity grade has distinct specifications for low- and high-temperature performance.

SAE viscosity selection should be based on the minimum and maximum service temperatures. Most of the commonly used gear lubricants are multigraded (i.e., 80W90, 80W140, and 85W140). These fluids meet both the low- and high-temperature requirements for the combined grades. For example, an 80W90 must have the low-temperature fluidity of a grade 80W as well as the high-temperature thickness of a 90 grade.

API Service Classifications

The API service classifications were developed to help manufacturers and end users select gear lubricants for a variety of operating conditions. The API service classes range from GL-1 to GL-6 and describe gear lubricants in terms of general type, severity of service, and application. Table 10.13 lists these classes.

With possible exception of API GL-1, each lubricant class is formulated using performance additives. The performance required for a particular service class de-

TABLE 10.12 Axle and Manual Transmission Lubricant Viscosity Classification

SAE J306 viscosity grade	Max. temperature for viscosity of 150,000 cP, °C		Viscosity at 100°C (212°F)	
	°C	°F	Min. cSt	Max. cSt
70W	−55	−67	4.1	
75W	−40	−40	4.1	
80W	−26	−15	7.0	
85W	−12	+10	11.0	
90			13.5	24.0
140			24.0	41.0
250			41.0	

TABLE 10.13 API Service Classifications

Classifications	Type	Typical application
GL-1	Straight mineral oil	Automotive manual transmissions (tractors and trucks)
GL-2	Usually contains fatty materials	Worm gear drives and industrial gear oils
GL-3	Contains mild EP additive	Manual transmissions and spiral bevel final drives (GL-3 not widely used)
GL-4	Equivalent to obsolete MIL-L-2105 specification usually satisfied by 50 percent GL-5 additive level	Manual transmissions and spiral bevel and hypoid gears where moderate service prevails
GL-5	Equivalent to present MIL-L-2105D specification primary field service recommendation of most passenger car and truck builders worldwide	Used for moderate and severe service in hypoid and all other types of gears; also may be used in manual transmissions
M2C105A/M2C154A	Same performance as Ford	Typically recommended in service conditions where more protection from scoring is desired. Passenger car hypoid axles with high offset

termines the type and amount of additive used. Most gear lubricants are designed for and used in truck applications. Only 1 percent of the vehicles in North America are heavy trucks and 75 percent are passenger cars. The size of truck differentials and the frequency of service results in a consumption of almost half of the gear oils produced. Passenger cars use only 7 percent of the total (Fig. 10.8).

The most commonly specified and available type of automotive gear lubricant in North America is API GL-5. In Europe and other parts of the world, API GL-4 is commonly used as API Gl-5. The API GL-5 specification does not address limited slip applications. Factory-fill formulations for limited slip differentials are

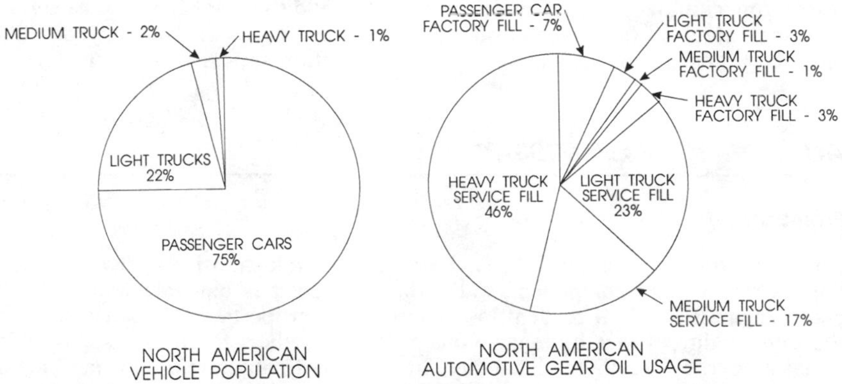

FIGURE 10.8 Vehicle population versus gear oil usage (1990).

TABLE 10.14 AGMA Lubricant Gear Specifications, Viscosity

AGMA lubricant no.	Viscosity, cSt @40°C (104°F) Min.	Max.	Equivalent ISO grade
1	41.4	50.6	46
2	62.2	74.8	68
3	90	110	100
4	135	165	150
5	198	242	220
6	288	352	320
7	414	506	460
8	612	748	680
8A	900	1100	1000

based on performance in the auto maker's individual test rig or vehicle. No standard industrywide test is available of a lubricant's ability to prevent chatter in a limited slip differential.

AGMA Lubricant Specifications

The American Gear Manufacturers Association (AGMA) has also developed specifications for gear lubricants to define physical properties and performance. A lubricant number has been defined to indicate the minimal and maximum viscosities (Table 10.14). A comparison of the various viscosity classification systems appears on Table 10.15. The SAE viscosity grades are also shown for engine oils. A 10W engine oil has viscosity which is equivalent to a 75W gear oil.

The AGMA has also established performance specifications for EP lubricants and oils to minimize rust and oxidation (R&O) (Table 10.16).

Formulations and Gear Oil Selection

A gear lubricant consists of a blend of a base fluid and additives (Table 10.17). The base fluid is the bulk of the mixture and defines the viscosity, acts as a heat transfer medium, and serves as a carrier for the additive package. The base fluid can be mineral oil or synthetic polymers such as polyolefins, polyglycols, and dibasic acid esters. Suggestions for gear oil selection appear in Table 10.18.

AUTOMATIC TRANSMISSION FLUIDS

Function

Automatic transmission fluids (ATFs) are very complex lubricants that may contain as many as 15 components. They have viscosities equivalent to SAE 20-grade engine oils with excellent low-temperature properties. ATFs use some of the same additives found in engine oils as well as others to provide special frictional properties and thermal stability. In addition to automatic transmission applications, ATFs are used in power steering systems, hydraulic systems for in-

TABLE 10.15 Comparative Viscosity Classifications

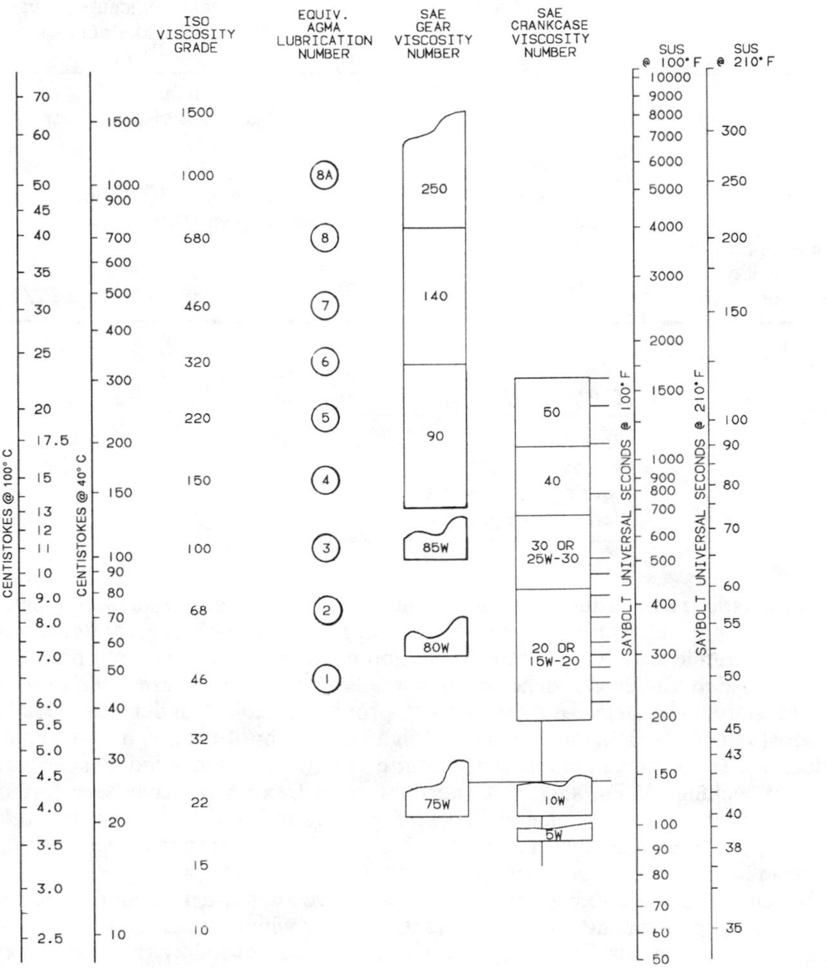

dustrial equipment, and air compressors and as a heat transfer fluid in many applications.

ATFs must perform five functions when used in the automatic transmission for passenger car, school bus, and light truck applications. The fluid must transfer hydrodynamic energy in the torque converter, transfer hydrostatic energy in the hydraulic control logic circuits, lubricate bearings, transmit sliding friction energy in lubricated bands and clutches, and transfer heat to maintain the transmission within proper operating temperatures. The oil must contain additives to prevent foaming, provide oxidation resistance, corrosion resistance, operate over a wide temperature range (−40 to 150°C, or −40 to 302°F) and be compatible with seal materials.

TABLE 10.16 AGMA Gear Lubricant Specifications, Performance

Property	R&O gear oils (3–10% fat)	EP gear lubricants (designated w/AGMA number and EP)
Viscosity index	90 minimum	90 minimum
Oxidation stability	1,2 1,500 h, minimum 3,4 740 h, minimum 5,6 500 h, minimum	Max. 10% viscosity increase in 02893
Rust protection	Pass—sea water	Pass—distilled water
Corrosion protection	Copper strip—1	Copper strip—1
Foam suppression	Yes	Yes
Demulsibility	Yes	Yes
EP properties		60-lb Timken 11-stage FZG

TABLE 10.17 Gear Oil Formulation

Base fluid	50–95%
Viscosity improver	0–35
Pour point depressant	0–2
Performance package	5–12

Automatic transmissions for automobiles and light trucks contain clutch plates and bands of cellulose fibers and resins. Other materials such as graphite, asbestos, and ceramic are added to improve durability and provide the proper frictional characteristics. Sintered bronze or semimetallic friction plates are often used in the transmission of heavy-duty vehicles to provide durability under heavy loads.

Most mineral and engine oils have a high coefficient of friction at low sliding velocities. Some clutch plate designs require low friction at low sliding speeds to prevent sticking. ATFs, such as General Motor's Dexron II, have been formulated to meet these requirements. These fluids are referred to as *friction modified.* Other clutch plate designs require a high friction at low speeds. These fluids (Ford M2C33-F) are called *nonfriction modified type F.*

Because excessive oxidation of ATFs can have such a detrimental effect on transmission performance, it is important to know when to change the fluid. A simple blotter spot test has been used for many years to follow the rate of oxidation of ATF and has been shown to be an effective means for rating fluids in transmissions or other operating equipment. In the test, one or two drops of used ATF are placed on blotter paper and the spot rated after 30 to 60 min. If the spot is wide and is red or light brown in color, the fluid is satisfactory. If the spot remains small and is dark, the fluid is oxidized and should be changed.

Viscosity—Temperature Properties

Since automatic transmissions must operate smoothly over a broad temperature range, the transmission fluid formulations must have special viscosity properties. A low-temperature viscosity limit of 50,000 cP at −40°C (−40°F) is required to prevent fluid cavitation at low temperatures. At 100°C (212°F) the minimum viscosity limits are 5.5 to 6.2 cts. Base stocks with low viscosities and a low wax

TABLE 10.18 Suggestions for Gear Oil Selection

Gear type	Characteristics and description	Gear oil selection
Spur	No end thrust Radial antifriction or journal bearings Simple and economical to manufacture Cast iron or steel—cast, ground, or hobbed Noisy and rough operation Single tooth contact with zero pitch angle 2000 ft/min maximum pitch line velocity 10:1 maximum single reduction ratio 4500 hp maximum with low-cost/hp Single tooth contact—rolling motion with sliding action at first and final contact	AGMA R&O gear oil Viscosity determined by manufacturer
Helical	Stepped spur gear Multiple tooth contact Smoother, higher-load capability Noise decreases with increased pitch angle and tooth contact End thrust Both radial and thrust bearings required 5000 ft/min. maximum pitch line velocity	AGMA R&O gear oil API GL-1 or GL-3 EP gear oil may be needed for thrust washers, gear case thermal expansion, or overloading
Herringbone	Double helical gear with no thrust load Expensive to manufacture Quiet and smooth for high-speed gearing 30,000 ft/min pitch line velocity for high speed Parallel shaft 50,000 hp maximum	Same as helical
Bevel	Spur gear on angle Intersecting shaft Bearing only on one side of gear Radial and thrust bearings Tapered teeth reduce strength at inner end, misalignment concentrates load at weak end 1000 ft/min maximum pitch line velocity Limited to low speed; rough and noisy operation	Same as helical

TABLE 10.18 Suggestions for Gear Oil Selection (*Continued*)

Gear type	Characteristics and description	Gear oil selection
Spiral bevel	Two or more teeth contact increases load capability Greater thrust loading 8000 ft/min maximum pitch line velocity 20,000 hp maximum	AGMA EP gear oil API GL-4 or GL-5 Low-speed and light-load units function with R&O oils
Worm	High-load capability High-speed reduction 70:1 maximum speed reduction ratio Smooth, quiet output Gear efficiency decrease with high-speed ratios Gliding straight line tooth contact Worm may be considered an endless rack Normally will not run backward 6000 ft/min maximum pitch line velocity 1500 hp maximum Wheel gear made of brass, low tooth loading Speed ratio determined by pitch of worm rather than wheel diameters	AGMA compounded gear oil EP gear oils with low copper corrosion and proper viscosity determined by manufacturer
Hypoid	Offset spiral bevel gear Smooth running Light weight and compact for loads used Long gear teeth make it strong and quiet Severe sliding action produces boundary layer lubrication All gears made of steel due to high tooth loading	API GL-5 Primarily automotive and truck use

content are needed to meet these viscosity requirements. Viscosity index (VI) improvers and pour depressants may also be added to the formulation. Other additives include dispersants, antioxidants, wear inhibitors, friction modifiers, corrosion inhibitors, and seal swell additives. As many as 15 different additions may be employed in the formulation.

Automatic Transmission Properties and Specifications

Properties and specifications of a typical friction modified transmission fluid (Dexron II) are compared to a nonfriction modified fluid (M2C33-F) in Table 10.19.

TABLE 10.19 Summary Comparison of Dexron II and M2C33-F Specifications

Test	Dexron II	M2C33-F
Seal tests		
Buna N	70 h at 149°C	168 h at 149°C
Polyacrylate	70 h at 149°C	
Silicone	8 h at 176°C	
Viscosity		
At 99°C (210°F)	5.5 cSt, min.	7.0 cSt, min.
At −17.8°C (0°F)		1400 Cp, max.
At −40°C(−40°F)	50,000 Cp, max.	55,000 Cp, max.
Foam	Special foam test procedure	ASTM D-892 procedure
Bench	Additional tests for flash point, fire point flash, fire, copper corrosion, rust and miscibility	Additional tests for flash point, fire point flash, fire, copper corrosion, rust and miscibility
Transmission oxidation	300 h at 163°C	300 h at 163°C
Transmission cycling	20,000 cycles on a dynamometer	8000 cycles in a vehicle
Friction	18,000 cycles in SAE 2 machine (low static curve required)	Six tests of 100 cycles each on SAE 2 machine (high static curve required)
Wear	50 h in Saginaw vane pump	Four ball wear tests

SPECIALTY LUBRICANTS

Many applications demand temperature ranges that exceed the range of typical mineral oils. Users are also demanding longer intervals between lubricant changes. These conditions dictate a synthetic or specialty lubricant. Other conditions, such as high vacuum, heavy loads, high speeds, presence of caustic and corrosion fluids, and the need for fire-resistant fluids, also may demand a synthetic lubricant.

Temperature Limitations for Synthetic Fluids

Lubricants are susceptible to failure at high temperature, especially in thin films and at long-term exposure. Most lubricants can operate at elevated temperatures for limited periods provided the ratio of volume to surface area is high. But oxidation, the primary form of lubricant degradation at high temperatures, eventually occurs even under these conditions.

Oxidation is a two-step molecular process. Primary oxidation products include alcohols and ketones. Next, secondary reaction products are formed which precede sludge formation. Once the primary oxidation products are formed at high temperatures, oxidation may proceed at normal operating temperatures. Increasing temperature rather than increasing shear load is the cause of most lubricant failures. Typically, specialty synthetic fluids are required for temperatures over 200°C (392°F). Table 10.20 gives temperature ranges for synthetic fluids.

TABLE 10.20 Temperature Range of Synthetic Fluids

	Typical mineral oil	SHF	Alkyl benzenes	Dibasic acid esters	Polyol esters	Polyglycols	Phosphate esters
Upper temperature range (intermittent service)							
°F	225/325	325/550	275/430	350/550	400/625	400/475	300/392
°C	107/163	163/288	135/221	177/288	204/329	204/246	149/200
Normal operating range (continuous service)							
°F	5/225	−50/325	−45/275	−40/350	−30/400	−30/400	−13/300
°C	−15/107	−45/163	−43/135	−40/177	−34/204	−34/204	−25/149
Lower operating temperature range (dependent on starting torque)							
°F	−35/5	−75/−50	−65/−45	−65/−40	−65/−30	−60/−30	−60/−13
°C	−37/−15	−59/−45	−54/−43	−54/−40	−54/−34	−51/−34	−51/−25

Properties and Major Applications for Synthetic Fluids

Formulations for synthetic fluids include phenyl silicones (−20 to 230°C) (−4 to 446°F), polyphenyl ether (5 to 285°C) (41 to 545°F), and fluorinated ether (−20 to 285°C, or −4 to 545°F); their properties are given in Table 10.21. Table 10.22 lists applications for synthetic lubricants. Over 285°C (545°F) resin-bonded solid-film lubricants generally replace oils and greases.

Among synthetics, expensive fluorinated ethers, fluorosilicones, and polyphenyl ethers have superior oxidation resistance. Silicate esters also have a wide temperature range; however, they break down slowly in high humidity. Silicones also have a wide temperature range, but they are unreceptive to most additives, including antioxidants. Shielded polysilicates are better lubricants than silicones, and they readily accept additives.

Synthetic Grease

High rotational speeds are most often encountered with precision bearings. In this application, grease lubricants are the usual choice because of design simplicity and ease of maintenance. In fact, stiffer, channeling-type greases are used to minimize heat generation due to churning. In addition, the stiffer greases counteract the inherent centrifugal and gravitational forces within the bearing assembly that tend to dislodge the lubricant from the intended surfaces.

High temperatures normally dictate the use of synthetic lubricants. While petroleum-based greases have a so-called speed factor (bearing bore in mm × rpm) of 600,000, their high-temperature limit is only from 93 to 120°C (200

TABLE 10.21 Properties of Synthetic Lubricants

Properties	MO	HMO	PAO	NPE	DBE	DB	PB	PAG	SE	PE	FC	S	FS
Viscosity index	F	E	G	G	E	F	F	G	E	F	F	E	E
Low pour point	P	G	G	G	G	G	G	G	E	G	G	G-E	G
Thermal stability	F	G	G	VG	G	G	F	F	G	G	E	G	G
Oxidative stability	F	G	VG	E	G	G	F	G	F	G	E	G	G
Hydrolytic stability	E	E	E	F	F	E	E	G	P	F	E	G	G
Paint compatibility	E	E	E	G	VG	E	VG	M	F	P	G	G	G
Fire resistance	L	L	L	L	L	L	L	L	L	H	H	G	VG
Lubricating ability	G	G	G	G	G	G	G	G	G	G	G	P-G	G
Low volatility	A	G	G	G	A	A	A	G	G	G	VG	G-E	G-E
Radiation resistance	H	H	H	A	A	H	H	A	A	L	H	G	G
Additive solubility	E	G	G	G	VG	E	E	G	F	G	P	P	P
Elastomer swell	L	N	N	M	H	M	L	L	L	H	L	L	L
Compatibility w/petroleum	—	—	G	F	G	E	G	P	F	G	P	P	P
Relative cost	1	2	5	5–10	3–5	5	4	5–10	30–45	5–10	20–30	12–20	20–30

E = excellent, VG = very good, G = good, F = fair, P = poor, A = average, H = high, L = low, N = nil.

MO = mineral oil, HMO = super refined petroleum, PAO = polyalphaolefin, NPE = neopolyol ester, DBE = dibasic acid ester, DB = dialkyl benzenes, PB = polybutenes, PAG = polyalkylene glycols, SE = silicate esters, PE = phosphate esters, FC = fluorinated compounds, S = silicones, FS = fluorosilicones.

to 248°F). Synthetic grease formulations and their speed factors and temperature ranges are found in Table 10.23.

Lubricants for High Vacuum

Conventional hydrocarbon liquid lubricants generally evaporate from lubricated surfaces quite quickly in a vacuum environment. The speed of evaporation depends upon the temperature and the molecular weight of the hydrocarbon. As a result, solid lubricants usually are preferred for vacuum conditions. The traditional vacuum lubricant is molybdenum disulfide in a bonded coating. MoS_2, which deteriorates in high humidity, actually improves its lubricating ability in vacuum environments. Diselenides are even better due to lower outgassing properties. One manufacturer quotes a temperature range of −200 to 230°C (−328°F to 446°F) for molybdenum disulfide dry film lubricant.

The addition of graphite extends the temperature range of 450°C (842°F). However, graphite should not be used alone in a vacuum. Graphite depends on condensable vapors to provide a lubricant film; thus, it exhibits poor lubrication properties in a vacuum.

TABLE 10.22 Major Applications for Synthetic Lubricants

Application	PAO	NPE	DBE	DB	PB	PAG	FC	PE
Engine oils and automotive lubricants								
SAE J-183s engine oils								
Multigrade	x	x	x					
Non-SAE J-183a								
Aircraft								
Piston	x							
Turbine		x	x					
Gas-fueled two-stroke			x		x	x		
Transmission and hydraulic fluids	x		x					
Gear oils	x		x		x			
Automotive grease	x		x					
Industrial lubricants								
General industrial								
MIL-spec. hydraulic	x							x
Hydraulic	x	x	x			x		x
Gear oils	x	x	x			x		
Other specified								
Turbine/circulating	x	x	x			x		x
Refrigeration				x				
Compressor	x	x	x		x	x	x	x
Other unspecified								
Vacuum pump								x
Bearing oils					x		x	
Textile oils					x			
Industrial engine oils								
Natural gas	x		x					
Metalworking oils								
Metal removing					x	x	x	
Metal forming		x	x		x	x	x	
Metal treating						x		
Industrial greases	x	x	x	x	x	x	x	

PAO = polyalphaolefin, NPE = neopolyol ester, DBE = dibasic acid ester, DB = dialkylbenzenes, PB = polybutenes, PAG = polyalkylene glycols, FC = fluorinated compounds, PE = phosphate esters

TABLE 10.23 Synthetic Grease Properties

	Conventional petroleum-based greases	Ester	Silicone	Synthetic hydrocarbons	Polyethers	Fluoroether
Speed factor (rpm × bearing bore mm)	600,000	400,000	200,000	400,000	500,000	400,000
Temperature range						
°F	– 40/250	– 100/300	– 100/400	– 75/300	– 50/350	– 40/550
°C	– 40/121	– 73/149	– 73/204	– 59/149	– 46/177	– 40/288

Although dry lubricants are the predominant vacuum lubricants, specialty greases are being made from certain synthetic hydrocarbons. Oils with unusually low vapor pressures can be produced by refining some higher-viscosity hydrocarbons. A fluoroether grease, for example, has a vapor pressure of 2×10^{13} torr. This grease is gelled with a fluorocarbon polymer and uses a viscous (1150 cSt at 38°C) fluorinated ether oil. It can be used from about 10 to 200°C (50 to 392°F); however, it is expensive. Thick silicone grease is a less expensive alternative; however, it cannot be used at the same levels of vacuum. Temperature range is 4 to 200°C (40 to 392°F).

P • A • R • T • 4

ECONOMIC AND MANUFACTURING CONSIDERATIONS

CHAPTER 11

ECONOMICS OF SELECTING SEAL TYPE FOR ROTATING SHAFT APPLICATIONS

INTRODUCTION

The seal user can choose from a variety of seal designs and types (packing, lip seals, mechanical seals) when deciding how to seal an application. Factors other than the initial seal cost should be considered when deciding which type of seal should be used. The costs that must be considered are the initial costs, the operating costs, and miscellaneous costs (Table 11.1). The true cost of sealing an application is the sum of these factors, which accrue during the life of the machinery that is sealed. The user should estimate and compare these total costs before deciding which seal type should be used.

TABLE 11.1 Seal Cost Factors

Initial costs	Operating costs	Miscellaneous costs
Purchase price Housing Inspection Assembly	Maintenance and replacement parts (seal, shaft, housing) Power consumed Loss of fluid Cost of spares stock	Production loss Standby units (capital and interest) Secondary damage Health Pollution

INITIAL COSTS

Purchase Price

The purchase price of packings, lip seals, and mechanical seals is highly dependent upon material. Choosing a seal with a material that is overspecified will penalize the user with unneeded extra cost. Simple, low-stress applications with mild lubrications can use inexpensive cotton packings. If the lubricants have aggressive additive packages, PTFE fiber packing is required. Increasing speeds or temperatures can result in more expensive packing materials such as graphite foil

TABLE 11.2 Relative Material Costs for Mechanical Seals

Stationary face carbons	Relative cost	Secondary seals O-ring	Relative cost	Rotating face	Relative cost
Plain	1.0	Nitrile	1	Stellited stainless	1
Metallized	1.1	Polyacrylate	2	Steel	
High temperature	2.6	Silicone	5	Ceramic coated	2
		Fluorocarbon	10	Copper alloy	
		PTFE	10	Tungsten carbide	7
		Perfluoro-elastomer	100		

or filaments. The cost of PTFE packing is approximately 3 times that of simple cotton fibers and graphite foils can be 5 times more.

Materials costs also vary with mechanical seals (Table 11.2). Lip seal costs also vary with material. A fluoroelastomer seal may cost 5 to 10 times more than a nitrile lip seal. Relative costs of various seal types appear in Table 11.3.

TABLE 11.3 Relative Purchase Costs for Various Seal Types, $

			Mechanical		
Diameter, mm	Soft packing	Elastomeric lip seals	Simple	Heavy-duty unbalanced	Heavy-duty balanced
25	0.4	0.6	10	60	75
100	1.0	1.0	40	95	105
200	2.0	2.50	75	110	250
250	3.00	5.5	100	300	500

Other Initial Costs

Lip seals often require carefully machined housing and plunge ground shafts for proper function. These tight specifications may require higher initial costs for the housing and the shaft than for either packings or mechanical seals. Initial shaft costs can be reduced if unitized seals or wear sleeves are used. The cost of the wear sleeves and the higher cost of the unitized seals must also be considered. Assembly of soft packings and some mechanical seals are often more complicated and costly than the simple lip seal. The difficulty of future replacement should also be considered when selecting a seal.

OPERATING COSTS

After the seal is installed, there are operating costs to consider and compare: labor costs for routine inspections, energy cost, fluid loss, maintenance, replacement parts, and cost of spare parts inventory. The cost of labor for routine seal inspection may be difficult to estimate since some inspection may be required for other components even if the seal is perfect. In most cases, seal inspection and replacement are required periodically and this labor should be changed to the cost of sealing the equipment. Seal power consumption is not usually a large percentage of the power required to drive the equipment. If the seal is large, power loss can amount to several kilowatts, and meaningful comparisons can be made between seal types. Fluid loss is becoming more important as environmental concerns increase. In some cases, the cost of the lost fluid is much less than the cost required to clean up the spill. Interest cost for spare seals and other replacement parts kept in inventory can be significant for high-cost mechanical seals when compared to low-cost packings. Maintenance and replacement part costs can be the most significant operating cost.

MISCELLANEOUS COSTS

In critical applications equipment availability, seal efficiency, or reliability may be an overriding consideration for seal type selection. Loss of production in a factory, loss of power generation in an electric power station, or loss of use of a ship or aircraft for a period would result in large losses while a failed seal is replaced. In nuclear or aerospace applications, reliability and effectiveness may be paramount. On a lesser scale, there is the inconvenience factor that predominates in domestic equipment and private cars where the consumer's bad experience may adversely affect reputation and future sales as well as causing expensive warranty claims. These consequential costs can be difficult to evaluate since failure probability is a key factor which is often unknown. A cost more easily evaluated is the capital (and interest) cost of standby equipment necessitated by the need to avoid the consequences of shutdowns through seal failure on a pump.

Less predictable is secondary damage that can result when leakage occurs. Leakage may cause bearing failure on a pump, which could in turn lead to destruction of the pump impeller and casing. Leakage might lead to fire. Corrosive leakage may cause structural damage to equipment. Damage to health may be caused by toxic, acidic, or caustic leakage. Leakage of pollutants is an increasing expense where effluent must be treated before discharge. It may result in the need for treatment plants larger than otherwise required. In some cases, leakage may lead to legal sanctions being imposed on a company. When costly secondary damage or a possible environmental hazard exists, the cost of a seal or sealing system may be of little consequence.

TOTAL LIFE COST: AN EXAMPLE

The total life cost concept can be applied to seals and is a useful method for comparing alternative seal types, especially where many pumps are involved. All costs quoted in the following example are relative. No dollar values are used since they will change with inflation. Six factors are identified as likely to have appreciable effects on the overall cost of the user:

Initial seal cost
Shaft wear
Length of working life
Leakage

Periodic adjustments
Power consumption

Each is discussed below:

1. *Initial seal cost assuming 250-mm seal diameter:* These are known relative costs and range from 3 for soft packings to about 500 for high-duty balance mechanical seals (Table 11.4).

TABLE 11.4 Total Life Costs (Relative Values)

			Mechanical seal		
				Heavy duty	
	Soft packing	Rubber lip seal	Basic	Un-balanced	Balanced
Initial relative cost	3	5.5	100	300	500
Estimated seal life, h	3000	2000	3000	5000	10,000
Estimated shaft life, h (relative cost to replace is 100)	5000	10,000	N/A	N/A	N/A
Estimated time between adjustments, h (relative cost is 10)	100	N/A	N/A	N/A	N/A
Relative labor cost to replace seal	100	100	100	100	100
Number of seals required	16	24	16	7	4
Total seal relative cost	48	132	1600	2100	2000
Labor for replacement	1600	2400	1600	700	400
Shaft replacement cost	900	400			
Adjustment cost	4990				
Total relative life costs	7538	2931	3200	2800	2400

2. *Shaft wear:* A specific value of 0.125 mm was chosen as the depth of wear which would necessitate shaft or sleeve replacement. Measured wear rates from rig tests enable the costs associated with shaft wear—labor and parts—to be estimated. Life of the shaft is 5000 h for soft packings and 10,000 h for rubber lip seals. No replacement costs will be incurred with mechanical seals. Shaft replacement relative cost is 100.
3. *Length of working life:* It is evident that when the seal ceases to perform satisfactorily because of uncontrollable leakage or excessive friction, the seal must be replaced. The cost of labor and materials were estimated to have a relative cost of 50. The life of basic mechanical seal is 3000 h, unbalanced high performance is 5000 h, balanced performance is 10,000 h, soft packing is 3000 h, and the lip seals is 2000 h.
4. *Leakage:* The most usual expense under this heading is the direct value of the lost fluid. Other expenses are the cost of heating to an elevated operating temperature and the cost of pressurizing to the operating pressures. Leakage costs were ignored for the example.
5. *Adjustments:* Soft packings often require periodic adjustment. Labor costs will be incurred to make the adjustment and to perform regular inspections to

check performance. A relative cost of $10 was assessed for adjusting soft packing. Adjustment is necessary every 100 h.

6. *Power consumption:* The work done against the friction of the stuffing box incurs a cost through consumption of energy. This will most often be the cost of the electricity consumed by the driving motor with some correction factor for the efficiency of the motor. In other situations, when the pump is driven by a diesel for instance, the cost of the power may be less obvious though still calculable. This cost was ignored for the example.

All the cost parameters were considered for a pump with a 5-year (50,000 h) life. Five seal types for the same application produced the total relative cost figures shown in Tables 11.3 and 11.4. It is seen that the soft packing seal type with the lowest initial purchase cost resulted in the highest total life cost. This cost was over 3 times that of the balance heavy-duty mechanical seal which had the highest initial purchase cost.

CHAPTER 12
MANUFACTURING AN ELASTOMERIC SEAL

INTRODUCTION

Seals and gaskets are made from a wide variety of materials, but most often, the material of choice is rubber (elastomer). This versatile material, when vulcanized (cured), designed into a suitable configuration, and applied properly, is unsurpassed at resolving the myriad of sealing problems confronting engineers today.[1] Quite often the statement "the seal leaked" or as in the case of the *Challenger* disaster, "the seal failed" is made. More often than not, the problem lies not with the seal itself but only with the type of sealing product selected and the manner in which is was applied. In this chapter, some manufacturing methods to be considered in the selection process will be reviewed.

There are over a thousand manufacturers of rubber sealing products in the world. From very small garage-type shops to very large international operations they manufacture seals in a great variety of types and sizes; seals so tiny that only under a magnifying glass can they be clearly seen to large roll-neck seals for the steel industry up to 15 ft in diameter are routinely manufactured to very exacting tolerances and standards, tolerances so fine that they challenge the capabilities of the most demanding precision mold shops.

MOLDING THE MODERN OIL SEAL

One thing the elastomeric seal manufacturers all have in common is molding equipment. Molding presses of less than 1-t (U.S.) capacity to presses of 1000-t (U.S.) clamp force are indicative not only of the size of sealing products manufactured but also the quantity. Rubber molding presses, unlike plastic molding presses, have heated decks and are usually compression-type presses rather than injection molding presses (Fig. 12.1). Some manufacturers have several different types of presses in their plants, including compression, transfer, and injection presses.

Elastomeric vulcanized rubber seals are manufactured by compressing uncured rubber compound between two hot mold plates for several minutes until full vulcanization occurs. Some compounds, especially if they are polyacrylate, silicone, or fluoroelastomers, are postcured. That is, they are removed from the

FIGURE 12.1(*a*) Various types of seal molding presses. (*a*) A line of 27-t Pentaject C-frame rubber injection molding presses used for molding door seal corner joints for cars and trucks.

mold and heated in ovens until full vulcanization occurs. After molding, the parts are deflashed, springs are installed if required, and the parts are lubricated, inspected, and stored or shipped to a customer.

THE SECRET RECIPE

A compound is prepared for use as a seal by mixing the raw polymer with other ingredients to obtain the desired physical and chemical properties required (Fig. 12.2). A proper balance must be made between good processability and good functional performance. Depending on the type of base rubber gum, the compounder will add ingredients to improve the elastomer's strength, fluid resistance, temperature resistance, cut resistance, rate of cure, friction characteristics, mold processability, shelf life, ozone resistance, adhesive bond strength, flash removal, tear strength, compression set, and tensile set, and will consider a host of other factors, including costs involved to purchase, store, mix, and manufacture the compound. It's obvious from the above discussion that no two manufacturers will have exactly the same compound even though the base polymer is the same. Each company prides itself on the qualities of its elastomeric com-

FIGURE 12.1(*b*) Various types of seal molding presses. (*b*) A top-drop 240-t single-station Pentaject press used for injection or compression molding large single-cavity seals or small multi-cavity seals.

pounds and takes great care not to release the secrets of its compounding ingredients to others. It takes years of development at considerable expense to arrive at the final recipe for a production seal compound.

COMPROMISES, COMPROMISES

The nitrile content in a compound can affect the temperature limitations of an oil seal. For example, high nitrile provides better performance at low temperature, and low nitrile content performs better at high temperature. Carbon black affects the strength. Should Dixon 1176 or Dixon 1175 graphite be used to reduce interface friction and mold release? A manufacturer must decide whether to use a peroxide or

FIGURE 12.1(*c*) Various types of seal molding presses. (*c*) A 150-t five-station rotary Pentaject press is used for injection or compression molding single- or multicavity seals.

sulfur cure system and must choose between calcium carbonate from East Coast oyster shells and calcium carbonate from Gulf Coast oyster shells; these are decisions that can only be made with years of education and experience.

Vulcanized natural rubber has been used in miscellaneous applications where its flexibility and resistance to chemicals and abrasion have been found to be invaluable. Until the 1930s natural rubber was the only material having "rubber-like" properties (i.e., great extensibility and resilience). In 1932 the commercial development of synthetic materials with similar properties started with the marketing of polychloroprene by DuPont, who later adopted the trade name Neoprene for that product. The Germans had already developed their own version of synthetic rubber but were having some difficulties in getting it out of the laboratory. They were certainly instrumental in the discovery and development of Buna rubber, known today as nitrile rubber, which was of a significant advantage to them in World War II. Nitrile is still the most versatile of all seal materials.

Since that time a large number of other synthetic materials have been devel-

FIGURE 12.2 A rubber take off mill is used for removing stock in strip form after mixing.

oped that have physical properties that are similar to natural rubber. The name *elastomer* is used to cover all rubberlike materials and is preferable since some of the synthetic rubbers bear little or no chemical similarity to natural rubber.

Since the introduction of synthetic rubber, the uses to which elastomers have been put and the conditions under which they can be applied have widened considerably. Some notable examples are their increasing use in the oil and chemical industries and the temperature range for which suitable products can be found.

During the same period the mechanical properties of elastomers have been studied, and engineering design with these materials is now on a firm basis.

PROCESSING ELASTOMERS

Vulcanization or Curing

Natural rubber in its raw state is fairly unstable to both heat and light and does not have useful mechanical properties. When mixed with sulfur and heated, a change occurs which is called *vulcanization.* Vulcanization is the name for the chemical reaction resulting in the establishment of crosslinks between long polymer chains. The amount of sulfur used can be varied and the properties of the end product vary accordingly. The higher the sulfur content, the harder the rubber. When the sulfur content is between 25 and 50 percent, the resulting rubber is called *ebonite.* Ebonite is unsuitable for use in seals.

The process of vulcanization is also known as curing the rubber. Although sulfur is the curing agent used for natural rubber and some of the synthetic rubbers, other curing agents, for example, metallic oxides and organic peroxides, are used for others. It is only in the cured state that elastomers have useful engineering properties. An indication of the changes in natural rubber brought about by vulcanization are:

Unvulcanized	Vulcanized
Low strength	High strength
Poor recovery after stretching	Good recovery after stretching
High plasticity	Low plasticity
Soluble in solvents	Insoluble in solvents
High freezing point	Low freezing point
Low softening point	High softening point

In addition to the curing agents, chemicals known as *accelerators* are used to speed up the rate of cure and influence the degree of vulcanization. The properties of natural rubbers depend very much upon the amount of sulfur and type of accelerator.

As a result of recent developments, there are now in commercial production elastomeric materials which are in fact thermoplastics. This means that synthetic rubbers which can be processed by melting are available, and the need for the vulcanization stage in the manufacture of seals can be eliminated. Though lacking the strong engineering properties of vulcanized elastomers, thermoplastic rubbers can reuse some of their waste rubber and can be processed on the more economical plastic injection presses, making them cost effective in less demanding applications.

Fillers

When fillers were first added to natural rubber, it was either to reduce product cost or produce a harder rubber. Today, different fillers (finely divided solids) are selected for a number of different reasons, the most important of which is reinforcement.

The most widely used and effective filler is carbon black. Other common fillers are calcium silicate, treated calcium carbonate, and finely divided silica. The ability of a filler to reinforce depends upon a number of factors such as particle size and surface characteristics. Examples of nonreinforcing fillers are talc, barytes, and magnesium carbonate.

The choice of fillers is important when considering certain chemical applications for elastomers. A filler such as calcium carbonate, which dissolves in acids, should not be used when the elastomer is intended for acid duty; otherwise the rubber will end up spongy. Siliceous fillers must be avoided when hydrofluoric acid is handled.

Antioxidants

Like many other polymers, elastomers require the addition of chemicals to inhibit oxidation and the effect of light and heat. In the case of elastomers they are also important in reducing the aging effect of mechanical flexing and therefore are of particular significance when the elastomer is used under dynamic conditions.

Antioxidants are complex organic chemicals and their development and selection is a specialized business. Without the use of antioxidants rubber will degrade. The result of this can be a hardening and cracking, or a softening, and even a sticky residue.

Softeners

Materials known in the industry as softeners, which range from mineral oils and waxes to plasticizers such as di-butyl-phthlalate, are added during compounding to aid mixing and to modify properties.

The Significance of Compounding

Compounding for manufacturing processes is a very complex business. Manufacturers consider processing know-how as confidential and do not normally disclose details. However, it is important for the user engineer to understand the significance of compounding and its effect on the properties of the elastomer. Many elastomers are compounded for general-purpose manufacturing use and may be unsuitable for a particular application. It may not be sufficient to refer to the elastomer by generic type—in some cases to get the best service it is necessary to refer to particular compounds. Some of the properties of elastomers which depend upon compounding are listed below:

1. Process characteristics (mixing, extruding, molding)
2. Strength
3. Modulus
4. Hardness
5. Compression strength and set resistance
6. Fatigue resistance
7. Heat resistance
8. Low temperature characteristics
9. Thermal expansion
10. Permeability
11. Abrasion resistance
12. Tear resistance
13. Electrical properties
14. Bonding to metals and fabrics
15. Suitability for use in the food and medical industries

TYPES OF ELASTOMERS USED IN THE MANUFACTURE OF SEALS

The elastomers available today are many and varied, and confusion can arise through the use of trade names, with and without suffixes, by the different manufacturers of similar products. Engineers have a habit of adopting trade names instead of using the generic name of the elastomer concerned. The information given in this section for the chemical compatibility and temperature range of various elastomers is not intended to be comprehensive but rather broadly informative. Initials in parentheses after the generic names are the internationally accepted abbreviations for those elastomers.

Natural Rubber (NR) and Synthetic Polyisoprene (IR)

This is resistant to most inorganic chemicals with the exception of strong oxidizing agents. It is satisfactory with alcohols and other polar organic chemicals, but is not to be used in mineral oil, aliphatic and aromatic hydrocarbon, or halogenated hydrocarbon applications. Service temperature can range from −30 to 100°C but 70°C is the normal upper limit.

Polybutadiene (BR) and Styrene Butadiene Copolymer (SBR)

These rubbers may be used by themselves or mixed with natural rubber. Chemical resistance is similar to natural rubber. Some grades have acceptable resistance to some hydraulic fluids.

Butyl Rubber (IIR)

This is a copolymer of isobutylene and isoprene. Butyl has better resistance to oxidizing acids than natural or polychloroprene rubbers. Butyl rubber has a lower permeability to gases and lower water absorption than other soft rubber compounds, and it is not suitable for use with mineral oils or aromatic hydrocarbons. It has good resistance to phosphate esters. Temperature range is −30 to 120°C.

Ethylene Propylene Rubbers (EPM and EPDM)

Ethylene propylene rubbers are available as copolymers and as terpolymers when dienes are used as the third monomer. The chemical resistance of ethylene propylene rubbers is similar to that of butyl rubber. Some formulations have excellent resistance to water and steam up to 150°C, and it is used for seals in central heating systems. At temperatures above 100°C it degrades in air, and at 150°C degradation is rapid. EP rubbers have good resistance to vegetable oils and phosphate esters but are not suitable for use with mineral oils.

Polychloroprene (CR)

The polychloroprenes are very good general-purpose rubbers. They have excellent weathering characteristics and are suitable for use with a wide range of inorganic chemicals. Polychloroprenes have fair resistance to mineral oils, fuel oils, and aliphatic hydrocarbons. Temperature service range varies from −30 to 100°C.

Acrylonitrile Butadiene Rubbers (NBR)

The name commonly used for NBR rubbers is nitrile. Nitrile rubbers are still used for a large percentage of seals. As with all rubbers of different generic types, there are a wide range of formulations. Nitrile rubbers are referred to according to acrylonitrile content and are classified as high, medium, and low nitrile. The effect of acrylonitrile on the properties of the rubber is illustrated below:

Acrylonitrile content	
High	Low
Tensile strength improves	Resilience and elasticity increases
Resistance to fuels and oils improves	Low-temperature flexibility improves

Nitrile rubbers are used for volume production of O-rings because their resistance to petrol, fuel oil, mineral oil, and ethylene glycol is very good. Nitrile rubbers are not suitable for use with hydraulic fluids of the ester type. The temperature range for nitrile rubber is −40 to 100°C. Hydrogenated nitriles extend the upper temperature range to 150°C.

Chlorosulfonated Polyethylene (CSM)

This rubber is noted for its excellent resistance to weathering and inorganic chemicals. It has reasonable resistance to oils but does not find much application for seals. Temperature range is −10 to 110°C.

Polyacrylic Rubbers (ACM)

Polyacrylic rubbers have excellent resistance to hot hydrocarbon oils. These rubbers have an advantage over nitrile rubber in that the upper service temperature is 150°C. Their resistance to water and acids is only moderate.

Fluorocarbon Rubbers (FPM)

Fluorocarbon rubbers have excellent resistance to a very wide range of chemicals including oils and some hydraulic fluids. Notable exceptions are methanol, acetone, acetic acid, and anhydrous ammonia. Resistance to steam is poor except for special grades which have been introduced in recent years. They can be used at temperatures from −30 to 180°C.

Perfluoro Elastomers

A small range of products sold by DuPont under the trade name *Kalrez* is resistant to a very wide range of chemicals and may be used up to 300°C. This elastomer does soften considerably at elevated temperatures. It has very poor compression set characteristics, which make it unsuitable for use when service conditions involve elevated temperatures and pressures and are cyclic. The material has found a wide application in mechanical seals in equipment that has difficult chemical duties. The material is very expensive and difficult to process. Finished products can only be obtained from DuPont.

Silicone Rubbers (SI)

Silicone rubbers have an advantage over most other rubbers in that in addition to being available in the conventional molded form, they are also available as liquid which can be cured at room temperature. These are available as one- or two-part materials. Two-part liquid silicone rubbers are used widely for sealing joints in large compressors operating at high pressure. Silicone rubbers can be used over the temperature range −50 to 250°C. Their oil resistance is reasonably good even though they do swell more than average. Silicone rubbers are used as seals in oxygen compressors. Silicone rubbers hydrolyze in steam. Many of the silicone rubber compounds have fairly low strength and most have poor tear resistance. The permeability of silicone rubbers to gases is high. They cannot be used with oils that have extreme pressure (EP) additive packages.

Polyurethane Rubbers—Ester Type (AV) and Ether Type (EU)

Polyurethane rubbers are characterized by their excellent resistance to wear. The oil resistance of these rubbers is good, but their general chemical resistance is limited. Maximum service temperature is about 80°C.

Epichlorhydrin Rubbers (CO Homopolymer and ECO Copolymer)

These rubbers have very good oil resistance, although general chemical resistance is limited. They are unsuitable for use with mineral acids and aromatic hydrocarbons. The permeability of these rubbers to gases is very low. They may be used up to 150°C for short periods.

Polysulfide Rubbers (TR)

Polysulphide rubbers are noted for their resistance to a very wide range of solvents, and they have excellent weathering properties. The temperature range is −30 to 100°C.

Never assume one manufacturer's compound is the same as another's even within the same elastomer group. Automotive companies invest years of their own test time evaluating various suppliers to make sure that they can expect equivalent performance from the vendors who are finally approved for a specific application. Even after exhaustive testing, slip-ups can occur and engineers, chemists, and/or compounders will discover they have not considered all of the variable interactions that can occur and sometimes do occur.

PREPARING (PREP) THE COMPOUND FOR MOLDING

When the raw rubber gum, or polymer, is first received, it is cut into a batch size for mixing and placed in a tote box with the other recipe ingredients. Usually the material is mixed in large capacity Banbury mixers and then dropped into a take-

FIGURE 12.3 Stearate bath is used for cooling and prevents mixed rubber from sticking to itself.

off mill where it is skived into strips for further use. As the strip comes off of the mill, it is fed through a stearate (soap) bath for cooling, coiling, and storage (Fig. 12.3). Later when needed, the strip will be fed through a tuber or extruder for slicing into round rubber rings called *prep* or preps. These are stored until required for molding. If the stock is to be injection molded, it is left in strip form for feeding into the injector screw system.[2] Occasionally the strip is made into pellets for feeding into an injection molding press.

MOLDING THE SEAL

When the mold plates are installed in the compression press and brought up to temperature (usually between 300 to 400°F and very closely controlled to a specific temperature), the prep is placed in loading trays for insertion into the molds (Fig. 12.4). If the prep is bonded to a metal insert or stamping, it too is loaded into the press at this time. The press is closed and the prep being squeezed between the upper and lower cavity is forced into every molding crevice at initial pressures of several thousand psi (but decaying quickly to several hundred psi). After the appropriate cure time at the correct temperature has occurred, the press opens and the parts are removed either by knockouts, air, or simply picking the parts out by hand. The seals are then transferred to other areas for additional trimming or deflashing if required. Some sealing products are "flashless" molded (i.e., flash located in an acceptable area and controlled to an agreed-upon customer acceptance level) and others require cold tumbling or knife trimming. O-rings are usually removed in sheets and cold tumbled (Fig. 12.5). Radial lip-type

FIGURE 12.4 Prep in containers await placement into loading trays for insertion into a compression molding press.

FIGURE 12.5 Sheets of molded products awaiting finishing.

seals are quite often molded flashless. Cold tumbling to remove flash can damage parts if it is not carefully done.

COMPRESSION INJECTION AND TRANSFER MOLDING

When molding an oil seal, there is no one best way. Compression molding is the oldest art and the one most practiced. Transfer molding and injection molding are also used extensively.[3] Whereas injection molding is used predominately in plastic molding, compression molding is the dominant method when molding rubber. A major reason is the high cost of a rubber injection press versus the low cost of a plastic injection press. Precision shot volume and cure control, shear rate, and scorch considerations, very high clamp forces, and extremely high rubber injection pressures (often over 20,000 psi) make a rubber injection molding press an expensive, complex, precision, highly sophisticated, computer-controlled machine of great strength and durability. The cost for such a machine is quite often beyond the normal capital investment capability of small- or medium-sized seal manufacturers who most often are not into high-volume runs. Besides, a compression machine, although considerably slower, can make parts of equal quality without the need to resort to highly skilled engineering talent to maintain and operate the equipment. It just does the job more slowly, and more art is required in the process. Some manufacturers use less-expensive, modified plastic injection presses for making seals usually for less demanding applications.

Molding an oil seal is a very difficult and challenging process. There are many variables that can be introduced, knowingly and unknowingly, that upset the functional quality of the product. Some process variables such as cure time and temperature rely on mold monitoring devices, but it is extremely difficult and expensive to monitor each and every cavity in a multicavity mold.

There are other molding variables equally important that are not quite so well known. The type of vacuum system, pressure, bumping, mold close timing, mold release, cavity pressure, stock viscosity, and scorch are very important and very difficult to monitor and control. Compare the compression mold process with the rubber injection mold process (Fig. 12.6).

Compression Molding

The following are the steps used in the compression mold process:

1. Mix chemicals into base polymer on Banbury or rubber mill.
2. Remove from mill in sheet or strip form.
3. Feed stock through a tuber to make a ring type of prep (Fig. 12.7), sometimes referred to as an I.D.-O.D. prep.
4. Place prep in mold on top of metal stamping that has been prepared with adhesive (Fig. 12.8).
5. Squeeze the two mold halves together in a heated press for an appropriate period of time.
6. Open the press and remove the product.

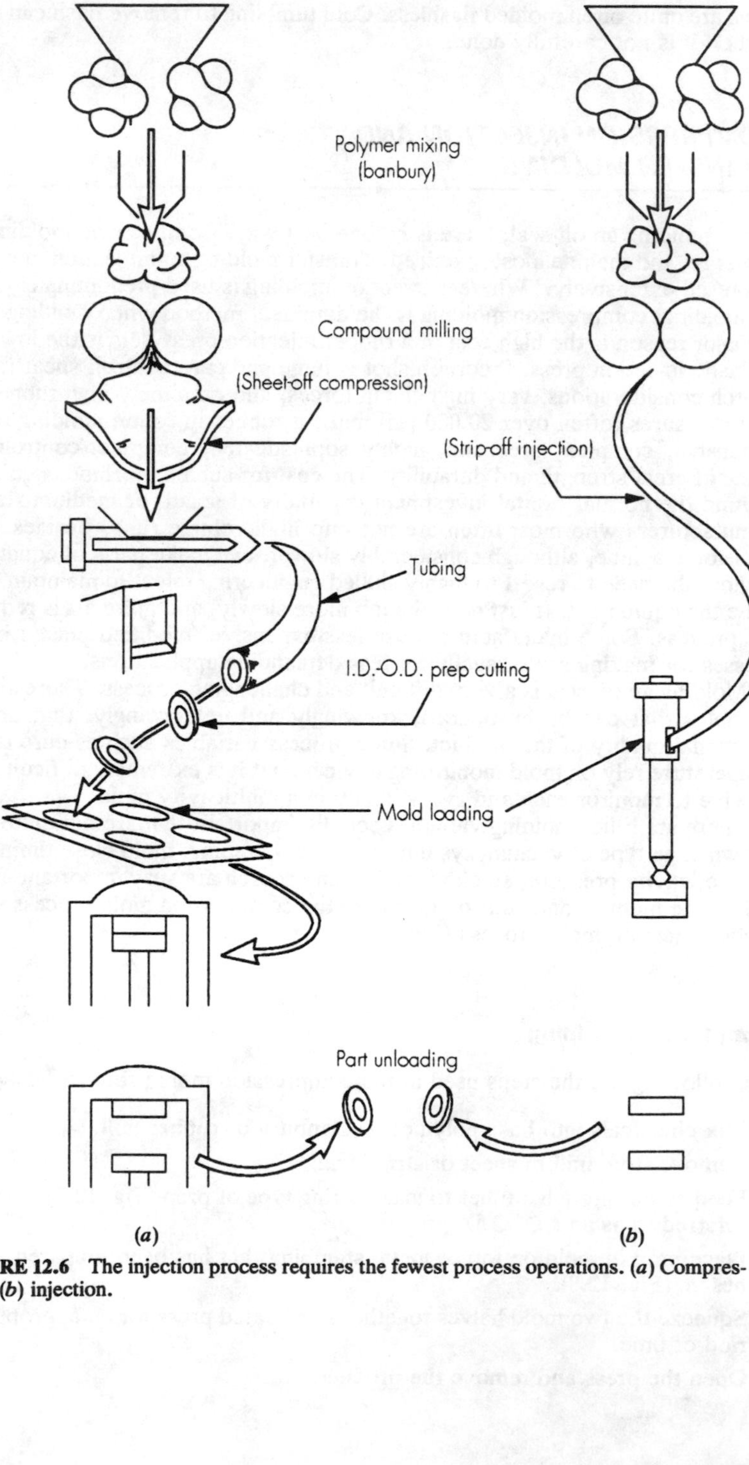

FIGURE 12.6 **The injection process requires the fewest process operations. (*a*) Compression; (*b*) injection.**

FIGURE 12.7 The tuber is an extruder used for making ring-type preps.

FIGURE 12.8 Molding a rubber to metal-bonded seal.

Making the prep for a compression mold can be very difficult since the prep weight will directly affect the pressure in the mold cavity and the functional quality of the part (Fig. 12.9). During the tubing, or extruding process, air gets forced into the rubber. The air thus entrapped in the prep can produce blisters on the finished oil seal lip in a hot oil environment. Water in the prep is another dangerous molding problem. If refrigerated prep is exposed to a hot, humid press room environment, steam can be introduced into the cavity during molding and can produce unfills, voids, and blisters. Water can also sometimes uncontrollably affect certain types of cure systems.

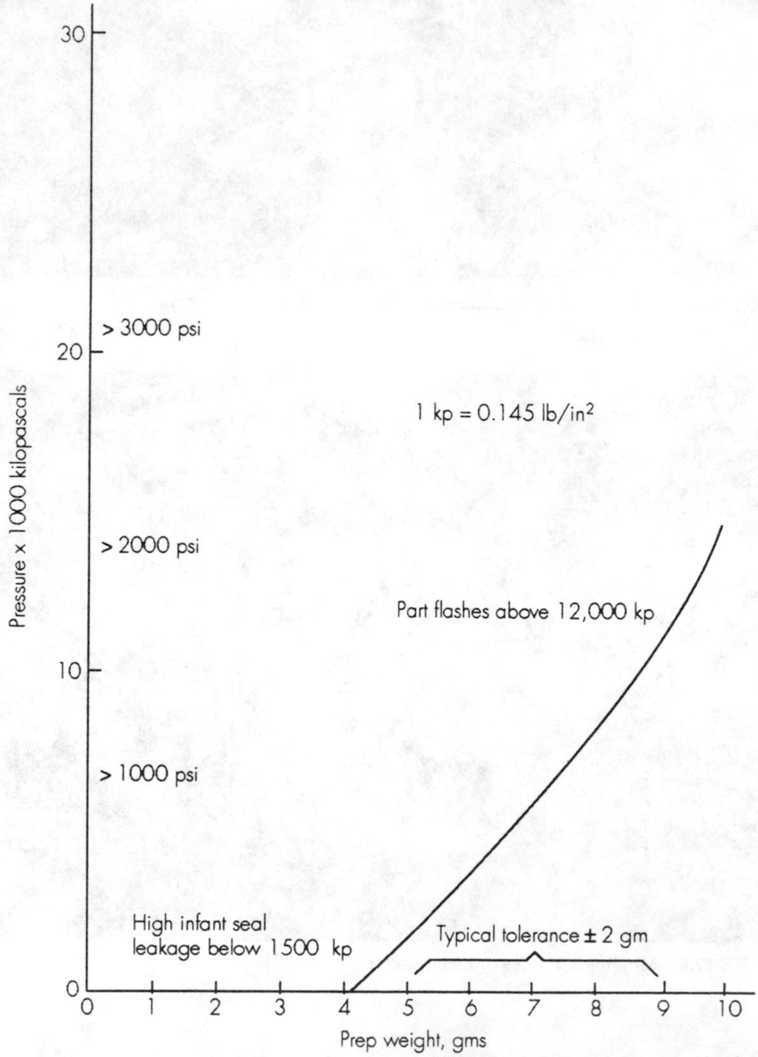

FIGURE 12.9 Compression prep weight affects cavity pressure and seal performance.

Most automotive seals are made in large compression presses with high-production multicavity molds due to the large number of oil seals required. Short cure times and high cure temperatures are used to speed the process. Quite often if a prep is placed on the hot mold for more than a few seconds, the skin of the prep begins vulcanizing, causing flow lines to appear in the finished product. These flow lines can sometimes screw the oil out of an application, much like a hydrodynamic seal. The flow lines do not always appear unless the seal is swelled in an oil environment.

Vacuum should be applied during molding to minimize the air trap problem. Air entrapment produces voids in the finished oil seal. However, the gases formed in the cavity during molding are difficult to remove. Press bumping is usually used. Although jogging the mold plates distributes the air more uniformly, it is usually too slight to even so much as open the mold to permit the air and gases to escape. If more than a slight bump is made, it can destroy the product geometry or cause additional flow marks to be formed once again.

Compression tooling must be tight so as to build up cavity pressure through resistance to stock flow. Tight tooling, stock viscosity, and proper prep weight all interact to determine or regulate the interior cavity pressure of the seal. Actually, this is almost impossible to do except over a very broad and nearly unacceptable range. With tight tooling, there is a momentary pressure peak and then a dramatic drop in pressure as the stock escapes through the restriction (Fig. 12.10). On multicavity setups, greater tolerances must be used in order to have all tools fit

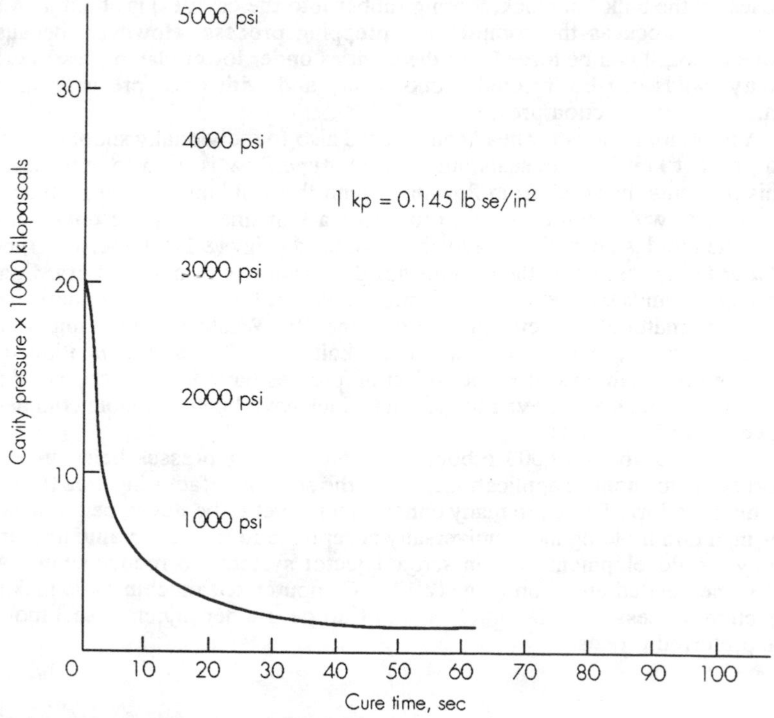

FIGURE 12.10 Pressure decays rapidly in compression molds.

into their respective holes and register. Loose register tools can produce parts that look good but operate with a high infant mortality. Lack of parallelism across the deck and plate warpage due to thermal distortion can also reduce the cavity pressure. The bond strength developed between the hot stamping and the curing rubber stock in the mold is also a function of the pressure generated. Pressure is controlled by the above-mentioned variables, which makes compression rubber molding a difficult engineering challenge. Some say it's a scientific art.

Injection Molding

The steps in the injection process are as follows:

1. Mix chemicals into base polymer in Banbury or rubber mill.
2. Remove from mill in strip form.
3. Feed stock through injection head and into a heated, closed, evacuated mold.
4. Open the mold and remove the part (see Fig. 12.6).

Injection molding differs from the compression molding process in that no preps or preforms are required, which eliminates most prepping variables.[4] Injection molding into multicavity molds using "cold" runners and "open mold" injection (similar to transfer molding, wherein a shot of rubber compound is injected into the center of the slightly open multicavity mold and then the press closes on the bulk hot stock, forcing rubber into the cavities) is about as wasteful of rubber stock as the compression prepping process. However, because the stock is hot, it can be forced into the cavities under lower clamp pressure. Single cavity molds can be injected economically and with great precision on multistation rotary injection presses.

A multinozzle injector head can be used also to individually shoot each cavity. To prevent knit lines in seals, an umbrella-type flow is used to enter the mold. This prevents the stock from flowing around the mold and having partially cured stock meet with uncured stock, producing a knit line. Pressure does not decay until the mold is opened if mold valves are used (Fig. 12.11). Otherwise there can be a drop in pressure as the injector nozzle withdraws from the cavity. Compression compounds can be used in injection molding, but the compounders prefer to tailor the materials for enhanced processability. Waste due to runners can become excessive, and runners can create knit lines. The above-mentioned problems and the high cost of rubber injection presses have limited the use of the injection molding system even though machines have been available commercially since 1964 (Table 12.1).

There are about 10,000 rubber injection molding presses being used today worldwide for nontire applications, but in the seal manufacturing area the number is still quite low. There are many enhancements yet to be developed to make rubber injection molding more universally acceptable to the seal manufacturer. Certainly the development of twin screw injector systems to reduce scorch should get some needed attention (Fig. 12.12).[5] Computerized machinery is making the injection process more favorable and in Europe, rubber injection seal molding is the preferred process.

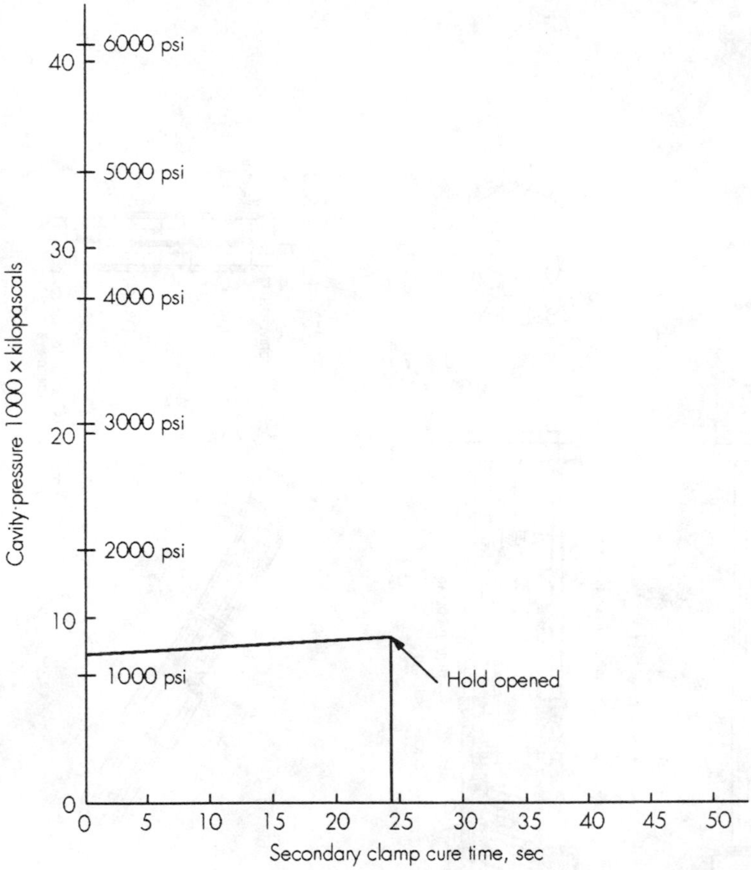

FIGURE 12.11 Injection mold values hold pressure in the cavity until the mold is opened.

TABLE 12.1

	Multicavity compression	Multicavity injection	Multinozzle or single cavity rotary injection
1. Rubber prep required	Yes	No	No
2. Vacuum in cavity during fill	No	Yes	Yes
3. High-volume production without knit lines	Yes	Yes (with cold runner tools)	Yes
4. Stock wastage	Moderate	Moderate	Low
5. Scrap parts	Average	Average	Very low
6. Functional quality	Extremely variable	Moderately variable	High quality
7. Tool cost	High	Average	Low to average
8. Setup cost	High	Average	Low to average
9. Press reliability	Average	Average	Average
10. Press utilization	Average	Average	High
11. Bumping required	Yes	Yes	No
12. Mold release	Yes	Yes	Occasionally
13. Cavity pressure	Extremely variable	Moderately variable	High pressure

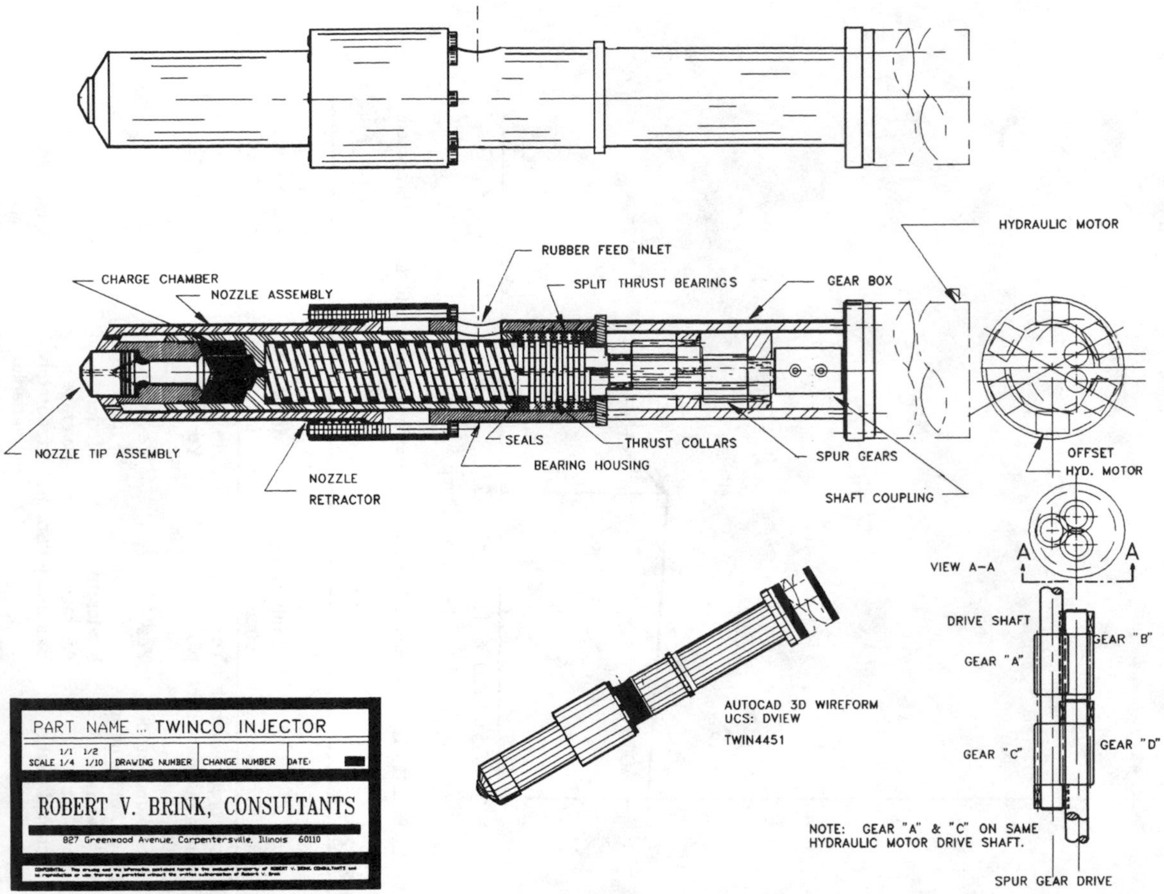

FIGURE 12.12 A corotating conjugate twin screw injector will move uncured compound through the injector barrel with virtually no shear, thereby minimizing the possibility of "scorch" and improving overall temperature control of the stock.

PRECISION MOLDS MAKE PRECISION SEALS

Molds for manufacturing elastomeric seals can be made from many different materials, but one of the best materials to use is an air quenched A-2-type steel, such as Carpenter 484, heat treated to approximately 60 Rockwell C Scale Hardness (Rc) and chrome plated to reduce wear and friction. Cold quenching the A-2 steel will also improve its mold growth resistance under elevated temperatures. The same process is used to make micrometers, calipers, and other types of precision tools. Usually several cycles down to −40°F (−40°C) are sufficient to provide years of satisfactory rubber molding service.

METAL CASES

Blanking and forming the metal stampings are usually performed in progressive dies if the volume permits. Radial lip-type seals use metal cases varying in thickness from 0.020 to 0.100 in and are normally made of SAE 1020 steel, although aluminum is also sometimes used. Water-based adhesives such as a Chemlok 607 will be applied to bond the rubber to the metal during the molding process, although some seal compounds are bonded to brass elements without adhesive.

Seals must be handled with great care throughout the manufacturing process to prevent damage to the sealing lip element. The slightest nick or imperfection can cause leakage in an application.

REFERENCES

1. American Chemical Society, *Rubber Technology,* 2d ed., Robert E. Krieger Publishing Company, Malabar, FL, 1981.
2. Brink, Robert V., "Optimising the Rubber Molding Process," *European Rubber Journal,* May 1981.
3. Brink, Robert V., "Molding the Modern Oil Seal," SAE Paper and presentation, Feb. 1976 at Winter Annual Meeting, Detroit, MI.
4. Harper, Charles A. *Handbook of Plastics and Elastomers,* McGraw-Hill, New York, 1975.
5. Brink, Robert V., "A Corotating Conjugate Twin Screw Feed Injector," Paper and presentation at the 136th meeting, Rubber Division, American Chemical Society, Detroit, MI, Oct. 17–20, 1989.

INDEX

ABOUT THE AUTHORS

Robert V. Brink, Editor in Chief, is President of Brink Consulting Services, providing professional design and development assistance to industrial clients. As a former research engineer, General Manager of Victor Seal Division, Dana Corp., Vice-President of Engineering, CR Industries, founder and President of Pentaject Corporation, an international manufacturer of large, computerized rubber injection press systems, Brink has created many technical innovations in the molding and sealing products industry. He has written many technical papers, including some of the sealing industry's basic papers on radial load devices, torque measuring devices, and other gages, transducers, and force sensing systems. Brink is currently providing engineering services to the Boeing Company.

Daniel E. Czernik is currently Vice-President of Technologies and Technical Planning for Fel-Pro, Inc., where he has been employed since 1969.

Leslie A. Horve has spent 23 years in the radial lip sealing industry, all with CR Industries where he is currently Vice-President of Industrial and Aerospace Technology.